Glycoprotein and proteoglycan techniques

LABORATORY TECHNIQUES IN BIOCHEMISTRY AND MOLECULAR BIOLOGY

Volume 16

Edited by

R.H. BURDON - *Department of Bioscience and Biotechnology, University of Strathclyde, Glasgow*

P.H. van KNIPPENBERG - *Department of Biochemistry, University of Leiden, Leiden*

ELSEVIER
AMSTERDAM · NEW YORK · OXFORD

GLYCOPROTEIN AND PROTEOGLYCAN TECHNIQUES

J.G. Beeley

Department of Biochemistry,
University of Glasgow,
Glasgow G12 8QQ, U.K.

1985
ELSEVIER
AMSTERDAM · NEW YORK · OXFORD

ISBN 0-444-80651-2 (pocket edition)
ISBN 0-444-80652-0 (library edition)
ISBN 0-7204-4200-1 (series)

Published by:
ELSEVIER SCIENCE PUBLISHERS B.V. (Biomedical Division)
PO BOX 211
1000 AE AMSTERDAM
THE NETHERLANDS

Sole distributors for the USA and Canada:
ELSEVIER SCIENCE PUBLISHING COMPANY, INC.
52 VANDERBILT AVENUE
NEW YORK, NY 10017
USA

Library of Congress Cataloging in Publication Data

Beeley, J. G.
Glycoprotein and proteoglycan techniques.

(Laboratory techniques in biochemistry and molecular biology ; v. 16)
Bibliography: p.
Includes index.
1. Glycoproteins--Analysis. 2. Proteoglycans--Analysis. I. Title. II. Series.
QP519.L2 vol. 16 574.19'2'028 s [574.19'285] 85-10396
[QP552G59]
ISBN 0-444-80652-0
ISBN 0-444-80651-2 (pbk.)

Printed in Belgium

Contents

CHAPTER 1

Introduction

There has been a remarkable expansion of interest in glycoproteins over the past two decades. From being a specialised area of structural research on the borderline between carbohydrate and protein chemistry the study of these molecules has become highly relevant to a wide range of biological phenomena. This change has come about because of developments in our knowledge of the distribution, biosynthesis, molecular organisation and functions of these molecules.

Glycoproteins (i.e. proteins containing covalently bound carbohydrate) are ubiquitous constituents of all living organisms with the possible exception of bacteria, in which they have to date only been unequivocally demonstrated in one genus (Sharon and Lis, 1982). Glycosylation is a very common modification of extracellular and integral membrane proteins of higher organisms. Covalently bound carbohydrate groups occur in glycoproteins which function as enzymes, antibodies, hormones, structural proteins, carrier proteins, mucins of epithelial secretions, membrane transport proteins and receptors.

The amount of carbohydrate present in glycoproteins can vary from less than 1 % to more than 85 % of the dry weight of these molecules. Proteoglycans are a class of highly glycosylated glycoproteins which are important constituents of the extracellular matrix of animal connective tissues. Some proteoglycans are closely associated with the surfaces of animal cells. The presence of glycoproteins and proteoglycans in, or attached to, the surfaces of animal cells (Hughes, 1976) has stimulated many enquiries into the role of these molecules in cellular adhesiveness, differentiation, in the control of growth and in disease processes such as neoplasia and the infection of tissues with bacteria and viruses.

The biosynthesis of glycoproteins occurs within the internal membrane systems of cells. Glycoprotein oligosaccharide units are assembled and modified as the molecules move through successive subcellular compartments *en route* for destinations outside the cell, as membrane components of the cell surface, or as components of the membranes or contents of cellular organelles. Studies of the biosynthesis of glycoproteins are therefore not only of inherent interest in showing the mechanisms by which these molecules are assembled but can provide insight into the biological problem of how macromolecules are directed to specific destinations. For example, the examination of glycoproteins synthesised by cells obtained from individuals with genetic disorders of proteoglycan catabolism ('mucopolysaccharidoses') has provided evidence that specific signals associated with the carbohydrate units of lysosomal enzymes are responsible for directing these glycoproteins to their destination as lysosomal contents (Neufeld and Ashwell, 1980). Studies of the biosynthesis and catabolism of proteoglycans and glycoproteins are also clearly of importance in understanding diseases of connective tissue, including the most widespread of all ailments, ageing.

The role of carbohydrate in the function of glycosylated proteins has been a continuing, and at times elusive, theme in research over many years. With such wide variation in both the protein components of these molecules and the carbohydrate units attached to them it would be surprising if several different functions had not evolved. In the proteoglycans and some glycoproteins (e.g. mucous glycoproteins) the physico-chemical properties associated with the carbohydrate units (such as visco-elastic behaviour, water retention and the exclusion of macromolecules from their solvent domain) are clearly of functional significance (Muir, 1983). The high negative charge on heparan sulphate present in the basement membrane of the renal glomerulus appears to have a physiological role in the retention of macromolecules in the bloodstream (Lemkin and Farquar, 1981). A role for oligosaccharide units of glycoproteins and proteoglycans in the protection of the peptide chains of these molecules from proteolytic cleavage has also been established. The carbohydrate units of glycoproteins can also have a pronounced effect on the folding of the peptide chain to which they are attached (Rose et al., 1984). However, there is now also sound evidence that the cells of animal tissues contain several different types of specific receptors which can recognise and respond to oligosaccharide units of glycoproteins by enhanced pinocytosis (Ashwell and Harford, 1982). The occurrence of these well-characterised receptors gives credibility to some of the many suggestions which have been made regarding the potential role of specific recognition of glycoproteins in biological processes.

The aim of this book is to describe techniques which can be used to answer some of the basic questions about glycosylated proteins. Methods are discussed for isolation, characterisation, compositional analysis, for determination of the primary structure of carbohydrate units and the nature of protein-carbohydrate linkages of glycoproteins and proteoglycans. An attempt has been made to keep in mind the diverse nature of glycosylated proteins and the many different types of problem, alluded to in preceding paragraphs, which readers may have to tackle. For example, quite different approaches are required for the isolation of mucins and membrane glycoproteins. The amount of sample available is often a limitation in the analysis of membrane glycoproteins or the products of viruses or cultured cells, and some emphasis has therefore been placed on the quantities of sample required for particular procedures and a chapter has been devoted to radioactive labelling techniques. Lectins have become important tools for the investigation of glycoproteins and these reagents are considered in a separate chapter. Emphasis has been placed on describing techniques which can be applied in most laboratories without requiring highly sophisticated equipment. However, when the best approach to a problem would be to use major equipment (e.g. high-resolution NMR) this has been indicated in terms of the nature of the sample required and the type of information obtainable, but without detailed description of instrumentation or the theoretical background.

A feature of this book is that proteoglycans and other glycoproteins are considered together. For historical reasons they have usually, although not always (Spiro, 1973), been treated separately and, on the whole, research workers have tended to concentrate on one or the other of them. The artificiality of considering proteoglycans separately from other glycosylated proteins has become obvious since it has been shown that a single peptide chain can carry carbohydrate units of both the 'proteoglycan type' and units characteristic of other glycoproteins. It is hoped that the inclusion of proteoglycans and glycoproteins will help to counteract the dichotomy which has arisen in the study of these molecules. The reader should be aware that there are also marked similarities between glycoproteins and glycolipids (Chapter 2). Some of the methods described here could quite easily be adapted to glycolipids.

Most research workers concentrate their efforts in a limited field of glycoconjugate research. The author's own interests, and limitations, may well be apparent in the selection of methods for this book, though an attempt has been made to cover a wide variety of different types of molecule.

Preparation of this book has been greatly aided and encouraged by a number of people. These include the secretarial staff of the Biochemistry Department at the University of Glasgow, who have typed the manuscript with great skill and perseverance. The illustrations have been prepared by Mr. Ian Ramsden and his staff in the Medical Illustration Unit. Dr. R. Eason kindly agreed to read and comment on Chapter 4 of the manuscript. Professor A. Kobata, Professor J. Montreuil and Professor N. Sharon generously provided copies of manuscripts prior to publication. In addition, a number of authors and publishers have given permission for the reproduction of original material subject to copyright which occurs, sometimes in slightly modified form, in several of the figures and tables in this book. Authorship of this material is indicated by the references given in the legends to figures and tables, and full details of the publications can be found in the references at the end of the book. Finally, I would like to thank the editors and publishers for their patience and encouragement.

CHAPTER 2

Glycoproteins and proteoglycans

2.1. Introduction

This chapter is intended to provide a brief guide to the molecular structures of glycosylated proteins. More extensive discussion of the structures of these molecules can be found in reviews of proteoglycans (Kennedy, 1979; Rodén, 1980) and other glycoproteins (Marshall, 1972; Spiro, 1973; Kornfeld and Kornfeld, 1976; Montreuil, 1980; Sharon and Lis, 1982).

Glycoproteins are proteins to which carbohydrates are covalently linked through glycosidic bonds (Spiro, 1973). Proteoglycans are a subclass of glycoproteins with distinctive features of carbohydrate structure (Spiro, 1973; Sharon and Lis, 1982). In describing methodology it is, however, convenient to differentiate between proteoglycans and other glycoproteins. For this reason the term glycoprotein will, subsequently in this book, be used to describe enzymically glycosylated proteins excluding proteoglycans.

Glycoproteins and proteoglycans have carbohydrate units which vary in size from monosaccharides to polysaccharides and there may be from one to some hundreds of carbohydrate units attached to a single polypeptide chain. Subsequent sections of this chapter contain descriptions of the components from which the carbohydrate units are built up (Section 2.2), types of protein-carbohydrate linkage (Section 2.3), the structural organisation of carbohydrate units (Section 2.4), polypeptide components (Section 2.5), molecular organisation of the different types of glycosylated protein (Sections 2.6 and 2.7), structural microheterogeneity (Section 2.8) and the relationship between glycoproteins, proteoglycans and other glycoconjugates (Section 2.11). Non-enzymatic glycosylation of proteins (Section 2.9) and chemically synthesised glycoproteins (Section 2.10) are also discussed.

2.2. Carbohydrate components

The structures of the monosaccharides which have been isolated from glycoproteins and proteoglycans are shown in Fig. 2.1. When the sugar residues are glycosidically linked they occur as six-membered pyranoside rings, with the exception of L-arabinofuranoside, which has been found in plant glycoproteins. The stereoisomers of monosaccharide given in Fig. 2.1 are those which have been identified in glycoproteins and proteoglycans.

The monosaccharide residues of glycoproteins often carry substituent groups. Hexosamines in glycoproteins and proteoglycans other than heparin and heparan sulphate are *N*-acetylated. Both heparin and heparan sulphate contain *N*-sulphated as well as *N*-acetylated hexosamine. Most proteoglycans and several glycoproteins have *O*-sulphate substituents. A few glycoproteins (lysosomal enzymes) have been found to have mannose residues esterified with *O*-phosphate (or β-*N*-acetylglucosaminyl-*O*-phosphate). Sialic acids can carry a wide range of substituents. Either the *N*-acetyl or *N*-glycolyl ($-COCH_2OH$) derivative of neuraminic acid can be present and there may also be a variety of *O*-acetyl and/or *O*-glycolyl groups.

The Haworth perspective formulae given in Fig. 2.1 show substituents and stereochemistry clearly but the conformations of sugar rings are not represented accurately. For consideration of the molecular shape and reactivity of the carbohydrate units of glycosylated proteins conformational formulae are preferable. The Haworth perspective formula and the conformational formula for the most stable $^1C_4$ structure in aqueous solution of β-*N*-acetyl-D-glucosamine-pyranose are illustrated in Fig. 2.2. Both Haworth and conformational representations of structures will be employed elsewhere in this book.

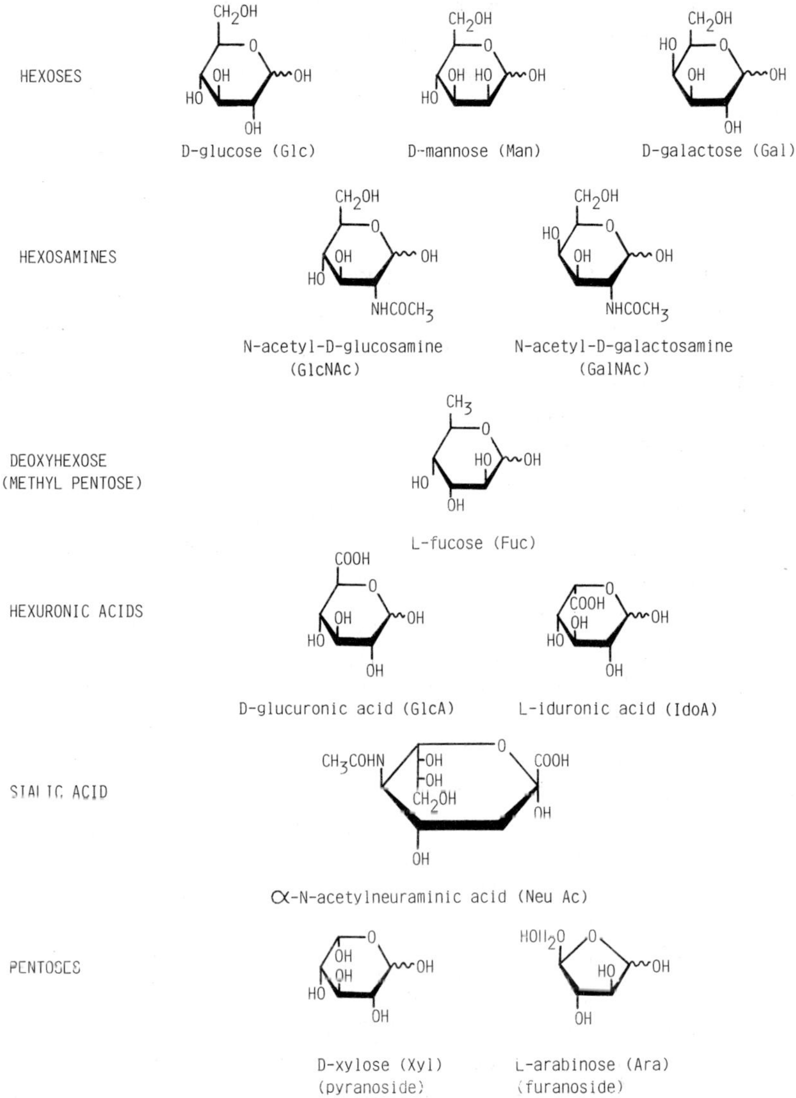

Fig. 2.1. Monosaccharides occurring in proteoglycans or other glycoproteins. Abbreviations used to represent monosaccharide residues in oligosaccharide structures are given in parenthesis.

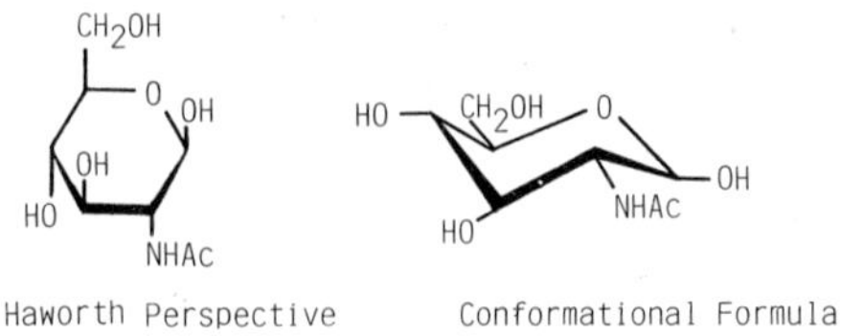

Fig. 2.2. Ways of representing the structure of a monosaccharide (β-*N*-acetyl-D-glucosamine). The conformational formula ($^4C_1$ chair form) shows that all OH groups and the *N*-acetyl substituent are equatorial to the ring in β-D-GlcNAc.

The ring structure of monosaccharide residues (furanoside or pyranoside) and both the anomeric configuration (α or β) and position of glycosidic linkages in oligosaccharides or polysaccharide chains can be represented by structural formulae, systematic names or in abbreviated form (Fig. 2.3). (The IUPAC-IUB Joint Commission recommendations on nomenclature are given in Eur. J. Biochem. *21* (1981) 455–477 and *126* (1982) 433–437.) It is often convenient to describe structures using the abbreviations for sugar residues given in Fig. 2.1 indicating the anomeric configurations and identifying the

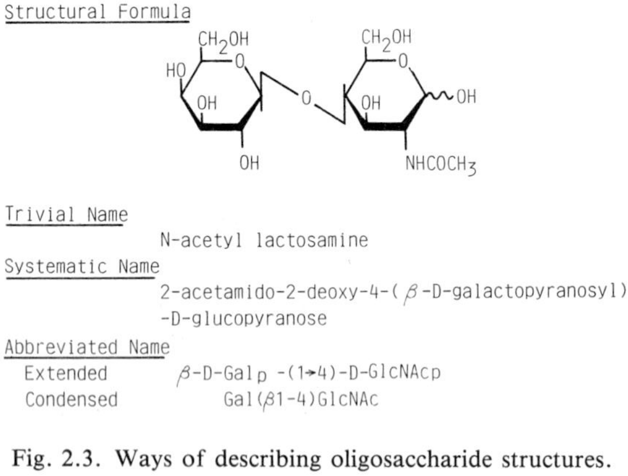

Fig. 2.3. Ways of describing oligosaccharide structures.

carbon atoms which are joined by glycosidic linkages as shown in the condensed abbreviated structure in Fig. 2.3. In this condensed system of nomenclature the common configuration and ring size are implied in the symbol. Thus, Gal means D-galactose pyranoside. When using the condensed system to describe branched oligosaccharides rounded brackets are used to indicate the branches but can otherwise be omitted. Thus Manα1–3(Manα1–6)Manβ1–4GlcNAc indicates a structure with two α-linked mannose residues attached to a third mannose which is β-linked to the reducing terminal *N*-acetylglucosamine. Squared brackets are used to define repeating units of condensed chains.

The occurrence of some of the monosaccharides is restricted to certain types of glycosylated protein. Glucuronic and iduronic acids are constituents of many proteoglycans but occur very rarely in other glycoproteins. Xylose is a constituent of the protein-carbohydrate linkage region of most proteoglycans and arabinose has been found only as a constituent of certain plant glycoproteins. Glucose occurs in collagen (and in intracellular biosynthetic intermediates of glycoproteins with *N*-linked carbohydrate units). Glucose residues can also become covalently linked to proteins by non-enzymatic reactions (Section 2.9).

2.3. *Protein-carbohydrate linkages*

In each carbohydrate unit of a proteoglycan or glycoprotein a single sugar residue is linked glycosidically to the side chain of an amino acid. The linkage is either *N*-glycosidic via the amide nitrogen of an asparagine residue or *O*-glycosidic via the hydroxyl of serine, threonine, hydroxylysine or hydroxyproline. Table 2.1 lists the types of sugar residues which participate in protein-carbohydrate linkages and gives some examples of the types of molecule in which they have been found. Methods for distinguishing between protein-carbohydrate linkages on the basis of differences in stability to alkali and other properties are discussed in Chapter 6.

TABLE 2.1
Types of protein-carbohydrate linkage

Linkage residues	Glycoside type	Example of molecules containing linkage
Xyl-Ser	O	Proteoglycans (except keratan sulphates)
GalNAc-Ser/Thr	O	Glycoproteins, especially mucous glycoproteins. Keratan sulphate type II.
Gal-Ser	O	Plant glycoproteins (extensin, potato lectin)
Gal-Ser/Thr	O	Earthworm cuticle collagen
Man-Ser/Thr	O	Yeast, fungal and *Nerei* cuticle glycoproteins
Gal-Hly	O	Collagen
Ara-Hpr	O	Plant glycoproteins (extensin, potato lectin)
Gal-Hpr	O	Plant glycoproteins
GlcNAc-Asn	N	Animal and plant glycoproteins (plasma glycoproteins, soyabean lectin). Keratan sulphate type I

2.4. Structures of carbohydrate units

If, in the construction of the carbohydrate units of glycoproteins, all of the possible combinations of anomeric linkages and positional isomerism were exploited an enormous range of oligosaccharide structures could be produced. It has been calculated, for example, that a trisaccharide having three different hexose units in the pyranoside form can have 1056 possible isomeric forms (Clamp, 1974). Fortunately for biochemists interested in glycoproteins there appear to be very considerable constraints on the structures actually encountered in nature. Certain patterns have been recognised in the carbohydrate units of glycosylated proteins, as indicated below.

2.4.1. Periodic and aperiodic structures

Two general structural motifs can be distinguished in glycosylated proteins; periodic structures with a repeating unit (such as a disaccharide) which form extended linear molecules, and aperiodic structures having relatively short and often highly branched oligosaccharide chains.

The glycosaminoglycan chains of proteoglycans have periodic linear structures. Identification of the repeating structure and the molecular weight defines the basic structure of this type of molecule. The types of monosaccharide and the types of linkage occurring in these molecules are quite restricted (Section 2.8). However, modifications of the repeating units occur at points along the chain as a result of variation in substitution (e.g., of sulphate) or secondary biosynthetic modification of sugar units (e.g., conversion of glucuronic to iduronic acid). At the non-reducing terminal end the periodic repeating polysaccharides are attached to protein via a short aperiodic 'linkage region' (Fig. 2.4).

<u>Linkage region of proteoglycans</u> containing Xyl-Ser linkage

Gal(β1-3)Gal(β1-4) Xyl(β1-2)Ser

Core structure of glycoproteins containing GlcNAc-Asn linkage

```
Man α1
     ↓
     3
     Man β1-4 GlcNAc β1-4 GlcNAc β1-Asn
     6
     ↑
Man α1
```

Fig. 2.4. Core structures of oligosaccharides chains adjacent to two types of protein-carbohydrate linkage. Glycosidic linkages at branch points are indicated by arrows, and the brackets used in the condensed description of linear chains are omitted.

The carbohydrate chains of most glycoproteins are of the aperiodic type. When unbranched chains occur they are usually short (1–5 monosaccharide residues). Many glycoproteins have branched structures of considerable complexity.

Some glycoproteins have carbohydrate units which have elements of repeating sequence together with branching and other characteristics of aperiodic structures. One example of this type of structure is found in the poly(glycosyl) (or 'lactosaminoglycan') chains of human erythrocyte membrane Band 3 (Krusius et al., 1978; Fukuda et al., 1984). This molecule has sections of repeating disaccharide units [3Gal(β1–4)GlcNAc(β1–] which carry branches. These structures carry blood group antigenic determinants. Fairly short repeating sequences (Type I and Type II chains) with peripheral substituents appear to occur in several partially characterised cell surface glycoproteins and in some mucins with blood group activity (Feizi, 1981).

2.4.2. Core structures

Certain sugar residues immediately adjacent to the protein carbohydrate linkage are referred to as the 'linkage region' or 'core structure' (Montreuil, 1980). These sequences appear to be exclusively associated with a particular type of protein-carbohydrate linkage. The core structures associated with the Xyl-Ser linkage found in proteoglycans and the GlcNAc-Asn linkage are shown in Fig. 2.4.

2.4.3. Peripheral residues

Sugar residues which occur at, or close to, the non-reducing terminal positions in the carbohydrate units of glycoproteins are referred to as 'peripheral residues'. Operationally this term is often used to describe residues which can be readily removed by digestion of the oligosaccharide chains with exoglycosidases. For glycoproteins which contain the GlcNAc-Asn linkage the term peripheral residues is also used to describe the terminal carbohydrate residues which are attached to the common inner core structure shown in Fig. 2.4. Short peripheral chains of 1–3 residues attached to invariant core structures have been termed 'antennae' (Montreuil, 1980).

Monosaccharides most commonly found as peripheral sugars in glycoproteins include sialic acid, fucose and galactose. Sialic acid and fucose always occupy non-reducing terminal positions.

2.4.4. Simple, complex and mixed types of carbohydrate units

The structures of a considerable number of the branched *N*-glycosidically linked carbohydrate units of glycoproteins have now been determined and the great majority of these structures seem to fall into three categories which contain the common core pentasaccharide shown in Fig. 2.4. Carbohydrate units of the 'simple type' (or 'high-mannose type') have a core structure substituted only with mannose residues. The 'complex type' of carbohydrate unit contains a core structure substituted with one or more of the sugars *N*-acetylglucosamine, galactose, fucose and sialic acid. These substituents can be monosaccharide residues or heterosaccharide chains. Within such chains disaccharide units with the structure Gal(β1–4)GlcNAc (*N*-acetyllactosamine) are often, but not always, present and 'complex-type' carbohydrate units have been described as the '*N*-acetyllactosaminic type' of structure (Montreuil, 1980). Complex units are described as di-, tri-, tetra- or penta-antennary structures according to whether the mannose residues of the core are substituted with two, three, four or five sugar chains (or GlcNAc residues).

A third family of carbohydrate moieties described as the 'hybrid type' (or 'mixed type') have been identified in which one of the two α-linked mannose residues of the core structure has substituents of the 'high-mannose' type while the other α-linked core mannose has substituents characteristic of the 'complex type' of unit. Examples of 'simple', 'complex' and 'hybrid' types of structure are shown in Fig. 2.5.

In some Asn-linked carbohydrate units the β-Man residue in the core is substituted with a GlcNAc residue between the two α-linked mannose residues. This type of structure is described as having an 'intersecting' of 'bisecting' GlcNAc residue.

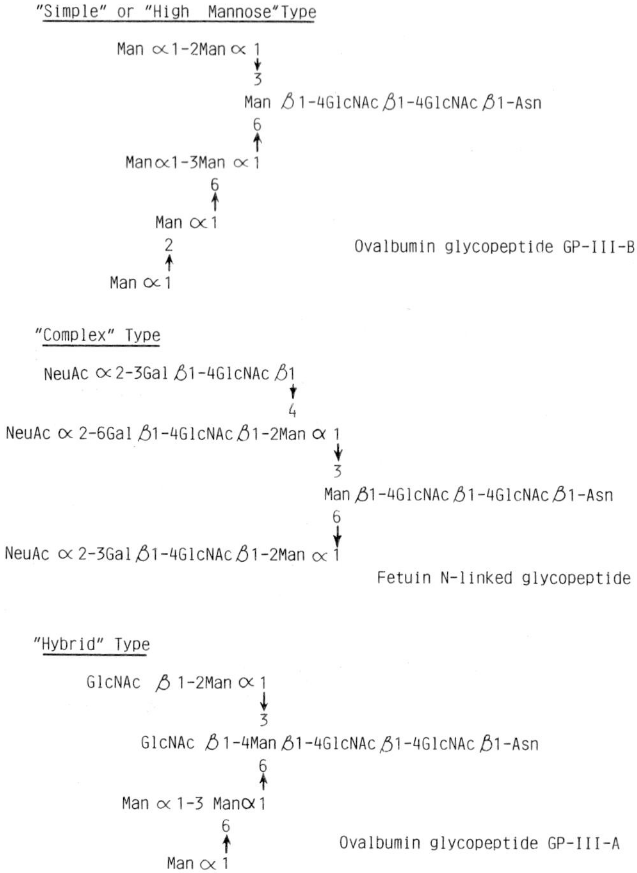

Fig. 2.5. Examples of simple, complex and hybrid types of asparagine-linked oligosaccharide units.

2.5. *Peptide moieties*

The carbohydrate groups of glycosylated proteins are attached at specific sites distributed along their peptide chains. It is therefore of interest to enquire how the addition of carbohydrate is related to the primary sequence and/or conformation of these molecules.

Amino acid sequences have been determined for a considerable number of glycoproteins having *N*-glycosidically linked carbohydrate groups. These glycoproteins, of varied functions, have carbohydrate attached to the side chain of asparagine residues in the sequence Asn-X-Ser/Thr, where X can be any amino acid residue except proline (Marshall, 1974). Not every sequence of this type in secreted proteins is glycosylated. In some proteins, such as bovine pancreatic ribonuclease, this tripeptide sequence is glycosylated in some, but not all, molecules.

No comparable regularity in amino acid sequence has been identified with relation to carbohydrate units *O*-glycosidically linked to GalNAc. In more highly glycosylated glycoproteins with this type of linkage, such as the mucous glycoproteins, glycosylated Ser and Thr residues are close together or adjacent. The amino acid proline occurs with high frequency in glycosylated regions of these polypeptide chains and this may influence the peptide chain conformation. Amino acid sequencing of the peptide units of proteoglycans suggests that in these molecules too the sites of glycosylation are closely spaced along the peptide chain.

The addition of carbohydrate to hydroxylysine residues in collagens seems to be associated with the amino acid sequence Gly-X-Hyl-Gly- (Marshall, 1974).

2.6. Secreted glycoproteins

The overwhelming majority of extracellular proteins of higher animals are glycoproteins. Among the proteins of human serum, for example, all but albumin and some of the lipoproteins are glycosylated during their biosynthesis. Secreted glycoproteins occur in the body fluids, in the extracellular matrix of connective and other tissues, and in the mucus coating of the digestive, respiratory and urino-genital tracts. The functions performed by these molecules include the lubrication and protection of epithelial surfaces, the transport of substances in the blood and a structural role in connective tissues.

Different types of molecular structures have evolved to perform these, and the many other, functions of glycoproteins. The brief consideration of some characteristics of mucous glycoproteins, serum glycoproteins and glycosylated fibrous proteins given below is intended to illustrate the different types of molecular organisation which can be found among secreted glycoproteins. These represent only a selection from some of the more extensively studied types of mammalian glycoprotein. Other structural adaptations of glycoproteins to specialised functions are already known (e.g. the antifreeze glycoproteins of arctic fish) and many more will undoubtedly be discovered as studies of these molecules are extended to the whole range of organisms producing secreted glycoproteins.

2.6.1. Mucous glycoproteins

These molecules are major organic constituents of mucus, a viscous, water-insoluble gel which forms a protective coating on epithelial surfaces. The physical and biological properties of purified mucous glycoproteins (or mucins) are dominated by their carbohydrate component, which can account for over 70% of their dry weight. Vertebrate mucous glycoproteins contain sugars which include galactose, fucose, *N*-acetylglucosamine, *N*-acetylgalactosamine and sialic acid but little or no mannose and no uronic acid. Analysis of carbohydrate composition can therefore distinguish these molecules from mannose-containing glycoproteins and proteoglycans which have uronic acid (Allen, 1983).

The carbohydrate units of mucous glycoproteins vary in size from disaccharides (ovine submaxillary mucin) to branched chains containing up to nineteen residues (porcine gastric mucin). Those oligosaccharides are usually closely spaced along much of the peptide chain. This close packing and the mutual repulsion of negative charges of sialic acid residues (and sometimes ester sulphate) produces a 'bottle-brush'-like structure with carbohydrate chains projecting around an extended peptide backbone. Blood-group-specific carbohydrate structures are present on mucous glycoproteins produced by individuals who are 'secretors'. Protein is linked to carbohydrate via the alkali-labile *O*-glycosidic linkage between *N*-acetylgalactosamine and serine or threonine. This protein-carbohydrate linkage is sometimes referred to as the 'mucin' type of linkage. The peptide chains of mucous glycoproteins have a high proportion of serine, threonine and proline. In the case of ovine submaxillary mucin all of the serine and threonine residues are glycosylated (Hill et al., 1977).

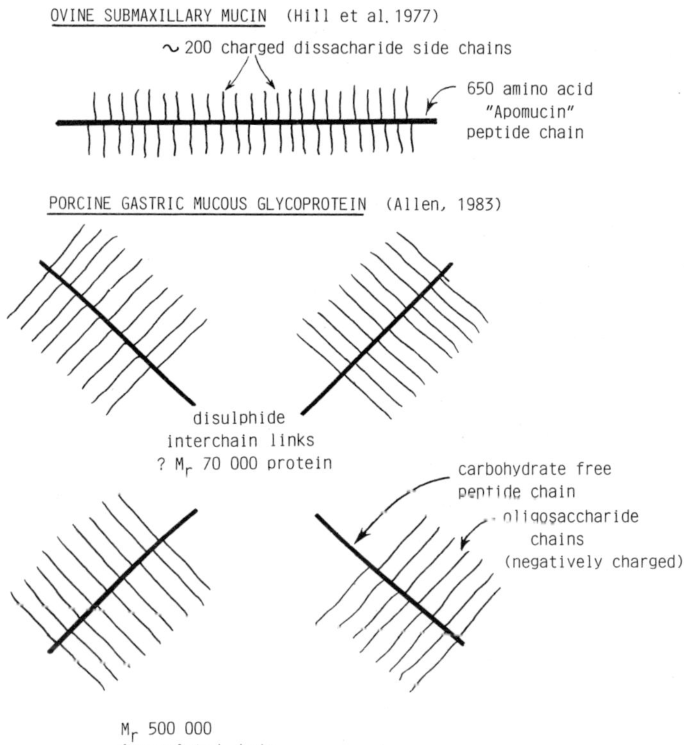

Fig. 2.6. Models for the structural organisation of mucous glycoproteins. Four glycosylated polypeptides in the gastric mucous glycoproteins are thought to be linked by disulphide bonds; a peptide of M_r 70 000 is also present but its role is not yet understood.

Two general patterns of molecular organisation have been identified in mucous glycoprotein and these are illustrated in Fig. 2.6. The simpler type of arrangement occurs in ovine submaxillary mucin, which has a single peptide chain of 650 amino acids carrying about 200 disaccharide chains, giving a monomer molecular weight of 154 000. The mucin self-associates by non-covalent protein-carbohydrate or protein-protein interactions in a concentration- and ionic-strength-dependent manner to form aggregates of molecular weight greater than $0.5 \cdot 10^6$ (Hill et al., 1977). More complex structures in which glycosylated peptide chains are linked through disulphide bridges have been proposed for other mucous glycoproteins. A structure suggested for pig gastric mucous glycoprotein (Allen, 1983) is shown in Fig. 2.6. The structural unit is made up of four glycosylated peptide chains each of molecular weight $0.5 \cdot 10^6$ and one non-glycosylated protein (molecular weight 70 000). Sections of the glycosylated chains which are devoid of carbohydrate appear to be involved in interchain disulphide bonds. The role of the non-glycosylated protein is unclear. Similar patterns of structural organisation but with differing numbers and sizes of subunits probably occur in other mucous glycoproteins (Allen, 1983).

The remarkable physical properties of mucous glycoproteins depend on the integrity of both their protein and carbohydrate moieties. Removal of carbohydrate from ovine submaxillary mucin by exoglycosidases produces an 'apomucin' which has little tendency to self-associate (Hill et al., 1977) and which is globular in shape rather than being thread-like (Rose et al., 1984). On the other hand, the viscosity of gastric mucin can be drastically modified by cleaving the peptide chain with proteases or by reduction of disulphide bonds.

2.6.2. Serum glycoproteins

The glycoproteins of mammalian blood serum (and plasma) have varied functions, acting as carriers, hormones, antibodies, blood clotting factors and components of the complement system. Their properties and activities arise primarily from their protein moieties.

However, the oligosaccharide units of serum glycoproteins may have a general role in the clearance of these molecules from the circulation, contribute to resistance to proteolysis, and have more specific functions in processes such as complement activation. Serum glycoproteins, unlike mucins, generally contain less carbohydrate than protein. Sugar residues which may be present in these molecules include *N*-acetylglucosamine, mannose, galactose, sialic acid, fucose and *N*-acetylgalactosamine. The type of protein-carbohydrate linkage found most commonly is the *N*-glycosidic linkage involving the side chain of asparagine. This GlcNAc-Asn bond is sometimes described as the 'serum type' of linkage. However, the GalNAc-Ser/Thr linkage does also occur in some serum glycoproteins. The *N*-linked carbohydrate units of mammalian serum glycoproteins usually have the 'complex type' of structure (Section 2.2.4), and one or more of these units may be attached to a single peptide chain. Where several carbohydrate units occur they may be widely spaced in the amino acid sequence, in contrast to the clustering of carbohydrate units in mucous glycoproteins. The carbohydrate units of serum glycoproteins occur at the interface between the surface of the folded protein and the surrounding solvent. The carbohydrate groups evidently project like antennae into the solvent, as they can be bound by cell surface receptors if they have the terminal sugar residues required for recognition. However, there is also evidence from X-ray crystallography (Deisenhofer, 1981; Sutton and Phillips, 1983) that non-covalent interactions occur between carbohydrate and amino acid residues at the surface of the protein. These protein-carbohydrate interactions may affect the accessibility of the peptide chain to macromolecules (e.g. proteases) and they may have some influence on protein conformation.

2.6.3. Structural glycoproteins

Collagens from a variety of connective tissues and basement membranes have been shown to contain carbohydrate units *O*-glycosidically linked to the side-chain hydroxyl of hydroxylysine. The linkage sugar is galactose and the carbohydrate units contain either a single galactose or the disaccharide Glc(β1–2)Gal. The number of oligosaccharide units per peptide chain varies with the source of collagen but the role of these units is unknown.

The hydroxyproline-rich extensins of plant cell walls are also glycoproteins. In those proteins, however, the side chain of hydroxyproline is linked to arabinoside chains, while serine side-chains are *O*-glycosidically linked to galactose residues.

2.6.4. Membrane glycoproteins

Membrane glycoproteins act as transport systems, light-transducing agents, antigens and receptors. These molecules are characteristically integral membrane components with peptide chains firmly associated with, and often spanning, the lipid bilayer. Carbohydrate units are located at only one side of the membrane, the extracellular surface of the plasma membrane and the cisternal surface of endoplasmic reticulum. The proportion of the molecule which is carbohydrate varies from below 10% to more than 60%. Carbohydrate units can

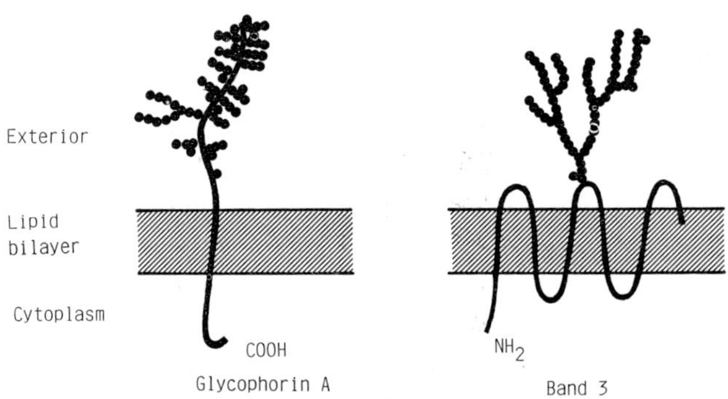

Fig. 2.7. Organisation of membrane glycoproteins. The general relationship of the peptide chains of two human erythrocyte membrane glycoproteins to the lipid bilayer is indicated. Folding of peptide chains, and carbohydrate structures, are not intended as precise representations. Transmembrane sections of peptides may be helical and, in band 3, may be associated. Monosaccharide residues are represented by filled circles.

be attached to protein by the GalNAc-Ser/Thr and/or GlcNAc-Asn linkages.

The number of membrane proteins for which there is complete information on primary sequence (of protein and carbohydrate) and membrane disposition is still small and the molecules which have been examined may not necessarily be representative. Fig. 2.7 shows the membrane orientation of two human erythrocyte membrane glycoproteins, glycophorin and band 3, which have been intensively studied. Human glycophorin A spans the bilayer once, has a highly glycosylated section of peptide close to the *N*-terminus which carries 1 *N*-linked and 15 *O*-linked carbohydrate units. In contrast, band 3, which functions as an anion transporter, has a peptide chain which spans the bilayer several times but glycosylation probably occurs at only one site (Drickamer, 1977). The carbohydrate unit has an unusual structure (see Section 2.4.1) and carries ABO blood group determinants. The cytoplasmic part of band 3 can participate in non-covalent interactions with cytoskeletal proteins of the erythrocyte.

2.7. *Proteoglycans and glycosaminoglycans*

The proteoglycans are a class of glycoproteins found in animal connective tissues. Their common structural characteristic is that they contain linear polysaccharide chains which have a repeating unit containing an amino sugar derivative. These carbohydrate chains are known as glycosaminoglycans. All of the glycosaminoglycans carry a high negative charge at neutral pH. Their repeating units (Table 2.2) contain uronic acid (except for keratan sulphate) or sulphate substituents (present in all but hyaluronate) or both types of charged group. This high charge has a marked effect on the physico-chemical properties of these molecules. Proteoglycans usually contain many glycosaminoglycan chains linked to a protein core. Protein and carbohydrate are covalently bonded by Xyl-Ser linkages, except for keratan sulphate chains, which have GalNAc-Ser/Thr or GlcNAc-Asn linkages. The repeating structure of the glycosaminoglycan chains is

TABLE 2.2

Periodic structure of proteoglycan carbohydrate units. These tetrasaccharides are intended to summarise types of periodic sequences found in proteoglycans. The substituents indicated in parenthesis are not always present (e.g. IdoA is not necessarily sulphated in dermatan sulphate) and the amount of iduronic acid present in the hybrid polymers heparin, heparan sulphate and dermatan sulphate is variable and may occur in blocks.

Hyaluronic acid	–4)GlcA(β1–3)GlcNAc(β1–4)GlcA(β1–3)GlcNAc(β1–
Chondroitin sulphates	–4)GlcA(β1–3)GalNAc(β1–4)GlcA(β1–3)GalNAc(β1– GalNAc: 4 or 6 OSO_3^-; GalNAc: 4 or 6 OSO_3^-
Keratan sulphates	–4)GlcNAc(β1–3)Gal(β1–4)GlcNAc(β1–3)Gal(β1– GlcNAc: 6 OSO_3^-; GlcNAc: 6 OSO_3^-; Gal: (6 OSO_3^-)
Heparin, heparan sulphate	GlcN: (6 OSO_3^-); GlcN: (6 OSO_3^-) –4)IdoA(α1–4)GlcN(α1–4)GlcA(β1–4)GlcN(α1– IdoA: (2 SO_3^-); GlcN: SO_3^- or Ac; GlcN: SO_3^- or Ac
Dermatan sulphate	–4)IdoA(α1–3)GalNAc(β1–4)GlcA(β1–3)GalNAc(β1– IdoA: (2 OSO_3^-); GalNAc: 4 OSO_3^-; GalNAc: 4 OSO_3^-

modified in the vicinity of the protein-carbohydrate linkage (see linkage region for Xyl-Ser linkages in Fig. 2.4).

Methods developed for the extraction of glycosaminoglycans from connective tissues have usually involved treatment with proteases or alkali, reagents which cleave peptide chains or Xyl-Ser protein-carbohydrate linkages, respectively. Thus the glycosaminoglycans are (with the possible exception of hyaluronate) degradation products of proteoglycans. Proteoglycans are usually named after their source and the predominant glycosaminoglycan derived from them (e.g. bovine nasal cartilage chondroitin 4-sulphate proteoglycan). Structural studies of intact proteoglycan molecules from several sources indicate that different types of glycosaminoglycan chain can be attached to a single polypeptide and that large numbers of oligosaccharide units of the 'glycoprotein' type may also be attached to the same core protein. In cartilage, it has been shown that proteoglycan chains are non-covalently associated with hyaluronate and 'link protein' to produce aggregates of very high molecular weight. This type of structure is illustrated in Fig. 2.8A, which shows the arrangement of carbohydrate chains in the major proteoglycan from vertebral cartilage (Hardingham et al., 1983). A single peptide chain (molecular

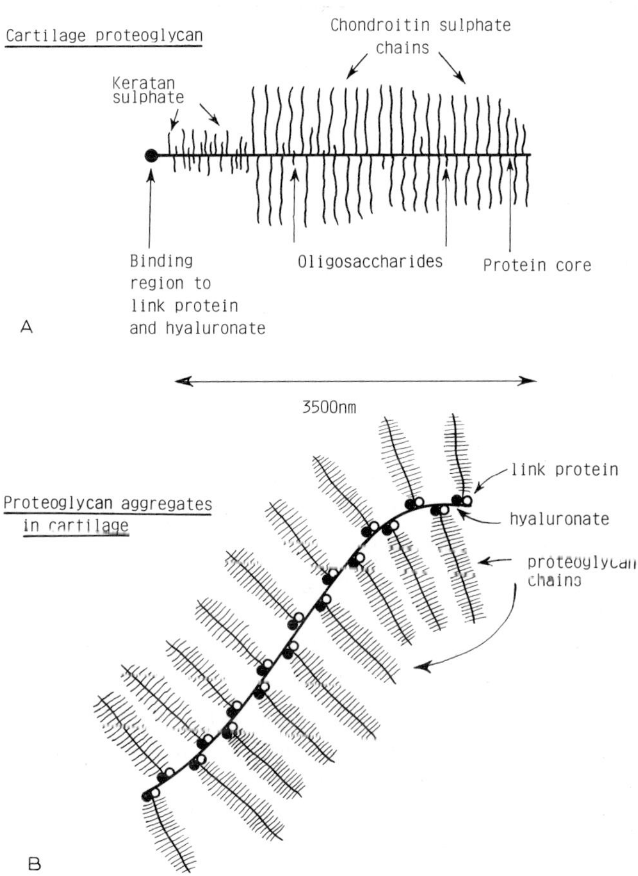

Fig. 2.8. Structural organisation of cartilage proteoglycan.

weight about 300 000) is densely substituted with chains of keratan sulphate, chondroitin sulphate and with short 'glycoprotein type' oligosaccharides. The different types of glycosaminoglycan chains are at least partially segregated along the length of polypeptide chain. One end of the peptide chain forms a globular structure that is relatively low in carbohydrate. This section of the molecule participates in intermolecular interactions with link proteins and hyaluronate forming the multimolecular aggregates illustrated in Fig. 2.8B.

2.8. Variability of carbohydrate structure

The polypeptide chains of glycoproteins (and presumably proteoglycans) are synthesised and processed to produce unique amino acid sequences of precisely defined length. In contrast, the biosynthesis and processing of the carbohydrate units of these molecules usually lead to a mixture of products showing appreciable structural variability (Montgomery, 1972; Beeley, 1974; Hatton et al., 1983). This can be detected by applying techniques which are sensitive to the distribution within the population of molecules of properties such as charge and size. Homogeneous glycoproteins and proteoglycans generally show behaviour characteristic of paucidispersity or polydispersity rather than the monodispersity characteristic of a pure protein containing only molecules of identical structure (Section 4.2). Structural variability can also be detected by the separation and structural analysis of fragments of glycoproteins or proteoglycans such as glycopeptides or glycosaminoglycans (Chapter 6).

The variation in carbohydrate primary structure between molecules of a glycoprotein is known as microheterogeneity and can take several forms. The carbohydrate units attached at a specific site on the peptide chain of a glycoprotein can vary in chain length with, for example, variation in the number of terminal sialic acid residues, or there may be variation in the number of branches (e.g. di-, tri-, tetra-antennary structures of *N*-linked chains), or alternative patterns of monosaccharide linkages may be present*. Two of the seven structures which can be found at the single protein-carbohydrate linkage site in ovalbumin are given in Fig. 2.5.

It has not yet been possible to make a comparable analysis of the glycosaminoglycan chains attached at specific loci on the protein cores of proteoglycans. However, it is clear that variations can exist within a population of glycosaminoglycan chains in the number of repeating units, the extent of sulphation, the distribution of sulphated residues along the polysaccharide chains and the number and distribution of iduronate residues (Kennedy, 1979; Rodén, 1980). This variation in primary sequence has also been described as microheterogeneity.

The occurrence of microheterogeneity in glycoproteins and proteoglycans creates problems in establishing criteria for assessing the homogeneity of preparations of these molecules (see Chapter 3) and in determination of the structures of their carbohydrate units (Chapter 6). The possible significance of glycoprotein heterogeneity has been discussed (Beeley, 1974; Hatton et al., 1983).

In addition to the microheterogeneity originating in the biosynthesis of glycoproteins and proteoglycans, enzymic or chemical degradation of these molecules can occur during their extraction and isolation. Measures for avoiding artifacts of this nature are discussed in Chapter 3. It should also be borne in mind, in carrying out experimental investigations of glycoprotein microheterogeneity (e.g. by isoelectric focusing), that some structural heterogeneity may also be present in the peptide chains of these molecules. Molecular species differing in charge can arise because of deamidation of asparagine or glutamine side-chains or because of secondary modifications such as phosphorylation, sulphation, etc. Amino acid substitutions have also

* Some authors restrict use of the term microheterogeneity to variation in the peripheral residues of glycoproteins (Montgomery, 1972).

been found in several glycoproteins, particularly where purification has been carried out on material pooled from several individuals.

2.9. *Nonenzymic glycoslation*

Mammalian extracellular proteins are exposed to appreciable concentrations of reducing sugars, particularly glucose, during their life-span. Reducing sugars can react nonenzymically with free amino groups of amino terminal residues or lysine side-chains. The initial product is a Schiff's base which can undergo Amadori rearrangement to form a stable ketoamine (Fig. 2.9). As a result of these reactions

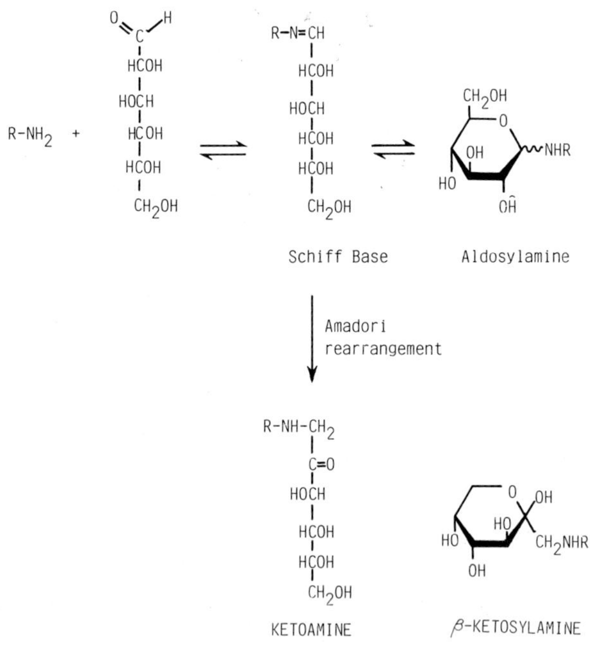

Fig. 2.9. Nonenzymic reaction of glucose with amino groups of proteins. R-NH_2 represents either an amino-terminal amino acid residue or the side chain of a lysine residue.

with glucose many proteins contain small quantities of stably bound monosaccharide. The amount bound depends on the period over which the protein is exposed to reducing sugar, and the proteins in which nonenzymic glycosylation has been detected, which includes haemoglobin, collagen, lens crystallin and serum albumin, tend to have quite long survival times. The extent of glycosylation also depends on sugar concentration. Nonenzymic glycosylation of haemoglobin and other proteins increases in uncontrolled diabetes melitus and there has been great interest in the measurement of glycosylated haemoglobins as a means of assessing control of hyperglycaemia.

The best-characterised nonenzymically glycosylated protein is haemoglobin Al_c. This modified haemoglobin has glucose bound (as ketoamine) to the amino-terminal valine residue of the β chain. Other glycosylated haemoglobins arise from reaction of amino-terminal residues with glucose 6-phosphate. Glucose is also linked as the ketoamine to lysine residues of haemoglobins.

Thorpe and Baynes (1982) have reviewed nonenzymic glycosylation and methods for detection and quantitation of this type of modification.

2.10. Semi-synthetic glycoproteins (neoglycoproteins)

It is possible to design and synthesise artificial glycoproteins by attaching carbohydrate units covalently to proteins by chemical methods. Either naturally occurring carbohydrate units isolated from fragmented glycoproteins, or oligosaccharides, or monosaccharide derivatives may be employed. Proteins to which simple sugar derivatives have been covalently bound have been called neoglycoproteins. Synthetic glycoproteins have been used to study the functional role of carbohydrate units and to raise antibodies directed against carbohydrate determinants.

The preparation and uses of synthetic glycoproteins have been reviewed (Lee and Lee, 1982) and will not be discussed further here.

2.11. *Relationships between glycoconjugates*

The glycoproteins (including proteoglycans) and glycolipids are collectively known as the glycoconjugates. Similarities in structure, location and functions exist between different categories of glycoconjugates. For example, common features are shared by glycolipids and membrane glycoproteins. Both types of molecule occur asymmetrically located in mammalian cell membranes and it has been suggested that their carbohydrate chains may participate in similar cellular processes. Both types of molecule (and mucous glycoproteins) can carry blood group determinants. There are structural similarities in terminal carbohydrate sequences of glycolipids and membrane glycoproteins which may indicate that the biosynthesis of these peripheral residues involves common glycosyl transferases (Rauvala and Finne, 1979). Common structural features occur in glycoproteins and proteoglycans including the oligosaccharide units attached to proteoglycans and the linkage regions of the keratan sulphates. There is much to be said therefore for avoiding rigid compartmentalisation of ideas about different members of the family of glycoconjugates.

CHAPTER 3

Isolation and fractionation

3.1. Introduction

The purification of glycoproteins or proteoglycans from tissues or cells involves their extraction in a soluble and preferably undegraded form, followed by the application of fractionation procedures to remove contaminants. Progress of the purification is monitored by carrying out appropriate assays to determine the specific activity or some similar parameter after each separation step. Finally, the homogeneity of the product is usually assessed on an analytical rather than a preparative scale by the application of further fractionation methods such as gel electrophoresis under denaturing and non-denaturing conditions.

The type of purificaton strategy employed can be varied according to the purpose of the investigation. If it is intended to study the biological activity or physico-chemical properties such as viscosity it is essential to avoid conditions of extraction or fractionation which modify these characteristics irreversibly. When the aim is to prepare material for investigation of primary structure it may be permissible to use methods which modify the folding of proteins (e.g. phenol extraction) but not conditions which could lead to the modification of the covalent structure. For example, exposure to alkali can cause cleavage of *O*-glycosidic bonds between protein and carbohydrate in glycoproteins and proteoglycans. Acidic conditions can lead to hydrolysis of susceptible glycosidic linkages between sugars, especially those involving sialic acid. However, where the interest is solely in the structure of the carbohydrate parts of glycoproteins or proteoglycans the use of degradative methods such as protease treatment can facilitate solubilisation of carbohydrate moieties.

Many glycoproteins occur in solution in fluids such as plasma, culture medium or egg white and do not require extraction. The presence of a high carbohydrate content is often associated with ready solubility. However, solubility difficulties are often encountered with highly glycosylated mucins as a consequence of intermolecular association as well as with glycosylated fibrous proteins. Membrane glycoproteins also require special extraction conditions for solubilisation (Section 3.3.2). Proteoglycans exist in connective tissues largely as macromolecular complexes which have unique solubility properties (Section 3.3.3).

The fractionation methods which can be applied to glycoproteins and proteoglycans (Section 3.4) include the traditional protein and carbohydrate techniques of selective precipitation, ion-exchange chromatography, zone electrophoresis and gel filtration. For molecules of high carbohydrate content the density may differ sufficiently from that of protein to allow buoyant density gradient ultracentrifugation to be employed as a preparative method (Section 3.4.5). Recently considerable use has been made of affinity chromatography on insolubilised lectins (Section 3.4.7 and Chapter 7) for the isolation of glycoproteins.

Purification using a succession of methods depending on a wide range of different molecular properties should be continued until further fractionation produces no change in specific activity and the preparation appears homogeneous when examined by sensitive analytical techniques (Chapter 4). The inherent variation in the carbohydrate structure of glycoproteins and proteoglycans can result in a single family of molecules having a fairly wide spread of size, charge or lectin affinity. Great care is required in distinguishing such microheterogeneity from the existence of molecular types differing in their peptide moieties. Fractionation methods which have proved especially useful for the separation of glycoprotein sub-populations which differ only in their carbohydrate moieties include isoelectric focusing (Section 3.4.4.3) and lectin affinity chromatography (Section 7.4).

Glycoprotein fractionation techniques have been reviewed (Green, 1981; Horowitz, 1977), isolation of mucin is discussed by Creeth (1978) and by Allen (1981). Methodology for the isolation of glycosaminoglycans has been described by Rodén et al. (1972) and the purification of glycosaminoglycans and proteoglycans was reviewed by Kennedy (1979).

3.2. *Choice of starting material*

It is possible to isolate glycoproteins and proteoglycans from a wide range of sources. However, careful choice of starting material can often simplify the problem of purification. As well as selecting material which is readily available and contains a high concentration of the glycoprotein, possible contamination with degradative enzymes (proteases and glycosidases) should be considered. The presence of a series of degradation products together with undegraded material greatly increases the difficulty of purification. For this reason sources which are contaminated with proteases, glycosidases or with bacteria should be avoided whenever possible. For example, the isolation of glycoproteins from whole saliva is complicated by the presence of bacterial enzymes capable of rapidly removing sialic acid from salivary mucin. Use of alternative sources of material such as animal submaxillary or submandibular glands or collection of the sterile secretions of individual salivary glands can facilitate purification of undegraded glycoproteins. The addition of sodium azide (0.01%) to inhibit bacterial growth in glycoprotein extracts is advisable.

Isolation of glycoproteins from a single gland still represents the product of several cell types. In some cases it may be possible to isolate glycoprotein from a single type of cell grown in culture. The quantity of material obtainable from cell culture will necessarily be limited but radioactive labelling techniques (Chapter 8) can be employed to increase the sensitivity of detection.

Genetic variation in glycoproteins can be minimised by examination of glycoprotein from a single (preferably homozygous) individual.

3.3. Extraction and solubilisation

Many glycoproteins, such as those which occur in serum, are freely soluble in dilute salt solutions and will remain in solution unless the protein moiety is denatured. Indeed some glycoproteins with high carbohydrate contents will remain in solution after the protein component has been denatured by heating; this property was the basis for some early preparative methods. The solubility of some glycoproteins (e.g. carcino-embryonic antigen) in protein precipitants such as perchloric acid has been made use of in some preparative methods. However, the strongly acidic conditions involved will promote hydrolysis of sialic acid residues and increase the heterogeneity of the preparation. In general it is advisable to avoid strong heating or acidic conditions which might modify covalent structure and alkaline conditions which can produce cleavage of *O*-glycosidic linkages in glycoproteins and proteoglycans (Chapter 5).

The extraction and solubilisation of the viscous mucins of epithelial secretions, membrane glycoproteins and proteoglycans will be considered in the following sections.

3.3.1. Mucous glycoproteins

The glycoproteins of epithelial secretions have visco-elastic properties which create difficulties in the application of purification techniques. However, these physico-chemical properties are fundamental to the biological function of mucins. The application of degradative methods which disrupt the structure of the mucins or their intermolecular interactions are undesirable when the objective is to examine the hydrodynamic properties of these molecules. More drastic methods are permissible when the aim is to determine the primary structure of protein or carbohydrate units.

Mucous glycoproteins can be isolated from mucous secretions or from mucus-producing glands or cells. Gastrointestinal mucus can be removed from the tissue surface by scraping after prior washing to remove extraneous material. Some contamination with material (mainly protein with some DNA) from damaged epithelial cells is liable to occur (Allen, 1981). Sources of mucus which is not contaminated with proteases, glycosidases or bacteria should be used whenever possible.

3.3.1.1. Centrifugation and homogenisation An initial separation of mucin into a clear supernatant sol and a more viscous gel phase can often be obtained by centrifugation. It may be possible to solubilise the gel by the application of shearing forces through homogenisation or prolonged stirring.

Snary and Allen (1972) have described the isolation of an in vitro labelled glycoprotein from scrapings of pig gastric mucosa. After dialysis against distilled water and centrifugation at 6 000 *g* for 20 min about 22% of the glycoprotein remained in solution. Of the water-insoluble glycoprotein gel 80% could be solubilised by other extraction procedures. Homogenisation of pig gastric mucin in 0.2 M NaCl (containing 0.01% sodium azide) using a Waring blender was effective in solubilising the gel phase of the mucin and the properties of the glycoprotein obtained were identical with those of the water-soluble glycoprotein (Robson et al., 1975). Bovine submaxillary mucin can be solublised from gland tissue by homogenisation in a Waring blender in 0.01 M NaCl (Hill et al., 1977).

The glycoprotein of sputum can be separated into sol and gel phases by centrifugation at 45 000 *g* for 1.5 h at 4°C (Roberts, 1974, 1976). Serum proteins were detected in the supernatant phase as well as glycoprotein of high molecular weight. Repeated washing of the gel phase was required to remove contaminants.

Bhushana Rao et al. (1973) obtained cervical mucus from cows in the oestrus period; at this time the mucus is abundant and of low viscosity. After lyophilisation the mucin was stirred overnight to 'fluidify' it. Effective solubilisation and reduction of the viscosity of mucus glycoproteins often requires the use of chaotropic agents or reduction of disulphide bonds.

3.3.1.2. Sonication Solubilisation of mucins can be obtained by a brief period of sonication (Ryley and Brogan, 1968). However, the possibility that free radicals generated by cavitation may degrade the carbohydrate and peptide chains make this procedure unattractive except perhaps for analytical procedures (Creeth, 1978).

3.3.1.3. Extraction with salts and urea Ovine submaxillary mucin can be solubilised from gland tissue in 0.01 M NaCl (Hill et al., 1977). However, in most cases it is necessary to employ high salt concentrations, chaotropic agents or protein denaturants such as urea or guanidinium chloride to extract mucins.

High concentrations of calcium and lithium salts have been used to extract gel-forming glycoproteins (Gibbons and Selwood, 1973). The salts of chaotropic anions (such as thiocyanates) are also effective solubilising agents (Bushana Rao et al., 1973). Strong chaotropes are liable to cause protein denaturation, particularly at high concentration. However, they may sometimes be employed at moderate concentrations. Khan et al. (1976) found that 0.2 M sodium thiocyanate was an effective solubilising agent for tracheal mucus and had less effect on the storage modulus of the gel than 6 M urea or 2 M guanidinium chloride.

Urea and guanidine salts have been used as dispersing agents for several types of mucus. In some cases solubilisation is time-dependent (Roberts, 1974). Extensive denaturation of the protein components of mucins occurs in concentrated solutions of these reagents. Removal of denaturant by dialysis may not always restore the conformation of mucins to their native state (Snary et al., 1974). These reagents must be used with caution if the goal is to study the non-covalent interactions of mucin but they will not affect the primary structures of protein or carbohydrate.

3.3.1.4. Phenol extraction Lyophilised cyst fluid has been extracted with phenol/water, 19:1 w/v, to solubilise protein. The residue is extracted with water and can be fractionally precipitated with ethanol. This procedure has been used to produce gram quantities of blood-group glycoproteins (Morgan, 1967).

With other secretions the phenol-insoluble material is not water-soluble and the carbohydrate groups can be solubilised only by enzyme treatment. Phenol treatment is likely to lead to protein denaturation. When DNA is present in the material to be extracted it may not be effectively removed by the phenol extraction procedure (Creeth, 1978).

3.3.1.5. Reducing agents Gelatinous mucins can, in many cases, be solubilised and made less viscous by the addition of reagents which reduce disulphide bonds. Dithiothreitol is highly effective but thiol-containing reagents such as *N*-acetylcysteine or mercaptoethanol can also be employed.

The effect of this treatment is to cleave disulphide bridges in proteins and glycoproteins present in the mucus. Disulphide linkages probably occur in mucins from gastrointestinal (Starkey et al., 1974), urino-genital (Gibbons and Selwood, 1973) and bronchial mucins (Roberts, 1976). The use of reducing agents can therefore lead to the isolation of mucin peptide chains differing from the intact molecule in molecular weight and physico-chemical properties. Some recent work has, however, cast some doubt on the role of disulphide linkages in bronchial mucin (Houdret et al., 1981).

Roberts (1976) carried out reduction of bronchial mucus glycoprotein with 0.1 M dithiothreitol at pH 8 for 1 h at room temperature either without denaturants or in the presence of 6 M urea or 6 M guanidinium chloride (under nitrogen). Sulphydryl groups were then alkylated with 0.3 M iodoacetamide. It is generally advisable to carry out reduction and alkylation under strongly denaturing conditions to ensure complete modification of the sulphydryl groups. There is a danger that the addition of thiol-reducing agents under non-denaturing conditions may lead to the activation of proteases present as contaminants in mucin preparations (Houdret et al., 1981).

Bushana Rao et al. (1973) reduced bovine cervical mucin under mild conditions (0.01 M dithiothreitol at pH 8.0 without denaturing agent) to decrease the viscosity sufficiently for gel chromatography to be undertaken. The pH was decreased to 6.5 to slow the reoxidation of sulphydryl groups. After removal of sialic acid a 4% solution of the cervical mucin was further reduced with 0.1 M dithiothreitol in 0.5 M Tris-HCl buffer, pH 8.2, containing 6 M guanidine-HCl for 90 min at 37°C. This material was then alkylated for 2 h at 37°C using 0.24 M iodoacetic acid in 1 M Tris buffer containing 4 M guanidine-HCl (Kandukuri et al., 1977). Radioactively labelled iodoacetate (or iodoacetamide) can be incorporated in this step.

3.3.1.6. Proteolysis Solubilisation of components of mucus can usually be accomplished by proteolysis. Digestion with pepsin was employed in the isolation of blood-group-specific glycoprotein from gastric mucosa and saliva (Kabat, 1956).

However, in addition to cleaving the unwanted protein components of secretions proteases are also likely to split those parts of the peptide chain not protected by heavy glycosylation. Snary and Allen (1972) digested pig gastric mucin with pepsin, papain, ficin and Pronase and found the latter enzyme to be most effective in solubilising the glycoprotein. Solubilised glycopeptides were separated from intact glycoprotein by density gradient ultracentrifugation in CsCl.

The fragments of glycoproteins obtained by proteolysis may be suitable for structural studies on their carbohydrate moieties but are obviously unsuitable for other purposes because of the degradation of the polypeptide chain.

3.3.2. Membrane glycoproteins

Membrane proteins can be classified on the basis of the conditions required to extract them from membrane (Marchesi et al., 1976). Proteins solubilised by variation in parameters such as pH and ionic strength are termed peripheral proteins because they are presumed to be associated with one or other of the membrane surfaces. In contrast, integral proteins are firmly associated with the hydrophobic lipid bilayer and their solubilisation requires disruption of the bilayer by detergents, strong chaotropes or organic solvents. Many glycoproteins fall into this latter class of polypeptide.

The strategy employed for the isolation of a particular membrane glycoprotein depends on whether the aim is the analysis of primary structure or the study of properties depending on protein conformation. Carbohydrate structure can be studied on glycopeptides released from cells or membrane preparations by proteolytic digestion. Extraction of membranes with strong chaotropes, organic solvents and certain ionic detergents is liable to cause denaturation and loss of function associated with the peptide chain. The extensive use of non-ionic detergents and bile salts for glycoprotein extraction has arisen because these reagents have relatively little effect on protein conformation and hence biological activity is often retained.

Many extracted integral membrane glycoproteins have a strong tendency to aggregate. One of the causes of such aggregation is interaction between the hydrophobic segments of the peptide chain which are normally buried in the lipid bilayer of the membrane. However, protein denaturation during extraction, for example by organic solvents, can increase aggregation by exposing further hydrophobic sections of the peptide chain.

Some extensively glycosylated membrane glycoproteins (e.g. glycophorin) become soluble in buffered salt solution after extraction from their lipid environment. However, in the majority of cases, especially when the carbohydrate content is not very high, it is necessary to have detergent present to maintain solubility. The detergent used may have an influence on the subsequent method of isolation. However, methods including gel filtration, affinity chromatography (including lectin affinity) and isoelectric focusing have been successfully applied in the presence of detergent. In addition SDS-gel electrophoresis has been widely applied to the characterisation of membrane glycoproteins.

Before attempting to extract integral membrane glycoproteins it is useful to remove some of the peripheral proteins from the membrane preparation. This can be accomplished by mild procedures such as extraction at high or low ionic strength or the use of chelating agents. The extraction of extrinsic membrane proteins has been reviewed by Tanner (1979).

General reviews of the extraction of glycoproteins from membranes include those of Hughes (1976), Juliano (1978) and Tanner (1979).

3.3.2.1. Detergents The solubilisation of membranes by detergents has been reviewed by Helenius and Simons (1975) and the characterisation of membrane protein in detergent solutions by Tanford and Reynolds (1976).

The detergents employed for the solubilisation of membrane glycoproteins can be divided into three categories. First the ionic (anionic and cationic) surfactants which have a long non-polar chain (e.g. sodium dodecyl sulphate); second, non-ionic surfactants (e.g. Triton X-100) and third, bile salts (e.g. sodium deoxycholate). Each type of detergent has particular properties which govern their interaction with lipids and proteins and hence their use in the solubilisation of membne glycoproteins.

In aqueous solutions of surfactants an equilibrium exists between monolayer, monomer and micellar forms. The concentration of free monomer is determined by the critical micellar concentration (CMC). Both temperature and (particularly for ionic detergents) the ionic strength influence the CMC. Generally ionic detergents such as sodium dodecyl sulphate have higher CMC values than non-ionic detergents at room temperature. The micellar size of non-ionic detergents such as Triton X-100 is, however, much larger than that of SDS. This has the important practical consequence that Triton X-100 cannot readily be removed from a protein by dialysis because the micelles are too large to pass through a dialysis membrane. The bile salts form small aggregates which can be separated from protein by dialysis.

The process by which membranes are solubilised by detergents can be considered in three stages (Helenius and Simons, 1975): (1) detergent binds to and is incorporated into the lipid bilayer, (2) a transition occurs from the lamellar membrane phase to mixed micelles of membrane phospholipid and detergent, (3) the size of mixed micelles decreases with increasing ratio of detergent to phospholipid. From this description of the process it is apparent that the ratio of detergent to membrane is a crucial factor in membrane solubilisation.

There are pronounced differences in the ways that classes of detergents interact with proteins. The high detergent monomer concentrations which can be obtained with ionic detergents such as SDS can induce cooperative binding of large amounts of detergent and unfolding and denaturation of most proteins. In contrast, soluble globular proteins usually have few, if any, sites at which non-ionic detergents such as Triton X-100 will bind. Non-ionic detergents have little tendency to denature proteins or to influence protein-protein interactions. They are therefore of great value in solubilising membrane glycoproteins where it is important to maintain protein function or when the isolation procedure employs techniques such as antibody precipitation or lectin affinity chromatography. Bile salts (e.g. sodium deoxycholate) are considerably less prone to denature proteins than is SDS but can in some cases cause denaturation (Lotan et al., 1977).

Binding of detergent to peripheral membrane proteins has not been extensively studied but would be expected to resemble detergent binding to soluble globular proteins. Solubilisation of integral membrane glycoproteins can generally be achieved using the denaturing detergent SDS. Detailed binding studies on a few glycoproteins indicate that the quantitative binding of SDS to membrane glycoproteins may differ from that of globular proteins (Grefrath and Reynolds, 1974). Studies of the binding of Triton X-100 to Semliki Forest virus glycoprotein show that large amounts of detergent are bound, presumably to the hydrophobic transmembrane region (Utermann and Simons, 1974).

3.3.2.2. Factors affecting detergent extraction of membrane glycoproteins The efficiency of extraction of membrane glycoproteins by non-ionic detergents depends on several experimental variables including the particular detergent chosen (Letarte-Muirhead et al., 1975), the ratio of detergent to membrane, temperature, time, number of extractions, the ionic strength, the nature of the ionic species present and pH. Letarte-Muirhead et al. (1975) examined the effects of a series of different detergents on the activity and solubilisation of a cell surface antigen and found marked differences in both parameters. Butters and Hughes (1974) examined the solubilisation of metabolically labelled glycoproteins from KB cell membranes (10 mg/ml) with concentrations of Triton X-100 between 0.01 and 5%. Optimal extraction (of about 70% of the labelled carbohydrate) was obtained with 0.5% Triton. Membrane glycoproteins were extracted more readily than labelled membrane protein as a whole. Approximately 65% of the membrane lipid was extracted under these conditions. Increased extraction of glycoproteins occurred at 37° compared with 4°C; however, the risk of proteolysis makes the use of a low temperature preferable.

Addition of detergents to membrane preparations can potentiate proteolytic activity and the addition of protease inhibitors (such as phenylmethylsulphonyl fluoride) prior to detergent extraction is an advisable precaution.

The contribution of the charge on membrane proteins to their solubilisation by non-ionic detergents has been demonstrated by Pratt and Cook (1979a,b), who studied the effects of pH on the solubility of components of human erythrocyte membranes. Ghosts (3 mg/ml) were extracted with Triton X-100 (1% v/v) in distilled water between pH 5.0 and 8.5. More than 80% of the sialoglycoprotein was solubilised by concentrations of Triton of 0.2% or greater in the range pH 6–7.5 but decreased extraction was obtained at higher and lower pH values. Following enzymatic removal of sialic acid the solubility of the membrane sialoglycoprotein was completely suppressed. Improved solubilisation was obtained using detergent in 50 mM phosphate. Still better extraction (91% of the hexosamine) of the glycoproteins from neuraminidase-treated erythrocytes was given by 1% v/v Triton in 50 mM borate buffer, pH 8.6. The authors suggest the use of non-ionic detergent in the presence of borate as a general procedure for the solubilisation of membrane glycoproteins deficient in sialic acid residues and applied the technique successfully to the extraction of macrophage membrane glycoproteins. They suggest that the charged borate-carbohydrate complex promotes solubilisation of glycoprotein in Triton micelles.

In addition to charge effects the solubilisation of membrane proteins by non-ionic detergents can also be influenced by intermolecular interaction between membrane components. Transmembrane glycoproteins may be anchored to the cytoskeleton of protein fibres that lie adjacent to the cell membrane and other membrane systems. The observation that only a fraction of the glycoprotein band 3 could be extracted from human erythrocyte membrane with buffered Triton X-100 (Sheetz, 1979), has been attributed to protein-protein interactions between band 3 and erythrocyte cytoskeletal proteins. It is obviously important to distinguish between the occurrence of such interactions and glycoprotein remaining associated with insoluble material because insufficient excess of detergent has been added to solubilise the lipid bilayer.

When establishing the best conditions for extraction of a membrane glycoprotein it is desirable to examine several detergents (including Triton X-100, Triton-borate and sodium deoxycholate) at a range of concentrations (say 0.1–2%). Extraction should be performed at a known ratio of detergent to membrane protein and at 0–4°C for 10–60 min. The extract is centrifuged at 100 000 *g* for 1 h to remove membranes and the amont of glycoprotein in the supernatant is determined. Repeated extraction can be employed to increase the yield. The effect of detergent on any assay for activity used should also be investigated.

Practical problems can arise because of batch-to batch variations in detergents with the presence of varying amounts of additives or water. For example, peroxides present in some batches of Triton X-100 can cause oxidation of methionine and free cysteine residues of proteins. Bile salts can, if necessary, be purified by crystallisation. Non-ionic detergents are more difficult to purify. Passage through a column of mixed-bed ion-exchanger removes ionised impurities (Pratt and Cook, 1979a).

The non-ionic detergents are often heterogeneous both in the chain length of the fatty acids and alcohols used in their synthesis and in their polyoxyethylene head groups (Helenius and Simons, 1975).

Sodium dodecyl sulphate is considerably more soluble than the potassium salt; the lithium salt which is still more soluble is now available commercially. The bile salts, which have ionised carboxyl groups, are more soluble than their corresponding acids. Thus sodium deoxycholate can be dissolved at a concentration of 1% w/v at pH 8.5 but on acidifying to about pH 7 a gel forms and at lower pH values deoxycholic acid is precipitated.

3.3.2.3. Chaotropes and protein denaturants Hatefi and Hanstein (1969) proposed the use of chaotropic anions, which favour the transfer of apolar groups to water, for the solubilisation of membranes. It is in the nature of such agents that they also tend to unfold and denature proteins. The protein-denaturants urea and guanidinium chloride, although probably not acting in the same way as chaotropic anions, are conveniently considered at the same time.

The chaotropic salt lithium di-iodosalicylate (LIS) has been used for the solubilisation of glycoproteins from the erythrocyte membrane (Marchesi, 1972) and from L cells (Hunt et al., 1975). Low concentrations (40 mM) of LIS have been used to extract peripheral proteins from erythrocyte membrane.

The effectiveness of LIS > guanidinium chloride > thiocyanate. In general urea and guanidinium chloride are not very effective for solubilisation of integral membrane proteins.

3.3.2.4. Organic solvents Various organic solvent systems have been employed to extract glycoproteins from erythrocytes. These include phenol-water mixtures (Springer et al., 1966; Howe et al., 1972; Marchesi, 1972), chloroform-methanol (Hamaguchi and Cleve, 1972), aqueous butanol (Maddy et al., 1972), aqueous pyridine (Blumenfeld and Zvilichovsky, 1973) and hexafluoroacetone (Juliano, 1972).

The chloroform-methanol and the LIS-phenol methods (Marchesi, 1972) give high yields of the erythrocyte sialo-glycoproteins in water-soluble form. Most of the other protein, and glycoprotein, components of the membrane are denatured and appear at the interface between aqueous and organic phases. Extraction with LIS-phenol has also been applied to solubilise glycoproteins from L cells (Hunt and Brown, 1975) and ascites cells (Rittenhouse et al., 1976). This type of extraction procedure with denaturing solvents seems likely to preferentially solubilise highly glycosylated components, but other glycoproteins may not survive.

Extraction with organic solvents is not usually advisable for studies of membrane protein function which require native protein. However, studies on the carbohydrate antigens of glycoproteins prepared in this way from cyst fluids have made an important contribution to our understanding of blood-group substances.

Extraction with organic solvents such as chloroform-methanol is sometimes applied to glycoprotein to remove traces of glycolipid which can be tenaciously bound.

3.3.2.5. Proteolysis Glycopeptides can be released from membrane glycoproteins by treatment with proteases. The size of proteolytic enzymes limits their action to the external surface of intact cells or vesicles. More specific proteases, such as trypsin, may remove only a selected range of glycopeptides from the cell surface, while treatment with non-specific enzymes such as Pronase removes much of the protein-bound carbohydrate of the membrane.

Winzler et al. (1967) digested intact human erythrocytes with trypsin and showed that one-third to one-half of the sialic acid of the cell was released as soluble glycopeptides. The glycopeptide fraction was purified and chemically characterised. Proteolysis with trypsin has been employed by Warren et al. (1974) to release labelled glycopeptides from the surface of cultured cells. The profile of glycopeptides from normal and transformed cells was examined by gel filtration following a second digestion with Pronase. Cummings and Kornfeld (1982b) employed Pronase to release glycopeptides from mouse lymphoma cells; the glycopeptides were then characterised by sequential lectin affinity chromatography (Section 7.4.5).

3.3.3. *Proteoglycans*

In the extraction of proteoglycans from connective tissue it is essential to maintain the primary structure intact. Conditions under which proteolytic cleavage of the protein core might take place and the use of alkaline extraction media which could degrade the protein-carbohydrate linkages should therefore be avoided.

Much of the proteoglycan of connective tissue is probably associated by non-covalent interactions with link proteins, hyaluronate, other proteoglycan molecules (Rosenberg et al., 1970) or collagen. The use of dissociative solvents (Sadjera and Hascall, 1969) which promote the dissociation of such interactions facilitate the extraction of soluble proteoglycan. The extent to which intermolecular interactions occur in the extract depends on the precise conditions which are employed. If the aim of the isolation procedure is to obtain free proteoglycan subunits from the tissue, mild non-dissociating conditions should be used. To isolate proteoglycans which are complexed in the tissue, extraction under highly dissociative conditions (e.g. with 4 M guanidinium chloride) may be appropriate. Other dissociative, non-denaturing extractants can also be used if proteoglycan complexes are being investigated.

Several studies (Brandt and Muir, 1971; Mayes et al., 1973) have shown that fractions of proteoglycan differing in the substitution of the protein core or the nature of their glycosaminoglycan chains can be obtained by sequential extraction or by using different extractants.

The conditions chosen for the extraction of proteoglycans are therefore an integral part of the strategy for their isolation.

3.3.3.1. Extraction conditions The surface area of connective tissue available for extraction can be increased by slicing, grinding or pulverising. Cartilage can be sliced with a Stanley surform (Sadjera and Hascall, 1969). Brandt and Muir (1971) pulverised cartilage by hammering in a steel die cooled in liquid N_2.

3.3.3.2. Effects of shearing forces and sonication Prolonged high-speed homogenisation of bovine nasal cartilage in water permits extraction of 80% of the proteoglycan (Malawista and Schubert, 1958; Sandson et al., 1966). However, material extracted in this way is degraded. Sadjera and Hascall (1969) showed that the average sedimentation coefficient of proteoglycan was decreased by the shearing forces during high-speed homogenisation. Proteases active at neutral pH have also been found in cartilage proteoglycans prepared by high-speed homogenisation (Seraffini-Fracassini et al., 1967).

Homogenisation under milder conditions may nevertheless be valuable in promoting the solubilisation of proteoglycan (Brandt and Muir, 1971).

Sonication may, in addition to shearing forces, give rise to free radicals which could potentially degrade protein or carbohydrate moieties of proteoglycans. Very marked reduction of *s* value occurs on sonication of proteoglycans (Sadjera and Hascall, 1969) and this technique should be avoided if intact proteoglycans are to be isolated.

3.3.3.3. Extractants Lucy et al. (1961) found that exposure of embryonic cartilage to water results in the release of lysosomal protease. Hence it is preferable to avoid hypotonic conditions for the extraction of proteoglycans, and protease inhibitors should be included during extraction (Oegema et al., 1975).

The effects of different concentrations of several salts on the extraction of proteoglycan from bovine nasal cartilage have been examined (Sadjera and Hascall, 1969; Hascall and Sadjera, 1970; Mason and Mayes, 1973; Rosenberg et al., 1970). The proportion of the tissue proteoglycan (usually measured as hexuronate) which can be extracted with 0.15 M KCl is about 20%. Maximal extraction (about 80% of the tissue hexuronate) was obtained with high concentrations of LiCl, $CaCl_2$, $MgCl_2$, $LaCl_3$ or guanidinium chloride. For the metal salts solubilisation increased to a maximum with increasing concentrations and declined at higher salt concentrations. Mason and Mayes (1973) found that there was a linear relationship between the concentration at which a particular metal salt is maximally effective in solubilising tissue proteoglycan and the enthalpy of hydration of the cation of the salt. Salt solutions which promote solubilisation, presumably as a consequence of dissociating intermolecular interaction between proteoglycans and other molecules in connective tissue, have been termed dissociative extractants (Sadjera and Hascall, 1969). High concentrations (4 M) of guanidinium chloride are very effective in solubilising proteoglycan. The action of this salt as a protein-denaturant is well known and its action on tissue proteoglycan complexes probably involves denaturation of link proteins with which proteoglycan is associated as well as denaturation of any secondary or tertiary structure which the proteoglycan subunit may contain. Urea is less effective in solubilising proteoglycan than guanidinium chloride but has a synergistic effect when added to salt solutions, decreasing the salt concentration producing maximal solubilisation. Sadjera and Hascall (1969) found that solubilisation in guanidinium chloride did not vary greatly in rate or extent between pH 6 and pH 9. Extraction at pH 5.8 resulted in optimal reaggregation after dilution of the guanidinium chloride (Hascall and Sadjera, 1970).

The time course of extraction of slices of bovine nasal cartilage with dissociative solvents (3 M $MgCl_2$ or 4 M guanidinium chloride) was examined by Sadjera and Hascall (1969). A period of 20 h at 25°C was found to be optimal but shorter extraction for other tissues should be considered. The presence of 4 M guanidinium chloride does not prevent proteolysis occurring in all tissues during extraction and it has become standard practice to add protease inhibitors (Oegema et al., 1975). An extraction mixture containing 4 M guanidinium chloride (ultrapure grade), 0.05 M sodium acetate, 0.01 M sodium EDTA, 0.1 M 6-aminohexanoic acid (Aldrich), 0.005 M benzamidine hydrochloride (Aldrich), pH 5.8, has been widely used.

The nature of the proteoglycan or proteoglycan complex which is extracted at lower ionic strengths (e.g. with 0.15 M KCl or 0.5 M guanidinium chloride) differs from that which is extracted under 'dissociating' conditions (e.g. into 2 M $MgCl_2$ or 4 M guanidinium chloride). Under non-dissociating conditions extraction of proteoglycan subunit predominates (Rosenberg et al., 1970; Emes and Pearce,

1975). Dissociating solvents such as $MgCl_2$ extract more proteoglycan in the form of high molecular weight aggregates which include link proteins (Emes and Pearce, 1975; Hascall and Sadjera, 1970). With 4 M guanidine the proteoglycan and link protein are extracted in dissociated form but can reassociate to form large aggregates on decreasing the concentration of guanidinium chloride. Sequential extractions under ‘non-dissociating’ and ‘dissociating’ conditions may produce fractions enriched respectively in proteoglycan subunit and proteoglycan aggregates (Royal et al., 1980).

However, there is abundant evidence that sequential extraction or extraction with different solvents can also lead to the fractionation of proteoglycans with different primary structures. Brandt and Muir (1971) extracted porcine articular cartilage repeatedly with 0.15 M sodium acetate, pH 6.8, and found that there was a progressive increase in glucosamine:galactosamine ratio and in protein content between proteoglycans isolated (by a precipitation method) from succeeding fractions. Mayes et al. (1973) extracted bovine nasal cartilage in media of different ionic strengths and found that the proteoglycan extracted with 0.15 M KCl was enriched in chondroitin sulphate but lower in keratan sulphate than higher ionic strength extracts. Variation in protein composition was also observed.

The amount and nature of protein extracted together with proteoglycan vary with the extractant. Larger quantities of collagen may occur in extracts of connective tissue with guanidinium chloride than with other extractants.

Much of the development of methods for extraction of proteoglycans has been conducted using bovine nasal cartilage. Other connective tissues differ considerably in the ease with which proteoglycan can be isolated. Emes and Pearce (1975) examined the extractability of proteoglycans from human intervertebral disc and found that the amount of proteoglycan (hexuronate extracted) seemed to be independent of the extractant used; 70–80% of the hexuronate was readily extracted even in 0.15 M KCl. The ease of extraction was attributed to the lower proportion of aggregate occurring in disc as compared to cartilage.

3.3.3.4. Reducing agents The viscosity of proteoglycan aggregates may be decreased by reduction. Reaggregation of proteoglycans, link proteins and hyaluronate can be inhibited by reduction and alkylation (Sadjera and Hascall, 1969).

3.3.3.5. Proteases Lysosomal proteases present in cartilage (Lucy et al., 1961) can cause degradation of the peptide moieties of proteoglycans and extraction conditions should be chosen to minimise contamination with these enzymes (Section 3.3.3.2). In some exceptional cases the proteoglycan may be resistant tooteolysis. Heparin proteoglycan isolated from peritoneal mast cells is resistant to proteases (Metcalfe et al., 1980) and heparin proteoglycan can be isolated from rat skin after treatment with Pronase (Robinson et al., 1978). However, extensive protease treatment of connective tissue usually leads to the release of glycosaminoglycans. Methods for the proteolytic digestion of tissue for the isolation of glycosaminoglycans are discussed in Section 6.4.2.

3.4. Isolation and fractionation

Once a glycoprotein or proteoglycan has been solubilised it becomes possible to apply techniques for purification and fractionation based on the solubility, size and shape, charge, density, absorption characteristics or affinity of the molecules concerned. Methods for the purification of proteins have been described elsewhere (Scopes, 1982) and this monograph will emphasise only those methods which are particularly appropriate for glycoproteins and proteoglycans.

The fractionation procedures employed may be limited by the conditions used to extract the glycoconjugate. Extraction with strong chaotropes or denaturing agents or ionic detergents such as SDS is liable to cause loss of native protein conformation. The presence of such solvents also prevents the protein-protein (or protein-carbohydrate) interactions which are required for affinity chromatography. Non-ionic detergents may bind to membrane glycoproteins in such a way as to interfere with gel filtration or rate-zonal centrifugation.

In addition to the removal of contaminating molecules, separation methods can be applied to fractionate glycoproteins or proteoglycans into sub-populations with the same peptide chains but differing by virtue of the microheterogeneity of their carbohydrate components.

3.4.1. Assays for glycoproteins and proteoglycans

To follow the purification of a glycosylated protein some type of assay for the biological, enzymatic, immunochemical or lectin-binding activity is often possible. In addition, assays for specific carbohydrate components can be applied to fractions. Commonly used procedures include phenol-H_2SO_4 for detection of neutral hexose, sialic acid estimation or hexosamine determination (Chapter 5). For proteoglycans uronic acid can be determined by the carbazole reaction (Section 5.6).

Purification can also be followed by use of SDS-gel electrophoresis coupled with staining for glycoproteins (Section 4.3.2).

Where radioactive label can be incorporated into the material under study this can prove useful in following the purification. [^{35}S]Sulphate labelling is useful for many proteoglycans and some glycoproteins and [^{3}H]- or [^{14}C]glucosamine has been used to label glycoproteins and proteoglycans (see Chapter 8).

3.4.2. Fractionation based on solubility

Several of the factors affecting the solubility of glycoproteins and proteoglycans have been discussed in relation to their initial solubilisation (Section 3.3) and selective solubilisation can be an important step in purification. Fractional precipitation with salts such as ammonium sulphate and with organic solvents are long-established methods of protein purification and have been widely applied to glycoproteins. Highly glycosylated glycoproteins may remain soluble in a relatively high concentration of ethanol (Björling, 1976), a fact exploited in the large-scale isolation of α_1-glycoprotein from serum by the Cohn fractionation scheme. Isolation of mucins by alcohol precipitation of a metal (e.g. K^+ or Ca^{++}) salt has been used quite widely and ethanol containing potassium acetate is an effective precipitant for glycosaminoglycans.

Glycoproteins with high carbohydrate content may be resistant to precipitation by common protein precipitants such as trichloroacetic acid or perchloric acid. Some protocols for glycoprotein isolation based on this property have been employed. However, exposure to strongly acidic conditions is liable to cause loss of sialic acid, thus inducing heterogeneity.

Macromolecular polyanions can be precipitated with certain ammonium salts. The nature of the interaction between cetyltrimethylammonium and cetylpyridinium salts with acidic polysaccharides has been described fully by Scott (1960). Quantitative precipitation of glycosaminoglycans released from connective tissue by papain digestion can be achieved by addition of an excess of anionic detergent salt. Precipitation is sensitive to the other ions present. In the presence of excess detergent, $MgCl_2$ and $CaCl_2$ tend to stabilise suspensions, whereas divalent anions (e.g. citrate, SO_4) have a strong coagulating action. After precipitation the polyanion can be recovered by dissolving at high salt concentrations. The quaternary salt can be removed subsequently by precipitation with sodium thiocyanate.

The solubility of the aliphatic ammonium salts–polyanion complex is strongly dependent on salt concentration. Polycarboxylates and polysulphated glycosaminoglycan complexes differ in their solubility and the charge density (i.e. charge per unit polymer length) also influences solubility. Hence the extraction of such complexes at different salt concentrations allows fractionation of charged polysaccharides of different types. In addition to glycosaminoglycans (and proteoglycans) highly charged glycoproteins are also precipitated as complexes with anionic detergents. Examples of the use of cetylpyridinium salts in the isolation of glycosaminoglycans and glycoproteins are given in Sections 6.7.3 and 3.4.7.5.

3.4.3. *Fractionation based on size and shape*

Gel filtration techniques employing beads of porous gels of polymerised dextran (Sephadex), polyacrylamide (Bio-Gel) or agarose (Sepharose, Bio-Gel) have found wide application to the purification of glycoproteins. Detailed descriptions of the use of these media can be found in the monography by Fischer (1969) in this series and in the excellent book on protein purification by Scopes (1982). Glycoproteins with significant carbohydrate content tend to elute earlier from gel filtration columns than globular proteins of the same molecular weight (Andrews, 1965; and Section 4.3.1). Microheterogeneity or polydispersity can lead to broadening of the peaks of glycoproteins and proteoglycans. It is generally advisable to carry out gel filtration at an ionic strength of 0.15 or greater to minimise any charge interactions between residual ionised groups on the gel filtration medium and the macromolecules being separated. High ionic strength also decreases charge interactions within polyanions (the electroviscous effect) and any electrostatic interaction between the glycoprotein and other molecules.

The viscosity of mucin preparations and the self-association of the molecules can give rise to difficulties in gel filtration. Only low flow rates may be obtainable and the concentration of sample which can be applied has to be limited. Information about the effect of ionic strength and pH on viscosity may be valuable in suggesting the most suitable conditions for column chromatography. Despite these difficulties gel filtration is one of the most important methods for the purification of mucous glycoproteins because their molecular weights (and Stokes radii) are often much greater than those of many contaminating proteins.

Gel filtration of membrane glycoproteins can be successfully carried out in the presence of ionic detergents such as SDS (Hamaguchi and Cleve, 1972) or deoxycholate (Letarte-Muirhead et al., 1975). The micelle size of non-ionic detergents such as Triton X-100 makes these less suitable for gel filtration.

The high molecular weight and asymmetric shape of proteoglycans result in the elution of these molecules in the excluded volume of all but the most porous gel filtration media. Useful fractionation of these molecules can be obtained under dissociative conditions by gel filtration on, for example, Sepharose CL-4B in buffer containing 4 M guanidine-HCl (Yanagishita and Hascall, 1983a,b).

Rate-zonal density gradient centrifugation, in which a sample is applied to the top of a density gradient of sucrose and centrifuged so that the sample sediments through the gradient, has been used to fractionate glycoproteins in the same way that it has been applied to proteins. The sedimentation rate depends on molecular size and asymmetry. Application of the technique to membrane glycoproteins may be affected by the detergent used for solubilisation. Letarte-Muirhead et al. (1975) found that the non-ionic detergent Lubrol PX decreased the rate at which the Thy-1 antigen glycoprotein sedimented on such gradients relative to the glycoprotein solubilised with deoxycholate.

3.4.4. Fractionation based on charge

Ion-exchange chromatography has been applied widely to glycoproteins using the methodology developed for protein separation (Scopes, 1982). Many glycoproteins carry a negative charge at neutral pH and will bind to DEAE-cellulose (or DEAE-Sephacel). Glycoproteins with high sialic acid content, or those containing sulphate, are usually strongly bound and elute at high salt concentrations. Charge heterogeneity due to variation in sialic acid content can give rise to broadening of peaks of glycoproteins eluted from DEAE-cellulose.

The high charge density of mucous glycoproteins gives rise to strong binding to anion-exchangers but the physical properties of mucins can lead to difficulties in obtaining satisfactory flow rates and poor recoveries of sample. Tracheal mucins have been fractionated on DEAE-cellulose columns in buffers containing 6 M urea (Gallacher et al., 1977).

Membrane glycoproteins such as band 3 of human erythrocytes can be purified by ion-exchange chromatography in the presence of non-ionic detergent (Section 3.4.8.3).

Ion-exchange chromatography is an important method for the purification of proteoglycans, particularly when they have been extracted from tissues or cell cultures which have a relatively low proportion of proteoglycan to total protein. In these circumstances it can be difficult to separate proteoglycans from contaminating protein by buoyant density centrifugation alone. Ion-exchange chromatography of proteoglycans is carried out on DEAE-cellulose in buffers containing 8 M urea (Antonopoulos et al., 1974; Yanagishita and Hascall, 1983) in 0.01 M sodium acetate, pH 6.0, with a gradient from 0.15 to 1.5 M NaCl. Glycoproteins and other contaminating proteins are weakly absorbed and are separated from proteoglycans which emerge at high salt concentrations (Section 3.4.8.9). Recoveries of small quantities of labelled proteoglycans in this system can be improved by the inclusion of the zwitterionic detergent CHAPS (Calbiochem) in the buffers (Section 3.4.8.8).

The use of cellulose-based ion-exchange media can lead to contamination of glycoprotein and proteoglycan preparations with small quantities of glucose arising from degradation of the cellulose. Sephacel products are less liable to produce this undesirable effect.

Chromatography on columns of hydroxylapatite has been used for the purification of human erythrocyte sialoglycoproteins (Liljas et al., 1976). It is possible to use this technique with samples solubilised in SDS (Warren et al., 1974).

3.4.5. Fractionation based on density differences

One of the major difficulties in the isolation of mucous glycoproteins is to separate them from proteins with which they are associated by non-covalent interactions. This problem can often be overcome by centrifugation in density gradients of caesium chloride or caesium bromide (Creeth, 1978; Allen, 1981). These gradients separate molecules of different buoyant density. The strong salt solution used to form the gradient can produce dissociation of ionic interactions. For the isolation of proteoglycans, protein denaturants such as guanidinium chloride can be added to the gradient, enabling the proteoglycan to be separated from protein (Sadjera and Hascall, 1969).

Buoyant density is inversely related to partial specific volume (Chapter 4). The value of the partial specific volume of proteins is about 0.70–0.75 ml/g, while values for non-sulphated polysaccharides are about 0.60–0.65 ml/g. Sulphated polysaccharides have even lower partial specific volumes, the value, for example, for chondroitin sulphate is 0.53 ml/g. Proteins band in caesium chloride density gradients at about 1.3 g/ml, while polysaccharides band at about 1.6–2.0 g/ml. Glycoproteins band at intermediate densities depending on their carbohydrate content and composition.

To carry out the method solid CsCl or CsBr is added to the sample to give an intermediate density. The sample is then centrifuged at high speed (100 000 *g*) for an extended period (24–96 h). Tube contents are fractionated by upward displacement and the densities determined by measurement of refractive index. Examples of the methodology applied to the isolation of mucins (Section 3.4.7.6) and proteoglycans (Section 3.4.7.7) are given.

3.4.6. *Affinity chromatography*

Affinity chromatography methods are based on the selective adsorption of a glycoprotein by another molecule which is attached to an insoluble support. The technique is widely applicable to proteins of differing functions (Lowe, 1979). While many types of interaction can be made use of, provided they have some specificity, the use of specific antibodies and of lectin-glycoprotein interactions are of special interest for glycoprotein isolation.

Lectin methods are discussed in Sections 7.4 and 3.4.7.4. They have found wide application, particularly in the isolation of membrane glycoproteins. Choice of appropriate detergents, the lectin used, and the method of coupling the lectin to its insoluble support are also important experimental factors. Elution of bound lectin is achieved by means of an appropriate sugar or glycoside.

Affinity chromatography of glycoproteins can also be achieved by the use of specific antibodies. The antibodies are coupled to an insoluble support and elution is by change in pH (acid or alkaline pH) or chaotropes. A pre-column of non-specific antibody (Section 3.4.7.4) can be used to avoid non-specific adsorption.

The use of monoclonal antibodies can provide a high concentration of bound molecules with identical specificities. High recoveries may be obtainable particularly where the affinity of binding is not too high.

3.4.7. *Examples of glycoprotein isolation*

Membrane glycoproteins

3.4.7.1. Glycophorin – chloroform-methanol extraction Human erythrocytes ghosts are homogenised and diluted to a protein concentration of 2 mg/ml with 10 mM Tris/0.1 mM EDTA-HCl buffer, pH 7.4. One volume of ghosts is mixed with 9 vols. of chloroform-methanol (2:1 vol/vol mixture) and stirred vigorously at room temperature for 30 min. After low-speed centrifugation the upper (aqueous) layer is carefully aspirated and centrifuged again to remove contaminating interphase material. The aqueous phase is concentrated by rotary evaporation at 37°C to give one-tenth vol. of clear solution.

The water-soluble extract contains periodate-Schiff staining glycoproteins (PAS I, II and III). Most other membrane proteins and glycoproteins are found in the precipitate at the interface and membrane lipid appears mainly in the organic phase. The extraction procedure has been adapted for the purification of the major sialoglycoprotein of the milk fat globule membrane (Snow et al., 1977).

Further purification of the extracted glycoproteins has been obtained by ethanol precipitation and by gel filtration on Sephadex G-100 columns equilibrated with 1% w/v SDS. Before chromatography samples were reduced (Snow et al., 1977) or reduced and alkylated

(Hamaguchi and Cleve, 1972) with mercaptoethanol and iodoacetate.

3.4.7.2. Glycophorin – LIS-phenol extraction (Marchesi, 1972)
Human erythrocyte membranes (25 mg/ml protein) are suspended in 0.3 M lithium di-iodosalicylate (LIS) and 0.05 M Tris-HCl buffer, pH 7.5, and stirred at room temperature for 10 min. Two volumes of water are added and the solution is stirred for 5 min at 4°C. After centrifuging at 45 000 *g* for 90 min at 4°C the supernatant is aspirated and mixed with an equal volume of (fresh) 50% w/v aqueous phenol. The mixture is stirred vigorously for 15 min at 4°C and then centrifuged at 4 000 *g* for 1 h at 4°C in a swinging-bucket rotor. The upper (aqueous) phase is removed, dialysed extensively against water to remove phenol, and lyophilised. This dried material is suspended in cold ethanol and stirred for 1–2 h in the cold and the glycoprotein is recovered by centrifugation. After three ethanol washes (to remove lipid) the pellet is dissolved in distilled water and dialysed against water overnight. A clear solution of the glycoprotein is obtained after centrifugation for 30 min at 4°C.

A similar procedure has been applied to the isolation of a membrane glycoprotein from L cells (Hunt and Brown, 1975). As with the chloroform-methanol extraction this procedure results in denaturation and precipitation of many membrane glycoproteins.

3.4.7.3. Band 3 extraction with Triton X-100 (Yu and Steck, 1975) Human erythrocyte ghosts are prepared by haemolysis and washing in 5 mM sodium phosphate, pH 8.0. The peripheral protein band 6 is removed by incubation of the ghosts on ice for 20 min in 20 volumes of 51 mM sodium phosphate, pH 8.0 (ionic strength 0.15). After centrifugation (20 min, 15 000 rpm) the supernatant is discarded and the pellet washed in 36 mM sodium phosphate, pH 7.5 (ionic strength 0.1). Band 6-depleted ghosts are extracted with 5 volumes of 0.5% v/v Triton X-100 in 36 mM sodium phosphate, pH 7.5 (ionic strength 0.1) for 30 min at 4°C. After centrifugation (20 min, 15 000 rpm) the supernatant Triton extract is collected.

Treatment of the Triton extract with the sulphydryl reagent *p*-chloromercuribenzoate (PCMB) is necessary to prevent copurification of band 4.1 with band 3 on ion-exchange chromatography.

A freshly prepared solution of 50 mM PCMB (0.3 ml) dissolved in 10 mM NaOH-0.1% Triton X-100 is added to 5 ml of the Triton extract and incubated at 4°C for 20 min. The mixture is immediately applied to a column (0.6 × 5 cm) of aminoethyl-cellulose (Bio-Rad) equilibrated with 1% Triton X-100 in 34 mM sodium phosphate, pH 8.0, and cooled with ice water. All buffers applied to the column are also cooled and are driven through the column by N_2 under pressure. After applying the sample the column is washed with 3 ml of 1% Triton in 44 mM sodium phosphate, pH 8.0 (ionic strength 0.13). Band 3 is eluted with 1% Triton X-100 in 82 mM sodium phosphate, pH 8.0 (ionic strength 0.24). The first 0.5 ml of eluate is discarded and the next 1 ml collected as the band 3 preparation. Finally the column is washed with 1% Triton in 170 mM sodium phosphate, pH 8.0 (ionic strength 0.5). Aggregation of band 3 occurs on storage in Triton X-100.

3.4.7.4. Purification of Thy-1 antigen by affinity chromatography (Letarte-Muirhead et al., 1975) Membranes isolated from approx. 10^{11} thymocytes (from 70–100 rat thymus glands) are resuspended in 25 ml of 10 mM Tris-HCl, pH 8.0, containing 0.01% (w/v) NaN_3, and 25 ml of 4% (w/v) sodium deoxycholate in the same buffer is added. After 60 min at 0°C the mixture is centrifuged for 6 750 000 *g* min, leaving solubilised Thy-1 antigen in the supernatant.

Gel filtration of solubilised membrane (50 ml) is carried out on columns of Sephadex G-200 (5 × 90 cm) using upward flow and eluting with deoxycholate-containing buffer. Fractions containing antigen activity are pooled.

Affinity chromatography is carried out on columns of lentil lectin covalently coupled to Sepharose 4B (3–14 mg lectin/ml beads). Before use the lectin column is washed with deoxycholate buffer, then with 0.5 M methyl α-D-glucopyranoside in deoxycholate buffer and finally in deoxycholate buffer. The sample is applied to a column

(2.5 ml beads with 14 mg lectin/ml) containing sufficient lectin to ensure that saturation does not occur. The column is washed with deoxycholate until the E_{280} of the eluate is zero and bound material is eluted with 0.5 M methyl α-D-glucopyranoside in deoxycholate buffer.

In this preparation only part of the Thy-1 antigen binds to lentil lectin. The activity which is not bound can be purified by antibody affinity chromatography.

Mucins

3.4.7.5. Ovine submaxillary mucin (Hill et al., 1977) Ovine submaxillary glands (obtained fresh and stored frozen) are suspended in 0.01 M NaCl (4 ml/g tissue) and homogenised in a Waring blender for 2 min. After centrifugation at 7 200 *g* for 20 min the supernatant is removed and the pellet re-extracted in 0.01 M NaCl (3 ml/g of original tissue) and centrifuged. Both supernatants are combined and adjusted to pH 4.7 by slow addition of 2 M acetic acid. The precipitate is removed by centrifugation at 7 200 *g* for 20 min and the supernatant applied to a column of sulphopropyl-Sephadex C-25 (4 ml packed volume/g original tissue) equilibrated with 0.05 M sodium acetate, pH 4.7. Fractions are collected and those containing sialic acid combined. Mucin is precipitated by addition of cetyltrimethylammonium bromide (0.1 ml of a 10% w/v solution/g original tissue) and centrifuged at 7 200 *g* for 20 min. The precipitate is redissolved in 4.5 M $CaCl_2$ (2 ml/g original tissue) and absolute ethanol added to a final concentration of 60%. After 60 min the solution is centrifuged at 27 000 *g* for 30 min, the precipitate discarded and the supernatant brought to 75% ethanol. After 60 min, the calcium salt of the mucin is collected by centrifuging at 27 000 *g* for 30 min. The precipitate is dispersed in 1 M NaCl (1 ml/g original tissue), using a Teflon homogeniser and stirred overnight. The solution is dialysed for 24 h against three changes of distilled water and made 0.01 M with sodium phosphate, pH 6.8.

The mucin solution is applied to a column of hydroxylapatite (4 ml packed volume/g original tissue) equilibrated and eluted with 0.01 M sodium phosphate, pH 6.8. Fractions containing sialic acid are pooled, dialysed and lyophilised, and stored at −20°C.

The final step in purification is the gel filtration of the mucin preparation after dansylation with 5-dimethylaminonaphthalene-1-sulphonyl (dansyl) chloride. This reagent allows the detection of protein impurities, which react with the reagent giving fluorescent dansyl derivatives, while the mucin, which has a blocked *N*-terminal residue and contains no lysine, does not react.

Mucin (10 mg/ml) in 0.5 M $NaHCO_3$, pH 9.8, is mixed with an equal volume of dansyl chloride in acetone (5 mg/ml) and incubated at 37°C for 20 min. The mixture is brought to pH 4 with formic acid and dialysed against 0.01 M sodium cacodylate, pH 6, containing 0.5 M NaCl. The dansylated mucin is applied to a column of Sepharose 4B equilibrated and eluted with the cacodylate-NaCl buffer. Fractions are analysed for fluorescence and sialic acid and those containing sialic acid but without fluorescence are pooled to give the purified mucin.

3.4.7.6. Glycoprotein from gastric mucus (Starkey et al., 1974; Allen, 1981) Gastric mucus obtained by scraping from the surface of pig stomach (Snary and Allen, 1972) is suspended in 0.2 M NaCl containing 0.01% sodium azide adjusted to pH 5.5 and homogenised for 1 min in a Waring blender. The solubilised mucin is centrifuged at 23 000 *g* for 10 min to remove cellular debris and the supernatant is fractionated by gel filtration on Sepharose 4B columns eluted with 0.2 M NaCl. Assay of fractions for carbohydrate (orcinol method) and protein (280 nm) indicates a high molecular weight component A appearing in the excluded volume. This material is then purified further by caesium chloride density gradient ultracentrifugation. A solution of mucin (1.2 mg/ml) is made up to 3.8% (w/w) with caesium chloride. The solution is centrifuged at 1.5×10^5 *g* at 5°C for 48 h. The contents of each centrifuge tube, collected by upwards displacement, are divided into six equal fractions with densities varying linearly from 1.38 g/ml (top) to 1.56 g/ml (bottom). Purified mucin is recovered from the two fractions of highest density.

Proteoglycans

3.4.7.7. Proteoglycan 'subunit' from bovine nasal cartilage (Hascall and Sadjera, 1969) Nasal septa are freed of non-cartilagenous tissue and rinsed with 0.9% NaCl solution at 4°C. The septa are then sliced with a Stanley Surform. Slices are about 0.5 mm thick. Cartilage slices are extracted with 15 volumes of 4 M guanidinium chloride containing 0.05 M sodium acetate, pH 5.8, at room temperature on a magnetic stirrer for 24 h. Fragments of cartilage are removed by vacuum filtration with the aid of 5% w/v Hyflo Super-cel. The filter cake is sucked dry and discarded without washing and the clarified extract dialysed against 9 volumes of 0.5 M acetate buffer, pH 5.8. Solid CsCl is added to the dialysed extract to give a density of 1.69 g per ml (1.19 g CsCl per g solution) and the solution is then centrifuged at 34 000 rpm in a Spinco type 40 rotor for 48 h at 18°C. Proteoglycan subunit is isolated from the lower two-fifths of the tubes.

3.4.7.8. Isolation of low buoyant density dermatan sulphate proteoglycan synthesised by cultured cells (Yanagishita and Hascall, 1983) Rat ovarian granulosa cells grown in culture are labelled with [^{35}S]sulphate, [^{3}H]glucosamine, [^{3}H]mannose or [^{3}H]serine as precursors. After labelling, the medium fraction is removed and solid guanidine-HCl (0.53 g/ml) and 0.01 vol. of 1 M *N*-ethylmaleimide in ethanol are added to bring the solution to about 4 M guanidine-HCl, 10 mM *N*-ethylmaleimide. The latter reagent is added to prevent disulphide interchange reactions. The sample is then run through a small column containing 8 ml Sephadex G-50 (fine), equilibrated and eluted with 8 M urea, 0.15 M NaCl, 0.05 M sodium acetate, pH 6.0. This step removes unincorporated isotope, guanidine-HCl and equilibrates the sample with 8 M urea-containing buffer. The detergent CHAPS is added to the breakthrough fractions to a final concentration of 0.5% (w/v). Then the sample is applied to a column (2 ml bed volume) of DEAE-Sephacel equilibrated with the urea-containing buffer containing 0.5% (w/v) CHAPS. After washing with 5 ml of buffer the column is eluted with a gradient from 0.15 to 1.5 M NaCl in the same solvent, using a total volume of 46 ml. The early fractions from this column contain weakly bound proteins and glycoproteins. Of the two peaks eluted by the salt gradient one contains a heparan sulphate proteoglycan and peak 2 contains crude dermatan sulphate proteoglycan. Peak 2 is then fractionated into two separate proteoglycan populations based on size using a Sepharose CL-4B column (0.7 × 100 cm) eluted with 4 M guanidine-HCl, 0.05 M Tris, 0.05 M sodium acetate, 0.2% (w/v) CHAPS. The two fractions of dermatan sulphate proteoglycan can be further characterised by caesium chloride density gradient centrifugation under dissociative conditions.

3.4.7.9. Purification of keratan sulphate proteoglycan from monkey cornea (Nakazawa et al., 1983) The keratan sulphate proteoglycan is extracted in 4 M guanidine-HCl containing protease inhibitors and purified by a combination of DEAE-cellulose chromatography, chondroitinase ABC digestion to remove chondroitin-dermatan sulphate proteoglycans and affinity chromatography on immobilised concanavalin A.

Proteoglycans are labelled in 15 corneas by incubation with $[^{35}S]$sulphate and $[2\text{-}^{3}H]$mannose. Labelled corneas are then extracted as described below and the extracts combined with extracts obtained from 300 unlabelled corneas. The 300 corneas are extracted in 200 ml of 4 M guanidine-HCl containing 0.01 M sodium EDTA, 0.01 M sodium acetate, 0.1 M 6-aminohexanoic acid, and 0.005 M benzamidine-HCl, pH 5.8, for 20 h at 4°C. After decanting the extract residual tissue is re-extracted with 100 ml of the same extraction cocktail and the extracts are combined, concentrated to 130 ml by ultrafiltration and dialysed against 8 M urea in 0.5 M Tris-HCl, pH 6.8. The extracts are chromatographed on a DEAE-cellulose column (2.5 × 25 cm) equilibrated with the buffered urea solution. Bound material is eluted with a linear gradient from 0 to 0.75 M NaCl (total 800 ml) in 8 M urea, 0.05 M Tris-HCl, pH 6.8, followed by 300 ml of 3.0 M NaCl to confirm the complete elution of proteoglycans. The proteoglycan peak detected by ^{35}S-labelling (Section 8.2) is pooled, dialysed and digested with chondroitinase ABC. The digest is dialysed against 8 M urea, 0.05 M Tris-HCl, pH 6.8, and rechromatographed on DEAE-cellulose as before. The keratan sulphate proteoglycan peak is concentrated and dialysed against 1 M NaCl, 0.05 M Tris-HCl, pH 7.0. This material is applied to a concanavalin A-Sepharose column (2.5 × 10 cm) equilibrated with 1 M NaCl, 0.05 M Tris-HCl, pH 7.0, washed with 300 ml of the same buffer and bound material is eluted with 300 ml of 1 M methyl mannoside, 1 M NaCl, 0.05 M Tris-HCl, pH 7.0. After dialysis and lyophilisation 50 mg of purified keratan sulphate proteoglycan is obtained (Nakazawa et al., 1983).

CHAPTER 4

Physico-chemical characterisation

4.1. Introduction

Having established that a proteoglycan or glycoprotein preparation is not heterogeneous, an essential step in its characterisation is to determine the molecular weight. Molecular weight analysis of non-glycosylated proteins is usually carried out by simple empirical methods such as SDS-polyacrylamide gel electrophoresis and gel filtration. Unfortunately these methods do not give reliable results when applied to highly glycosylated glycoproteins. Nevertheless, empirical methods are useful in characterising the molecular weights of glycoproteins with low carbohydrate content, glycosaminoglycans, glycopeptides and oligosaccharides (Sections 4.3.1 and 4.3.2). It is also possible to analyse the molecular weight of the peptide chains of enzymically or chemically deglycosylated glycoproteins by SDS-polyacrylamide gel electrophoresis. The best general procedure for determination of the molecular weights of intact proteoglycans and highly glycosylated glycoproteins is sedimentation equilibrium ultracentrifugation (Section 4.3.3). Molecular weights of glycoproteins and proteoglycans can also be measured by combining information from sedimentation velocity ultracentrifugation with measurements of diffusion or viscosity, by light scattering or by osmotic pressure measurement (Section 4.3.4).

Difficulties can arise in the physico-chemical characterisation of highly glycosylated glycoproteins and proteoglycans as a result of the molecular heterogeneity of their carbohydrate units, their asymmetric shape, molecular interactions and high charge (Section 4.2). Information about the size heterogeneity of glycoconjugate preparations can often be obtained during molecular weight analysis and evidence of molecular interactions may also be gained. Heterogeneity of charge can be examined by isoelectric focusing of glycoproteins or by gel electrophoresis (Section 4.4). Different types of glycosaminoglycan can also be distinguished by electrophoretic methods (Section 4.4.2).

The gross shape of glycoconjugates in solution can be assessed from their scattering and hydrodynamic properties. Electron microscopy has also contributed valuable information about the molecular shapes of glycoproteins and proteoglycans. Studies of chiral properties such as circular dichroism and the spectroscopic properties of complexes with dyes have yielded some information about the conformation of proteoglycans in solution. Recently nuclear magnetic resonance techniques have been applied to examine the conformation and flexibility of glycoproteins and proteoglycans in solution and studies have been carried out on proteoglycan complexes in situ in connective tissue. The detailed atomic structures of a few glycoproteins have been determined by X-ray crystallography. X-ray diffraction techniques have been applied to stretched films of glycosaminoglycans and some studies on proteoglycans have been reported. Methods for investigating the shape, interactions, conformation and flexibility of glycoproteins and proteoglycans are outlined in Section 4.5.

In describing experimental approaches to the physico-chemical characterisation of proteoglycans and other glycoproteins emphasis will be placed on methods involving equipment available in most laboratories. Use of major instruments (e.g. the analytical ultracentrifuge) will not be described in detail because information in the form of manufacturers manuals and local expertise is usually available where this type of equipment is located.

Applications of physical techniques to glycoproteins (including proteoglycans) have been reviewed by Bettelheim (1977) and in more detail by Gibbons (1972). Problems associated particularly with epithelial mucins are dealt with by Gibbons (1972) and membrane glycoproteins are reviewed by Juliano (1978). The physico-chemical properties of proteoglycans and glycosaminoglycans are detailed by Kennedy (1979).

4.2. *Problems in characterising glycoproteins and proteoglycans*

The characterisation of glycoproteins with low carbohydrate content (<5% by weight) can be approached in the same way as for non-glycosylated proteins. However, when larger amounts of carbohydrate are present difficulties may arise because of polydispersity, highly asymmetric shape, molecular interactions and high charge density (Gibbons, 1972).

Pure preparations of proteins are generally monodisperse with respect to molecular weight (i.e. all molecules in the population are identical in size). Proteins are also usually monodisperse in other physical properties such as charge, although sensitive techniques such as isoelectric focusing may reveal paucidispersity (a few molecular species) resulting from differences in the extent of deamidation of asparagine and glutamine residues. Glycoproteins often show paucidispersity of charge on isoelectric focusing due to variation between molecules in sialic acid content. They may also show evidence of polydispersity of molecular weight giving, for example, broad peaks in gel filtration elution profiles and broad bands on SDS-polyacrylamide gel electrophoresis. Molecules which differ slightly in size and charge, but which still form part of a single statistical population, are described as polydisperse. The physical properties, such as charge or molecular weight, of a polydisperse preparation have a unimodal

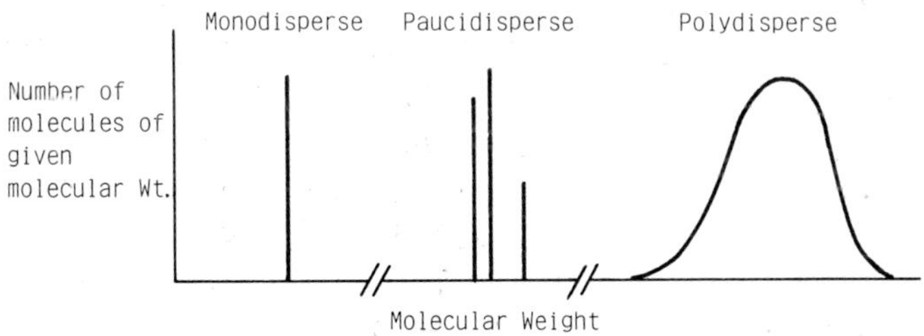

Fig. 4.1. Distribution of molecules in monodisperse, paucidisperse and polydisperse preparations.

distribution (Gibbons, 1972; Fig. 4.1). Proteoglycans and glycoproteins with high carbohydrate content are polydisperse in molecular weight and charge.

For a monodisperse system the molecular weights determined by different types of measurement will be equal. However, for polydisperse systems the values obtained will depend on the type of molecular weight average measured. The most widely encountered of these are number average and weight average. Where n_i = number of molecules of molecular weight M_i and c_i is the concentration by weight of the ith species then (Svedberg and Pederson, 1940):

$$\text{Number average} = \frac{\sum M_i n_i}{\sum n_i}$$

$$\text{Weight average} = \frac{\sum M_i^2 n_i}{\sum M_i n_i}$$

Osmotic pressure measurements (or other colligative properties) depend on the number of molecules present and give number average molecular weights. Most other physical measurements depend on the weight of material and give rise to weight average molecular weights. The ratio between different molecular weight averages gives an indication of polydispersity and for this reason it is useful to obtain more than one molecular weight average for polydisperse glycoproteins or proteoglycans.

Glycoproteins with high carbohydrate content tend to be asymmetric molecules with high frictional ratios, high intrinsic viscosities and large second virial coefficients. This last property necessitates careful measurement of physico-chemical properties at low concentrations so that valid extrapolation to zero concentration can be obtained.

The shapes of the carbohydrate chains of glycoproteins and proteoglycans in solution probably range from highly flexible random coils through stiffened coils occupying large volumes of solvent to more extended rod-shaped structures. In some cases the peptide chain may be flexible with carbohydrate attached at frequent intervals giving rise to a 'bottle brush' type of structure. Alternatively, the carbohydrate units may extend into the solvent from a globular protein structure. Frequently physical measurements give only a crude indication of the structure of glycoproteins in solution.

The extended, solvent-filling, structure of some glycoproteins and proteoglycans leads to the possibility of gels forming through the overlap of molecular domains or because of entanglement of threadlike chains. Such entanglement can give rise to visco-elastic behaviour and may impede physical studies.

Strong intermolecular association may also occur in some glycoproteins as a result of protein-carbohydrate, or carbohydrate-carbohydrate interactions (Hill et al., 1977). Such interactions lead to marked concentration-dependence of physical properties.

Most proteoglycans and glycosaminoglycans and some glycoproteins have a high charge density. Because of the flexible nature of the carbohydrate chains the presence of charged groups can make the conformation of such molecules highly dependent on pH and ionic strength. In practice it is usually advantageous to study the physico-chemical properties of glycosaminoglycans, proteoglycans and glycoproteins with high charge in solutions of high ionic strength to minimise such charge effects.

4.3. *Molecular weight determination*

4.3.1. *Gel filtration*

Gel filtration is a simple, inexpensive, empirical method of molecular weight determination. The technique requires calibration of the chromatographic system with standards whose molecular weights have been established by a primary method (e.g. sedimentation equilibrium ultracentrifugation) and which are similar in shape and hydration to the sample under investigation. Gel filtration has been used quite widely for the determination of the molecular weights of glycosaminoglycans and oligosaccharides and glycopeptides derived from glycoproteins. The technique can also be used to obtain information about the polydispersity of these molecules. However, for glycoproteins containing more than a few percent of carbohydrate serious errors can arise in molecular weight analysis by non-denaturing gel filtration. For glycoproteins containing 5–50% carbohydrate the best empirical approach is to perform gel filtration of the reduced alkylated glycoprotein in 6 M guanidine hydrochloride.

The theory and practice of gel filtration and the application of the technique to the determination of protein molecular weight have been considered in a previous volume in this series (Fischer, 1969) and in the review by Ackers (1975). This section will therefore give only a summary of the method and an account of the application of gel filtration to glycoproteins and proteoglycans.

In gel filtration experiments measurements are made of the volume of solvent V_e required to elute a macromolecule from a column filled with beads of a porous gel. Molecules which are so large that they do not enter the gel are unretarded and elute in the void volume V_o. Smaller molecules are eluted at larger volumes approaching the limit of V_t, the total column volume. The partition coefficient, K_{av}, describing the distribution of a substance between the moving phase (of liquid outside the gel beads) and the stationary phase (which includes the volume of liquid inside the gel and the gel itself), can be defined as

$$K_{av} = \frac{V_e - V_o}{V_t - V_o}$$

Alternatively the partition coefficient K_d for the distribution of a solute between the liquid outside the gel and the volume of liquid inside the gel V_i (excluding the volume of the gel matrix) can be written as

$$K_d = \frac{V_e - V_o}{V_i}$$

Laurent and Killander (1964) showed that K_{av} could be related to the apparent Stokes radius, r_s, of globular proteins or of polysaccharides according to the equation

$$K_{av} = \exp[-\pi L(r_s + r_r)^2]$$

where L is the concentration of the stationary phase and r_r is the radius of the (rod-shaped) molecules of the stationary phase. There is a constant relationship between Stokes radius and molecular weight for molecules of similar shape and hydration.

Empirical relationships have been established between K_{av}, K_d or other gel filtration parameters, and the molecular weight of globular proteins (Fischer, 1969) and glycosaminoglycans (Wasteson, 1971a,b). Calibration curves can be obtained for globular proteins by plotting K_{av} against log *M;* interpolation of the K_{av} values for other globular proteins allows their molecular weights to be estimated (Fischer, 1969).

Glycoproteins and glycopeptides

Andrews (1965) demonstrated that columns of Sephadex can be calibrated with proteins of known molecular weight to obtain a

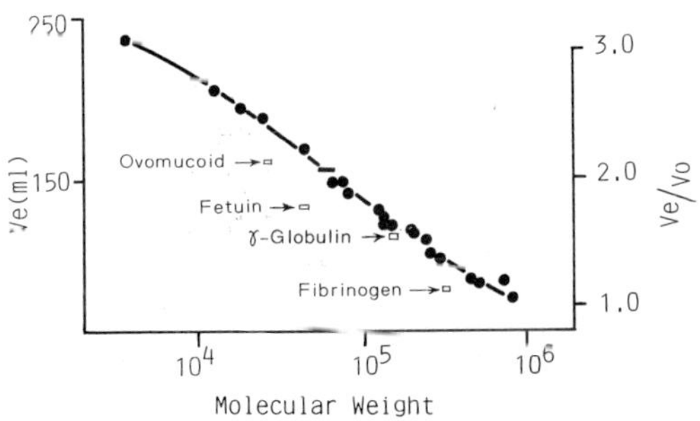

Fig. 4.2. Behaviour of glycoproteins on gel filtration. The logarithmic plot of protein molecular weight against elution volume from Sephadex G-200 is redrawn from Andrews (1965). The highly glycosylated glycoproteins ovomucoid and fetuin (as well as the highly asymmetrical glycoprotein fibrinogen) fall well below the calibration curve.

standard curve of V_e against log M. While the data for most globular proteins fall on this curve (Fig. 4.2), the elution volumes of glycoproteins of known molecular weight generally lie below the calibration curve. This could lead to considerable error in using a gel filtration system calibrated with proteins to estimate the molecular weights of glycoproteins. For this reason gel filtration under non-denaturing conditions is not a satisfactory method of molecular weight analysis of glycoproteins containing more than a few percent of carbohydrate.

The reason for the anomalous behaviour of glycoproteins is that their asymmetric, branched, structure gives them an apparent Stokes radius greater than that of a globular protein. For a valid calibration curve to be obtained for use with glycoproteins it would be necessary to use standard glycoproteins with a similar carbohydrate content and of similar shape to the unknown.

The best approach to the molecular weight analysis of glycoproteins by gel filtration is to employ the random-coil-producing solvent guanidinium chloride (Leach et al., 1980b). Columns (1.6 × 100 cm) are packed with 5% agarose (Pharmacia or Bio-Rad) equilibrated and eluted with 6 M guanidinium chloride (Sigma, Grade I) adjusted to pH 6.0. A flow rate of 1.4 g h^{-1} cm^{-2} is employed and fractions of about 1 g are collected. For sufficiently precise measurement of the column eluate it is necessary to collect the eluate into weighed test tubes and determine the total eluate weight. Samples of standard proteins and the glycoprotein sample (at least 1 mg) are dissolved in 6 M guanidinium chloride and disulphide bonds are reduced and carboxymethylated (Mann and Fish, 1972). Proteins are detected in the eluate by measuring absorbance at 280 nm. The void volume of the column is determined by adding Blue dextran (1 mg) to samples. DNP-glycine can be employed as a marker for the internal volume (V_i) of the column. Gel chromatography data can be treated for molecular weight determination by plotting K_d against log M or by other standard procedures (Mann and Fish, 1972).

When this procedure was applied to 16 glycoproteins of known physico-chemical properties satisfactory molecular weight values were obtained in most cases (Fig. 4.3). It appears that the oligosac-

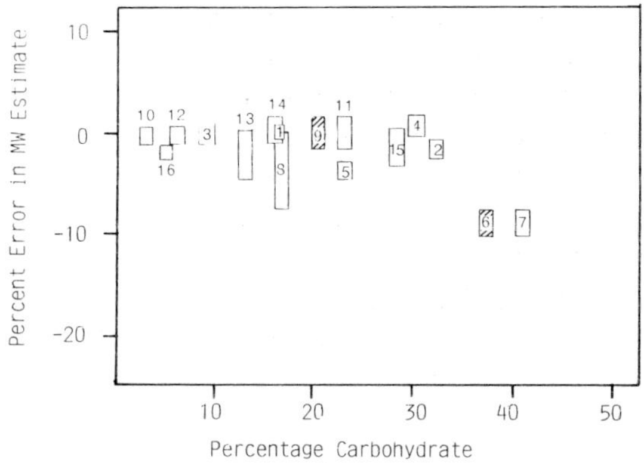

Fig. 4.3. Molecular weight analysis of glycoproteins by gel filtration in 6 M guanidinium chloride (Leach et al., 1980b). Gel chromatography of reduced, carboxymethylated glycoproteins was carried out on columns packed with 6% agarose gel beads. Molecular weights determined in this system were compared with values obtained by sedimentation equilibrium analysis or calculation from the primary sequence. The size of each rectangle represents the confidence limits of the molecular weights determined by these primary methods. Shaded rectangles represent desialated glycoproteins. Samples were: (1) thryrotropic hormone; (2) human chorionic gonadotropin, α-chain; (3) α-lactalbumin; (4) hCG β-chain; (5) ovomucoid; (6) asialo α_1-acid glycoprotein; (7) α_1-acid glycoprotein; (8) horseradish peroxidase; (9) asialofetuin; (10) ovalbumin; (11) fetuin; (12) IgG, H chain; (13) cellobiohydrolase C; (14) glucose oxidase; (15) Tamm-Horsfall glycoprotein; (16) transferrin.

charide branches produce an effect on the unfolded polypeptide chain giving an effective hydrodynamic radius which is the same as that of a linear polypeptide of the same total mass (Leach et al., 1980b). However, there is no theoretical reason indicating that this should always be the case and the molecular weight of the highly glycosylated α_1-glycoprotein determined in this way deviates significantly from its true value. While this method is probably the most reliable empirical method for estimating glycoprotein molecular weight the results obtained should not be regarded as better than a first approximation.

Agarose gel filtration in high concentrations of guanidine hydrochloride is time-consuming and resolution is not high. Improvements in both of these factors can be obtained by making use of an HPLC column packing of rigid spherical silica gel bonded with hydrophilic substituents (Ui, 1981). Columns of TSK-GEL-G3000 SW can be run with HPLC equipment to separate glycoproteins in 6 M guanidinium chloride with a run time of 50 min, compared with the several days necessary for techniques using soft gels.

Gel filtration under non-denaturing conditions has been used to estimate the molecular weight of glycopeptides (Bhatti and Clamp, 1968). Methodology is described in Section 6.7.1. Bio-Gel P4 gel filtration and HPLC methods have also been employed for molecular weight analysis of glycopeptides and oligosaccharides (Section 6.7.4.2).

Glycosaminoglycans

The molecular weight of the glycosaminoglycan chains of proteoglycans can be determined by gel filtration provided that molecular weight standards of the glycosaminoglycan are available. Wasteson (1971a) has described the isolation and characterisation of molecular weight standards of chondroitin sulphate. The following procedure for determination of the molecular weight of chondroitin sulphate (or other glycosaminoglycans for which molecular weight standards are available) can be applied to small samples (100 μg) and analysis of the elution profile can give information about the polydispersity of the preparation (Wasteson, 1971b).

Columns of Sephadex G-200 (0.8 × 63.5 cm) are packed and equilibrated with 0.2 M NaCl at 4°C. Samples (100–200 μg) of chondroitin sulphate are applied to the column, which is eluted with 0.2 M NaCl at a flow rate of about 3 ml cm^{-2} h^{-1}. Effluent fractions (0.9 ml) are collected and analysed for uronic acid or ^{35}S radioactivity. The void volume of the column is determined using Blue Dextran 2000. The elution volume, V_e, for a glycosaminoglycan sample is the volume corresponding to the peak appearing on the elution profile. A graph of log M against K_{av} is obtained for the standards and the molecular weights of unknown samples can be determined by interpolation. It is possible to make use of the elution profile to calculate both M_w and M_n and thus obtain an estimate of the polydispersity of the preparation (Wasteson, 1971b). Columns of Sepharose 6B (1.3 × 79 cm) can be employed for fractionation of glycosaminoglycan chains too large for effective separation on Sephadex G-200.

Gel filtration is a simple method requiring small amounts of sample, although it is less sensitive than gel electrophoresis (Section 4.3.2). The major limitation of the gel filtration method is in obtaining the appropriate standards.

A high-speed gel-permeation chromatography system for the separation of glycosaminoglycans on columns packed with Fractogel TSK HW-55(S) has been described by Irimura et al., (1983). Elution was carried out with 0.2 M NaCl at 55°C and was complete in 50 min. A calibration curve relating log M to retention time was obtained for a series of different glycosaminoglycans and glycopeptides (hyaluronate, chondroitin 6-sulphate, heparan sulphate, chondroitin 4-sulphate, heparin and thyroglobulin glycopeptides). The wide molecular weight range and the apparent applicability of this system to a variety of different glycosaminoglycans suggest that its further development could be of great value for rapid estimation of glycosaminoglycan molecular weight.

4.3.2. Electrophoretic methods

The use of SDS-polyacrylamide gel electrophoresis in the characterisation of glycoproteins and the determination of the molecular weights of glycosaminoglycans by polyacrylamide gel electrophoresis will be discussed in this section. Electrophoretic separation of different types of glycosaminoglycan will be described in Section 4.4.2.

Glycoproteins

Because of its simplicity, speed and high sensitivity SDS-polyacrylamide gel electrophoresis is now the most widely employed method for estimating the molecular weights of the individual peptide chains of proteins. The technique of SDS-gel electrophoresis is very valuable in the characterisation of intact glycoproteins particularly as a method for assessing homogeneity and for the examination of membrane glycoproteins which are solubilised by the anionic detergent sodium dodecylsulphate. A qualitative indication of the polydispersity of glycoproteins can also be obtained from the width of the stained bands after electrophoresis. However, SDS-gel electrophoresis is subject to considerable and unpredictable errors as a means of molecular weight determination for native glycoproteins with more than a few percent of carbohydrate.

It is generally found that glycoproteins migrate more slowly on SDS-gel electrophoresis than standard polypeptides of the same mass (Glossman and Neville, 1971; Schubert, 1970; Segrest et al., 1971; Leach et al., 1980b; Fig. 4.4). Segrest and Jackson (1972) reported that the empirically observed molecular weights of glycoproteins approached their true values more closely in gels with a high percentage of acrylamide than in low-percentage gels. This has been con-

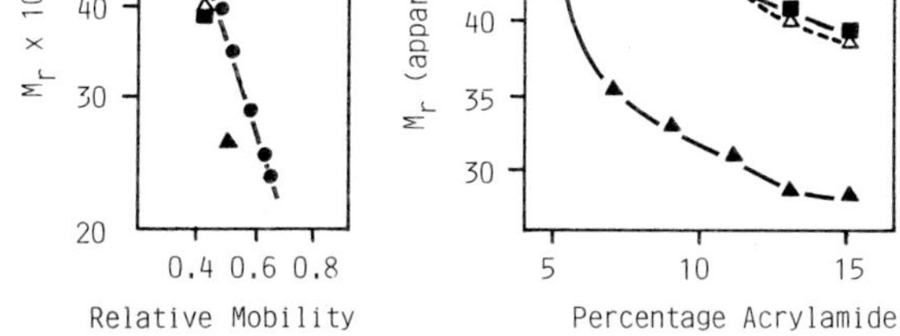

Fig. 4.4. Behaviour of glycoproteins on SDS-polyacrylamide gel electrophoresis (Leach et al., 1980a). (a) Electrophoretic mobilities on 7% gels for fetuin (□), horseradish peroxidase (△), α_1-acid glycoprotein (■), and ovomucoid (▲) relative to a series of polypeptide standards (●). (b) Dependence of apparent molecular weights of glycoproteins on acrylamide concentration for fetuin (□), horseradish peroxidase (△), α_1-acid glycoprotein (■) and ovomucoid (▲).

firmed by Leach et al. (1980b) but as their results show (Fig. 4.4) the dependence of apparent molecular weight on acrylamide concentration differs for each glycoprotein. The best approach to estimating the approximate molecular weights of intact glycoproteins is therefore to determine their mobilities relative to standard proteins in gels at high (e.g. 15%) polyacrylamide concentrations. Preferably values for the apparent molecular weight of glycoproteins should be obtained in gels of different polyacrylamide concentration and plotted as shown in Fig. 4.4b to give an indication of the dependence of apparent molecular weight on acrylamide concentration.

The anomalous behaviour of glycoproteins on SDS-gel electrophoresis results from both aberrant hydrodynamic behaviour and from decreased binding of sodium dodecyl sulphate compared with polypeptides of equal mass (Leach et al., 1980a).

Another approach to estimating the molecular weights of glycoproteins is to remove the carbohydrate units and then analyse the molecular weight of the peptide chain by SDS-gel electrophoresis. The molecular weight of the intact glycoprotein can then be calculated if the proportion of carbohydrate in the molecule is known. Methods for the removal of carbohydrate without peptide-chain cleavage include the use of endoglycosidases, the combined action of exoglycosidases or chemical deglycosylation (Section 6.11). It is essential that enzymes are free of proteases.

The molecular weight analysis of deglycosylated peptide chains has the great advantage that highly polydisperse glycoproteins can be converted to monodisperse peptide chains. Trimble and Maley (1977a) showed that *Saccharomyces cerivisiae* invertase could be converted from its native polydisperse form (apparent molecular weight 90 000–160 000) to peptide chains of molecular weight 60 000 which migrated as sharp bands on SDS-gel electrophoresis, by removal of the carbohydrate units with β-endo-*N*-acetyl glucosaminidase H. The same endoglycosidase, which cleaves carbohydrate units of the high-mannose type, has been used to produce deglycosylated peptide chains of other enzymes for molecular weight analysis by SDS-gel electrophoresis (Trimble and Maley, 1977b). Endo-β-*N*-acetylglucosaminidase F can be used to cleave both high-mannose and complex units from glycoproteins (Elder and Alexander, 1982). Chemical deglycosylation with anhydrous trifluoromethanesulphonic acid, which removes both *N*- and *O*-linked carbohydrate units, has been employed for the molecular weight analysis of the peptide chain of human C$\overline{1}$ inhibitor (Harrison, 1983). This type of approach to the analysis of the molecular weight of the peptide units of highly glycosylated proteins has great potential for future development. It is also possible to obtain polypeptide chains free of *N*-linked carbohydrate units by the use of inhibitors of glycosylation (Section 8.2.8).

Analysis of the molecular weights of glycoproteins is carried out by flat-bed gel electrophoresis using the discontinuous buffer system of Laemmli (1970). Very small amounts of sample (0.5 μg) can be analysed by using the microelectrophoresis apparatus (with running gel length of 33 mm) described by Amos (1976), which can readily be constructed in a workshop. For most purposes it is more convenient to use a 'semi-micro' apparatus constructed to the same design but having glass plates to contain the gel 70 mm long × 200 mm wide with Teflon spacers giving a gel thickness of 1.5 mm. Gels are prepared from stock solutions of 30% acrylamide (electrophoretic grade) containing 0.8% *N,N'*methylenebisacrylamide. Running gels (7.5–15% acrylamide) contain a final concentration of 0.375 M Tris-HCl, pH 8.8, and 0.1% w/v SDS. Polymerisation is produced by the addition of 0.025% v/v tetramethylethylene diamine and ammonium persulphate. The running gel is overlaid with water to give a flat surface. Stacking gels (5% acrylamide) contain 0.126 M Tris-HCl, pH 6.3, and 0.1% SDS. Electrode buffers (pH 8.3) contain 0.025 M Tris, 0.192 M glycine and 0.1% SDS. Samples for electrophoresis are made up to a final concentration of 0.0625 M Tris-HCl, pH 6.8, 1% SDS, 10% w/v sucrose, 50 mM dithiothreitol and 100 ng pyronin Y and are heated for 2 min at 100°C prior to application to sample slots on the gel. Purified glycoproteins, 2–20 μg, can be applied in sample volumes not exceeding 30 μl. When using the periodate-Shiff stain for glycoproteins a minimum of 5 μg sample should be applied. Suitable standard protein molecular weight markers are available commercially as kits (e.g. Pharmacia high and low molecular weight calibration kits) or they can be made up from mixtures of purified proteins. Electrophoresis is carried out at constant current (15 mA) until the tracker dye reaches the end of the gel (about 1 h). The dye front is marked by means of a needle dipped in Indian ink.

Staining and fixing of proteins (Fairbanks et al., 1971) can be achieved simultaneously by soaking gels overnight in a solution containing 0.05% (w/v) Coomassie Brilliant Blue in acetic acid:isopvopanol:water (10:25:65 by volume) followed by destaining for 6 h in acetic acid:isopropanol:water (10:10:80 by volume) and then in 10% acetic acid.

Staining of glycoproteins with the periodic acid - Schiff's (PAS) reagent can be done on gels fixed overnight in 10% acetic acid – 15% isopropanol followed by 10% acetic acid –25% isopropanol. Thorough removal of SDS is essential; failure to do this can result in staining of non-glycosylated proteins. PAS staining is carried out by soaking the gels in the following solutions at room temperature with continuous gentle shaking.

1. 0.5% periodic acid for 2 h;
2. 0.5% sodium arsenite in 5% acetic acid for 30 min;
3. 0.1% sodium arsenite in 5% acetic acid for two 20 min washes;
4. 5% acetic acid for 10 min;
5. Schiff's reagent (British Drug Houses) overnight (no shaking required);
6. 0.6% sodium metabisulphite in 0.01 M HCl, repeated until rinse solutions fail to turn pink after addition of formaldehyde. Glycoproteins give pink bands which should appear within a few minutes of the addition of Schiff's reagent.

Some soluble glycoproteins may not be adequately fixed in isopropanol. Leach et al. (1980a) suggested fixation in 5% phosphotungstic acid in 2 M HCl for 1.5 h followed by extensive washing in 7% methanol – 15% acetic acid. Periodate oxidation was performed with 1% periodic acid in 7% trichloroacetic acid for 1 h. Gels were then treated with 0.5% sodium metabisulphite in 0.1 M HCl for 1 h followed by staining with Schiff's reagent.

A more sensitive fluorescent method for the detection of glycoproteins has been described by Eckhart et al. (1976). The aldehyde function generated by periodate oxidation is condensed with dansyl hydrazine.

For molecular weight determinations the distance moved from the top of the separating gel to the centre of the stained bands of proteins and glycoprotein standards is measured. Relative mobility is obtained by dividing the distance moved by the proteins by the distance moved by the tracker dye. A plot of relative mobility of standard proteins against log molecular weight should give a straight line (Fig. 4.4a). Molecular weights of unknown samples are determined by interpolation.

Proteoglycans

Electrophoresis in polyacrylamide gels provides a rapid and sensitive method for estimating the molecular weights of glycosaminoglycans. As with other empirical methods, samples of the same type of glycosaminoglycans of known molecular weight are required as standards.

This method has been described by Matthews and Decker (1971) and further details are given by Rodén et al. (1972) and by Matthews (1976). Apparatus for carrying out gel electrophoresis in 5 × 75 mm tubes at a constant temperature (e.g. Polyanalyst electrophoresis apparatus, Buchler Instruments) is required. Samples containing 10–20 μg of glycosaminoglycan are applied to polyacrylamide gels (3–10% acrylamide depending on molecular weight of the sample) made up in 0.05 M phosphate buffer, pH 7.5. Electrophoresis is carried out at 27.5 ± 0.5 V for 120 min with water circulation in the apparatus at 25°C. Gels are removed and stained with 0.5% Alcian blue SGX (Fisher Chemical Co.) in 3% acetic acid for 60 min followed by destaining overnight in 7% acetic acid. The migration distance from the top of the gel to the peak of stain intensity is measured. At least two reference samples of known molecular weight of the same type of glycosaminoglycan should be included.

The distance of migration for the standards is plotted against log molecular weight and the unknown molecular weight is determined from this graph. As the electrophoretic mobility depends on the charge as well as the size of the glycosaminoglycan the standards should preferably have a similar ratio of sulphate to hexosamine to that of the sample.

Similar methodology can be applied to proteoglycans if electrophoresis is performed in large-pore composite polyacrylamide – agarose gels (McDevitt and Muir, 1971). The average pore size is sufficient to allow penetration of all but the largest articular cartilage proteoglycans, making it possible to distinguish aggregated from disaggregated molecules. Mobilities of proteoglycans were found to be related inversely to hydrodynamic size as determined by gel chromatography. In contrast, the relative mobilities of separate chains of different types of glycosaminoglycans were shown to be a function of their charge density (compare Section 4.4). Cylindrical gels are prepared containing 0.5–1.5% acrylamide and 0.6% agarose. After electrophoresis proteoglycans are stained with 0.2% Toluidine blue in 0.1% acetic acid. Alcian blue should not be used as it stains the matrix strongly.

4.3.3. Sedimentation equilibrium ultracentrifugation

Sedimentation equilibrium ultracentrifugation remains the most important primary method for determination of the molecular weights of glycoproteins and proteoglycans. This method has the great advantage that it is independent of molecular shape and it is applicable to milligram or microgram quantities. While it is necessary for the sample to be homogeneous, the technique is suitable for the molecular weight analysis of polydisperse glycoconjugates.

Difficulties can be encountered in using this method to determine the molecular weights of glycoconjugates which undergo self-association. However, the occurrence of association is indicated by data obtained in sedimentation equilibrium experiments and the technique can be used to characterise some associating systems. Molecular weight determination can be carried out in solvents which break down intermolecular interactions such as solutions containing high concentrations of salt, urea, guanidine hydrochloride or detergent. The major limitation to the accuracy of molecular weight analysis by sedimentation equilibrium ultracentrifugation usually lies in the precision of determination of the partial specific volume ($\bar{v}$) of the glycoprotein or proteoglycan in the solvent used for molecular weight determination.

The remainder of this section will give a brief introduction to the sedimentation equilibrium analysis of glycoconjugates. It is not, however, possible to give a rigorous theoretical treatment of the subject or details of operation of the ultracentrifuge. Good reviews of the theoretical principles of ultracentrifugation (Birnie and Rickwood, 1978), practical methodology (Rickwood, 1984) and specific problems associated with glycoconjugates (Gibbons, 1972; Bettelheim, 1977) are available.

The basic equipment required for sedimentation equilibrium ultracentrifugation of glycoconjugates is a Beckman model E analytical ultracentrifuge equipped with either interference optics or an ultraviolet scanner. It is desirable, but not essential, to have equipment for measurement of the partial specific volume of the glycoconjugate such as a Paar mechanical oscillator.

The principle of the method is that a solution of the glycoconjugate is centrifuged in a sector-shaped cell until equilibrium is reached between sedimentation and diffusion. From thermodynamic considerations it can be shown that for an ideal (non-interacting) two-component system the distribution of a homogeneous macromolecule can be described by the following expression (Aune, 1978):

$$\frac{\mathrm{d}\ln m_2}{\mathrm{d}r^2} = \frac{[M_2(1-\bar{v}\rho)\omega^2]}{2RT} \qquad (1)$$

where r is the distance from the centre of rotation; m_2 is the molality of the solute; M_2 the molecular weight and $\bar{v}$ the partial specific volume of the solute; ρ is the solvent density and ω the angular velocity. Experimentally the molalities employed are such that

$$\mathrm{dln}m_2 \simeq \mathrm{dln}C_2$$

Hence a plot of $\ln C_2$ against r^2 for an ideal system is linear with a slope equal to the right-hand side of equation (1) from which M_2 can be determined. The quantity $\ln C_2$ can be determined from the fringe pattern obtained with interference optics (Creeth and Pain, 1967; Aune, 1978) or by the use of an ultraviolet scanner. For glycoproteins or proteoglycans with a low content of aromatic amino acids absorption can be measured in the region of 220 nm.

For glycoproteins and proteoglycans non-linearity of the graph of $\ln C$ versus r^2 can be indicative of heterogeneity, polydispersity or of molecular interactions. Polydispersity produces upward curvature of the line. However, linear plots of $\ln C$ against r^2 can sometimes be obtained, as the result of upward curvature due to polydispersity being cancelled out by downward curvature arising from non-ideality (Creeth and Knight, 1968). Further information about association or polydispersity can be obtained by computing molecular weight averages at different positions in the ultracentrifuge cell (Aune, 1978). For a polydisperse glycoconjugate the molecular weight average for the original unredistributed sample is obtained by integration over the whole cell (Gibbons, 1972).

The time required for equilibrium to be established by the low-speed techniques is usually several hours, longer times being required for substances of low molecular weight. However, satisfactory results can be obtained even with glycoprotein fragments of molecular weight below 5000 (Beeley, 1976a). Molecular weight determinations can be carried out more rapidly (but with some sacrifice of precision) using the high-speed equilibrium method of Yphantis (1964). The centrifuge is operated at about three times the usual speed so that a steep concentration gradient is formed and the concentration of sample remaining at the top of the cell, near the meniscus, can be neglected relative to the initial concentration. For this reason this is sometimes referred to as the 'meniscus depletion method'. Very low concentrations of sample can be employed.

The solvent for sedimentation equilibrium ultracentrifugation of glycoproteins and proteoglycans is usually a buffered salts solution. Before analysis the sample is dialysed to equilibrium against the solvent. To minimise electroviscous effects, a salt concentration of 0.15 M or greater is employed with charged glycoconjugates. If intermolecular interactions occur these may be decreased or prevented by the use of high salt concentrations (e.g. 2 M NaCl) or high concentrations of urea (8 M) or guanidinium chloride (6 M) or detergent (e.g. 1% sodium dodecyl sulphate). However, it then becomes necessary to take into account interactions involving this third component. In particular it is necessary to determine the value of apparent specific volume of the macromolecule in solvent containing the third component.

Determination of $\bar{v}$

Measurement of molecular weight by the-equilibrium method requires accurate estimation of the partial specific volume $\bar{v}$. This is defined as the change in volume which occurs when one gram of solute is added to an indefinitely large volume of solvent. For a substance which behaves ideally this quantity would be equal to the reciprocal of the density of the dry material. However, for a real glycoprotein the quantity will include any volume changes occurring as a result of interactions of peptide and carbohydrate moieties with solvent.

When sufficient quantities of glycoprotein (100–500 mg) are available $\bar{v}$ can be determined by pycnometry using the relation

$$\bar{v} = \frac{1}{\rho_s}\left(1 - \frac{\Delta\rho}{c}\right)$$

where ρ_s is the solvent density and $\Delta\rho$ the difference in density between solution at concentration c (g/ml) and solvent. Lyophilised glycoprotein should be dried to constant weight at 70–80°C over P_2O_5 in a good vacuum to ensure that it is anhydrous. Even brief exposure to air can lead to uptake of moisture. A weighed quantity is then made up accurately to a known volume and the resulting solution is weighed to determine its density. The quantity $\Delta\rho$ is obtained by subtracting the density of the pure solvent. Generally solute concentrations of about 1% w/v are employed. Strictly, measurements should be made at several different concentrations and the $\bar{v}$ values obtained extrapolated to zero concentration (Gibbons, 1972). For cartilage proteoglycan Woodward et al. (1972) showed that there was significant concentration-dependence of $\bar{v}$.

The partial specific volumes of smaller samples (about 10 mg) of glycoconjugates can be measured by the use of a mechanical oscillator technique (Kratky et al., 1973; Elder, 1979) using commercially available equipment (A Paar, Gratz, Austria). Other techniques for the measurement of $\bar{v}$ have also been described (Krupke and Beams, 1972; Sakura and Reithel, 1972; Hunter, 1978). Where only small quantities of material are available $\bar{v}$ can be estimated from sedimentation equilibrium measurements in solutions whose density is altered by the addition of different proportions of D_2O to H_2O (Edelstein and Schachman, 1967). Deuteration of the exchangeable protons on the hydroxyl and *N*-acetyl groups of the glycoprotein as well as exchangeable groups on the protein must be allowed for.

The partial specific volume of glycoproteins has been found to vary significantly with the carbohydrate content. Creeth et al. (1974) showed that glycoproteins varying between 7.1 and 98% in peptide content had $\bar{v}$ values ranging between 0.608 and 0.73 ml/g. Muir and Hardingham (1975) show that for polysaccharides $\bar{v}$ is 0.60 to 0.65 ml/g and the presence of sulphate residues gives glycosaminoglycans even lower values of $\bar{v}$.

When the complete amino acid and carbohydrate composition of a glycoprotein is known in detail it is possible to calculate a value for $\bar{v}$ assuming that the partial specific volume of the macromolecule is the sum of the partial specific volumes of its constituent amino acid and sugar residues. The partial specific volume of the macromolecule is given by $\bar{v}_i w_i$ where $\bar{v}_i$ and w_i are respectively the partial specific volume and the weight fraction of the *i*th residue of sugar or amino acid (Schachman, 1968). Values of $\bar{v}$ for carbohydrate residues of glycoproteins and some proteoglycan repeating units obtained from crystallographic data are listed in Table 4.1 and values for amino acids are tabulated by Zamyatnin (1972). For ionised residues such as sialic acid the partial specific volume of the appropriate salt must be used (Gibbolns, 1972; Perkins et al., 1981).

TABLE 4.1

Partial specific volumes of some constituents of glycoproteins and proteoglycans (Perkins et al., 1981). Values for mono- and disaccharides are derived from crystallographic data. In deriving $\bar{v}$ for fucose and sialic acid it was assumed that these residues occurred in non-reducing terminal positions and that NeuAc was present as the sodium salt.

	$\bar{v}$ (ml/g)
Glc	0.622
Gal	0.622
Man	0.607
GlcNAc	0.684
GalNAc	0.684
Fuc	0.671
NeuAc	0.581
Hyaluronic acid	0.669
Sodium hyaluronate	0.588
Chondroitin sulphate	0.662
Disodium chondroitin sulphate	0.534
Na^+	−0.288
SO_4^{2-}	0.310

In general the agreement between such calculated values and those determined experimentally is surprisingly good (Gibbons, 1972) and this approach has been adopted in many sedimentation equilibrium studies of glycoproteins. Nevertheless it is preferable, whenever possible, to dctermine $\bar{v}$ experimentally.

Solvent containing extra components such as urea, guanidine hydrochloride or other salts

In carrying out molecular weight determinations it is often desirable to add large quantities of a third component (in addition to glycoconjugate and water) such as urea, guanidine hydrochloride or other salts to prevent intermolecular interactions or promote solubility. The possibility of interactions occurring between solute and third component gives rise to substantial theoretical difficulties. The problem can be circumvented by an approach developed by Casassa and Eisenberg (1964). In principle this involves considering the situation obtaining when water (component 1), glycoconjugate (component 2) and third component (urea, salt etc.) are dialysed to equilibrium against water and third component. Any interaction between glycoconjugate and third component will lead to an excess or deficiency of third component inside the dialysis bag. Molecular weight analysis requires knowledge of the apparent specific volume φ_2' of the macromolecule. This can be obtained by measuring $\Delta^*\rho$, the difference in density between the solution dialysed to equilibrium and its dialysate, and C, the concentration of glycoconjugate in the indiffusable portion. The latter concentration can be determined by ultraviolet absorption or by quantitation of hexose, sialic acid, etc. The background to this procedure is given fully by Casassa and Eisenberg (1964) and experimental procedures are discussed by Lee et al. (1979). This type of method has been applied, for example, to sedimentation equilibrium studies on cartilage proteoglycans and glycosaminoglycans (Woodward et al., 1972).

The dialysis equilibrium technique is not practicable for low molecular weight substances or where very little material is available. In such cases it may be necessary to calculate φ_2'. A method for the calculation of the apparent partial specific volumes of proteins in 6 M guanidinium chloride has been described by Lee and Timasheff (1979). This approach has been applied to some glycoproteins using the (untested) assumption that each residue of carbohydrate binds one molecule of guanidinium hydrochloride.

Solvent systems containing detergents

Many membrane glycoproteins require the addition of detergent to maintain them in solution or to prevent aggregation. Sedimentation equilibrium methods have been developed to determine the molecular weights of proteins in either ionic or non-ionic detergents. The difficulty arising in both cases is that detergent interacts with the macromolecular component and equations describing the condition of sedimentation equilibrium contain a term $(1-\varphi'\rho)$ where φ' is the effective partial specific volume of the detergent-protein complex.

One approach to assessing φ' requires knowledge of the amount of detergent bound and the individual partial specific volumes of protein and detergent (Grefrath and Reynolds, 1974). It is necessary to have labelled detergent, or a good chemical assay of detergent concentration so that binding of detergent to glycoprotein can be measured by, for example, equilibrium dialysis. The quantity $(1-\varphi'\rho)$ can, to a good approximation, be divided into independent contributions from glycoprotein and bound detergent, i.e.

$$(1-\varphi'\rho)=(1-\bar{v}_{GP}\rho)+\delta_{Det}(1-\bar{v}_{Det}\rho)$$

where δ_{Det} is the amount of detergent bound per gram of protein. The partial specific volume of the glycoprotein, $\bar{v}_{GP}$, can be calculated or measured by densimetry as described earlier. A value of $\bar{v}_{Det}$ for sodium dodecyl sulphate of 0.87 ml/g has been reported (Tanford et al., 1974). This method has been applied to human erythrocyte glycophorin A solubilised in sodium dodecyl sulphate (Grefrath and Reynolds, 1974). In this study sodium dodecyl sulphate was added to prevent aggregation rather than to solubilise the glycoprotein. Aggregation was observed when the bound detergent was less than 1.8 g/g glycoprotein but at higher concentrations of detergent an unambiguous monomer molecular weight of 29 000 was obtained.

A second approach which can be applied to membrane proteins solubilised in detergent does not require knowledge of detergent binding. The contribution of detergent to the sedimentation potential is blanked out by carrying out centrifugation in solutions containing different proportions of D_2O (Reynolds and Tanford, 1976).

Self-associating systems

Many glycoproteins and proteoglycans are subject to self-association to form aggregates which may be of extremely high molecular weight. This behaviour can be indicated by curvature of the plot of $\ln C_2$ against r^2. Where the aim of the experiment is to determine the 'subunit' molecular weight of a glycoprotein it may be possible to promote dissociation by the use of low concentrations of sample, by the addition of high concentrations of urea, guanidine hydrochloride or salt (Hill et al., 1977) or by adding a detergent such as SDS.

However, information about the intermolecular interactions involved in self-association can also be obtained from sedimentation equilibrium data. Stone and Reynolds (1975) have described procedures for dealing with such self-associating systems. Using this approach Hill et al. (1977) were able to determine that the association constant for ovine submaxillary mucin monomers in equilibrium with dimers was 6×10^4 litres/mole.

Woodward et al. (1972) have treated the self-association of chondroitin sulphate and proteoglycan preparations by a procedure based on that of Roark and Yphantis (1969).

4.3.4. Other methods

Several other physico-chemical techniques have been employed to determine the molecular weights of glycoconjugates. These include light-scattering and osmotic pressure measurements and the combination of information obtained from sedimentation velocity ultracentrifugation and diffusion or viscosity. The principles underlying these methods and their applications to glycoproteins and proteoglycans have been reviewed (Gibbons, 1972; Bettelheim, 1977). In recent years the use of these methods apart from scattering techniques has been relatively infrequent compared to sedimentation equilibrium ultracentrifugation and empirical methods.

Examples of the application of light-scattering techniques to the analysis of the molecular weight of proteoglycans can be found in the publications of Pasternack et al. (1974) and Sheehan et al. (1978). Neutron-scattering of a glycosylated proteoglycan fragment has been studied by Perkins et al. (1981).

The most definitive and accurate method for determination of the (subunit) molecular weight of glycoproteins is to analyse their primary structure.

4.4. Characterisation of charge

Glycoproteins

Glycoproteins containing sialic acid are generally paucidisperse or polydisperse with respect to charge due to the existence of molecular species differing in the number of sialic acid residues they contain. The most useful technique for examination of charge heterogeneity is isoelectric focusing. Species varying by a single charge per molecule can usually be resolved without great difficulty on an analytical scale and preparative separations can also be obtained. Paucidispersity arising from sialic acid variation can be distinguished from other types of charge heterogeneity by studying the effects of removing sialic acid from the glycoprotein by treatment with neuraminidase.

Analytical isoelectric focusing of glycoproteins is carried out in slabs of polyacrylamide gel using ampholytes which give a shallow pH gradient in the region of the isoelectric point of the glycoprotein. Suitable apparatus is available commercially (e.g. LKB Multiphor, Pharmacia Flat Bed Electrophoresis System, Bio-Rad Horizontal Electrophoresis Cell) and the preparation and use of these systems have been fully described in an earlier volume in this series (Righetti, 1983). Less sophisticated apparatus which can readily be constructed also gives excellent results (François-Gerard et al., 1979; Awdeh et al.,

1968). Ampholytes can be obtained from LKB, Pharmacia or Bio-Rad. After focusing, the glycoprotein is fixed in the gel and stained for 2 h in 0.2% Coomassie Brilliant Blue G-250 dissolved in water/ethanol/acetic acid (10:9:1 by vol) and destained in the same solvent (François-Gerard et al., 1979). Some variation in procedure may be necessary for glycoproteins which are not completely fixed in this solvent (see Section 4.3.2). The isoelectric points of the glycoprotein species can be determined if the pH gradient along the gel is measured with a microelectrode (Beeley et al., 1972) prior to staining.

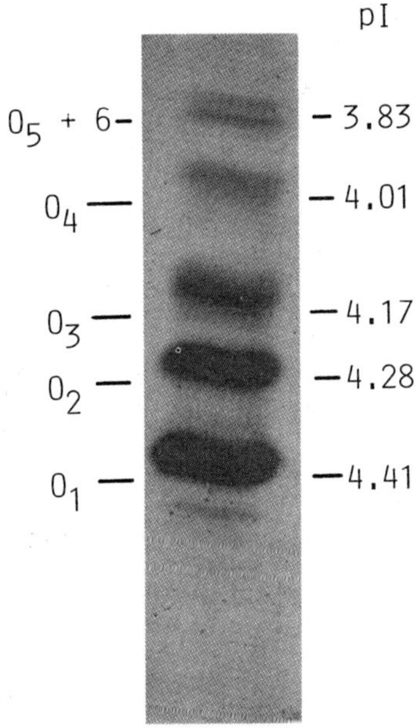

Fig. 4.5. Detection of charge heterogeneity of glycoproteins by isoelectric focusing. Charged variants of hens' egg ovomucoid were separated by isoelectric focusing in a gradient produced by pH 3–6 Ampholine (Beeley and McCairns, 1972).

Glycoproteins containing sialic acid give a series of from two to ten or more closely spaced bands differing in isoelectric point by small increments usually falling within the range 0.05–0.5 pH unit. The pattern of bands obtained on isoelectric focusing of hens' egg ovomucoid is shown as an example in Fig. 4.5. Following complete removal of sialic acid from ovomucoid by treatment with neuraminidase only bands corresponding to the two most basic components 0_1 and 0_2 are present, indicating that there is residual charge heterogeneity not accounted for by sialic acid.

In removing sialic acid from glycoproteins to examine charge microheterogeneity it is essential to ensure that digestion is complete and that no protease activity is present. The desialylated product can be examined for peptide chain cleavage by SDS-gel electrophoresis. For the removal of sialic acid from ovomucoid the glycoprotein (16 mg) dissolved in 500 μl of 0.05 M sodium acetate buffer, pH 5.5, containing 0.15 M NaCl and 7 mM $CaCl_2$ was incubated with 160 mUnits of *Vibrio cholerae* neuraminidase (Behringwerke), at 37°C for 24 h under toluene. Liberation of sialic acid was determined colourimetrically by the method of Aminoff (1959).

It is also possible to separate species of glycoprotein differing in sialic acid content by preparative isoelectric focusing or by ion-exchange chromatography (Beeley, 1971). The interconversion of species on removal of sialic acid can then be demonstrated directly.

Isoelectric focusing is not suitable as a method for characterising charge heterogeneity of glycosaminoglycans or proteoglycans. Artefactual banding patterns are obtained due to complex formation between polyanions such as heparin and ampholytes (Righetti, 1983).

Glycosaminoglycans

Glycosaminoglycans can be separated and identified rapidly, in microgram amounts, by zone electrophoresis on cellulose acetate strips or agarose gels. The amounts of glycosaminoglycans present in a mixture can be estimated quantitatively by densitometric scanning of the stained bands following electrophoretic separation. Because of their simplicity and sensitivity electrophoretic methods are widely used in the characterisation of glycosaminoglycans.

In supporting media which do not have an appreciable molecular sieving effect the electrophoretic mobility of glycosaminoglycans depends primarily on the charge per repeating disaccharide unit (Breen et al., 1976). The number and types of ionising group per disaccharide unit are listed in Table 4.2. The state of ionisation of these groups depends on their pK_a values and the buffer pH. At neutral pH the carboxyl and sulphate substituents are fully ionised and consequently heparin has the highest charge density of the glycosaminoglycans, while hyaluronate and unsulphated chondroitin have only one charge per disaccharide unit. Electrophoresis at pH values intermediate between the pK_as of uronic acids and sulphate half-ester can differentiate between glycosaminoglycans containing carboxyl groups and those which are sulphated. At low pH (e.g. in 0.1 M HCl) the charge a glycosaminoglycan has depends only on the sulphate groups and migration distance is related to sulphate content (Wessler, 1971).

TABLE 4.2
Ionising groups per disaccharide unit of glycosaminoglycans (Breen et al., 1976)

	COO^-	$-SO_3^-$	Total
Hyaluronic acid	1	0	1
Keratan sulphate	0	1.2–1.3	1.2–1.3
Chondroitin	1	0	1
Chondroitin 4-sulphate	1	1	2
Chondroitin 6-sulphate	1	1	2
Dermatan sulphate	1	1	2
Heparan sulphate	1	1	2
Heparin	1	2.3	3.3

Separation of glycosaminoglycans of the same net charge can be obtained by electrophoresis in solutions containing divalent cations (e.g. Ba^{2+}, Ca^{2+}, Zn^{2+}, Cu^{2+}, Cd^{2+}, organic diamines) which bind to them and modify their mobility. Electrophoreses in solutions of barium acetate have been found to be particularly useful. Wessler (1968) showed that the order of migration of GAGs on cellulose acetate strips in barium acetate differed from that obtained in buffers containing monovalent cations at the same pH. This system also showed considerably less band spreading due to variation in sulphation within populations of glycosaminoglycan molecules. Another useful system (Seno et al., 1970) employs calcium acetate and is capable of resolving chondroitin 4-sulphate, chondroitin 6-sulphate and dermatan sulphate.

It is not possible to separate all of the known glycosaminoglycans in a single buffer system. To obtain a complete analysis of the components present in a mixture it is usually necessary to carry out electrophoresis in two or three different systems. Alternatively, procedures have been devised for carrying out electrophoresis successively in two buffers run in the same direction (Morris et al., 1981) or by two-dimensional electrophoresis (Hata and Nagai, 1972; Sato and Gyorkey, 1974; Dietrich et al., 1977). A separation of all glycosaminoglycans on a single cellulose acetate strip has been obtained by a combination of electrophoresis and precipitation with ethanol (Cappelletti et al., 1979). For characterising the glycosaminoglycans in a mixture the combined use of electrophoretic separation and digestion with specific glycosidases (Section 6.9.4.1) is very valuable.

Following separation glycosaminoglycans are stained with Alcian blue or Toluidine blue and can be quantitated by densitometry or by elution of bands.

Procedures

Samples are prepared by exhaustive digestion of proteoglycan with papain or pronase followed by deproteinisation and precipitation of glycosaminoglycans (Section 6.4.2). Apparatus suitable for cellulose acetate electrophoresis is available commercially from Beckman (Microzone) and Gilson. The following electrophoretic procedure is based on that of Murata (1980).

Samples (0.25 μl) containing 2–10 μg of glycosaminoglycan are applied to an origin marked on strips of cellulose acetate (Oxoid, Separex or Sephrapore III) which have been moistened with one of the following buffers; A, 0.1 M formic acid – pyridine (pH 3.1); B,

0.1 M calcium acetate; or C, 0.1 M barium acetate. Electrophoresis is carried out in buffer A at 0.5 mA/cm for 60 min, in buffer B at 0.5 mA/m for 18 min and in buffer C at 5 V/cm for 180 min (Murata, 1980). Electrophoresis strips are stained in 0.5% Alcian blue 8-GX or 0.5% Toluidine blue in 0.5% acetic acid for 2 h followed by destaining in 0.5% acetic acid until the background is clear.

In buffer A the order of migration of standard glycosaminoglycans is heparin > chondroitin sulphates > heparan sulphate > dermatan sulphate > keratan sulphate > hyaluronate. Heparan sulphate is heterogeneous in charge and may form a broad band overlapping chondroitin and dermatan sulphates (Rodén et al., 1972). Buffer B separates chondroitin 4- and 6-sulphates and dermatan sulphate. Heparin sulphate and keratan sulphate migrate close to the chondroitin sulphates and may interfere in their detection. In buffer C, polysaccharides are fractionated into three groups of increasing mobility: heparan sulphate-heparin, dermatan sulphate-hyaluronic acid and the chondroitin sulphates.

Identification of glycosaminoglycans should not be made solely on the basis of electrophoretic mobility. Samples should be digested with specific enzymes and re-electrophoresed to determine whether degradation has occurred (section 6.9.4.1).

Quantitation of glycosaminoglycans can be obtained by clearing the cellulose acetate strips and densitometric scanning as described by Breen et al. (1976).

Cellulose acetate electrophoresis can be carried out rapidly with simple equipment and using small quantities of sample. One limitation of the method is that a quite concentrated (1–5 mg/ml) sample, free from salts and major impurities is required. For larger volumes of sample, or where impurities are known to be present, agarose gel electrophoresis is a useful alternative (Dietrich et al., 1977; Funderburgh and Chandler, 1978). A second limitation of both cellulose acetate and agarose systems is the overlapping of glycosaminoglycan bands which arises from polydispersity of charge and molecular weight. This type of problem is minimised by use of the divalent-cation-containing systems which show relatively little band broadening due to charge polydispersity. If necessary improved resolution may be obtained by using a more complex system such as the cellulose acetate system of Cappelletti et al. (1979) or the agarose method of Bianchi et al. (1980) in which bands are sharpened by introducing a discontinuity into the electrophoresis buffer.

Because of variations in the relative amounts and characteristics (such as the extent of sulphation, existence of hybrid structures and molecular weight heterogeneity) of glycosaminoglycans from different sources it is necessary to adapt methodology to the particular sample being analysed. Some examples of analytical procedures which have been applied to the separation of glycosaminoglycans from tissues are given in Table 4.3.

TABLE 4.3

Some methods for the electrophoretic characterisation of glycosaminoglycans from particular sources

Cellulose acetate	Human tissue	a. 0.05 M $LiCl_2$/0.01 M HCl, pH 2 b. 0.05 M phosphate, pH 7.2 c. 0.2 M $ZnSO_4$, pH 5.1	Breen et al., (1976)
	Pigeon aorta	0.3 M cadmium acetate, pH 4.1	Curwen and Smith (1977)
	Synovial fluid and articular cartilage	0.1 M phosphate, pH 2.0	Smith et al., (1980)
Agarose	Cultured cells	0.3 M barium acetate, pH 5.7	Fundenberg and Chandler (1980)

4.5. *Shape, interactions and flexibility*

Information can be obtained about the molecular shape and interactions of glycoproteins and proteoglycans by several physico-chemical techniques in addition to those described in earlier sections of this chapter. The types of parameter measured by hydrodynamic, scattering, optical, electron microscopic and NMR techniques are summarised in Table 4.4.

Hydrodynamic methods in general give a fairly crude assessment of molecular shape which can be interpreted in terms of models of flexible, stiffened or 'worm-like' coils, rods, elipsoids or spheres (Gibbons, 1972; Bettelheim, 1977; Tanaka, 1978). In some circumstances it is also possible to assess chain flexibility from viscosity measurements (Morris and Ross-Murphy, 1981). Information about molecular interactions can be obtained from viscosity (Allen et al., 1976; Pasternack, 1974; Morris and Ross-Murphy, 1981) and from sedimentation behaviour (Sheehan et al., 1978).

Scattering properties can give information about molecular size, shape (radius of gyration) and macromolecule-solvent interactions (second virial coefficient). For example, light-scattering measurements have been used to examine changes in the shape and solvent interactions of proteoglycans in guanidinium chloride solutions (Pasternack et al., 1974) and the self-association of laryngeal cartilage proteoglycan in 0.15 M sodium chloride (Sheehan et al., 1978). Neutron scattering (Perkins et al., 1981) appears to have considerable potential for studying the shape of glycosylated proteins. An interesting feature of this technique is that the scattering due to the carbohydrate component can be selectively blanked out by the use as solvent of D_2O-H_2O mixtures of equivalent scattering density to the carbohydrate. The scattering due to the entire molecule can therefore be compared with that due only to the peptide moiety.

Electron microscopy has proved to be very valuable for examining the shape of proteoglycans and their aggregates (Rosenberg et al., 1975; Kimura et al., 1978). A mixture of the sample with cytochrome *c* is prepared in the form of a thinly spread film which is then stained with uranyl acetate. Electron microscopic techniques can also provide information about the shape of glycoproteins (Slayter, 1981). Heavy metal shadowing is usually employed and has been particularly useful in examining the shape of large extended macromolecules such as lubricin (the boundary lubricant of articular cartilage), fibrinogen and epiglycanin. The influence of the carbohydrate chains of ovine

TABLE 4.4

Methods for assessing molecular shape, interactions and flexibility of glycoproteins and proteoglycans

Properties	Parameter measured	Reference
HYDRODYNAMIC		
Sedimentation velocity	sedimentation coefficient, frictional ratio	Bettelheim (1977)[RPG], Gibbons (1972))[RGP], Sheehan et al. (1978)[PG], Rose et al. (1984)[GP]
Diffusion	diffusion coefficient	Gibbons (1972)[RGP], Swan et al. (1981)[GP]
Viscosity	intrinsic viscosity, shear and concentration-dependence of viscosity	Morris and Ross-Murphy (1981)[RGP], Woodward et al. (1972)[PG], Matthews and Decker (1977)[PG], Tanaka (1978)[PG], Allen et al. (1976)[GP]
COLLIGATIVE PROPERTIES		
Osmotic pressure	M_n, second virial coefficient	Bettelheim (1978)[RPG], Wasteson (1971)[PG]
SCATTERING PROPERTIES		
Light scattering	M_w, radius of gyration, second virial coefficient	Pasternack et al. (1974)[PG], Matthews and Decker (1977)[PG], Sheehan et al. (1978)[PG], Gibbons (1972)[RGP]
Quasi-elastic light scattering	Stokes radius	Perkins et al. (1981)[PG]
X-ray scattering	M, radius of gyration	Cleland (1977)[PG]
Neutron scattering	M, radius of gyration	Perkins et al. (1981)[PG]
OPTICAL PROPERTIES		
Circular dichroism Optical rotary dispersion	Spectra depend on primary structure and conformation of protein and carbohydrate moieties	Morris et al. (1981)[GP], Stone (1976)[PG], Morris et al. (1975)[PG], Franks and Morris (1978)[GP]
Refractive index	Refractive index increment	Pasternack et al. (1974)[PG], Sheehan et al. (1978)[PG]

Properties	Parameter measured	Reference
OTHER PROPERTIES		
Electron microscopy	Average length, width, shape	Rosenberg et al. (1975)[PG], Kimura et al. (1978)[PG], Rose et al. (1984)[GP], Slayter (1981)[RGP]
X-ray diffraction from fibres or oriented films	Spacing of repeating units interpreted by model fitting to give secondary structure	Winter (1982)[RPG]
X-ray diffraction from single crystals	Diffraction pattern used to calculate electron density and hence 3-D structure	Diesenhoffer (1981)[GP], Wilson et al. (1981)[GP]
Nuclear magnetic resonance	Chemical shifts, line width, relaxation times, nuclear Overhauser enhancement	Darke et al. (1975)[PG], Brewer and Kaiser (1975)[PG], Egmond et al. (1979)[GP], Dill and Allerhand (1979)[GP], Rademachar et al. (1983)[GP]

References relating particularly to glycoproteins or proteoglycans are given the superscripts GP and PG, respectively. Review articles are indicated by RGP or RPG when they relate primarily to one or other category of molecule.

submaxillary mucin on the shape and intermolecular interactions of subunits has been dramatically demonstrated by electron microscopy following enzymic removal of the carbohydrate chains (Rose et al., 1984); the intact glycoprotein has an extended thread-like appearance but the deglycosylated peptide chain is globular. It is, of course, necessary to show that the shapes of molecules observed by electron microscopy are consistent with their physico-chemical properties in solution. Examples of such correlations can be found in papers by Rose et al., (1984) and Swann et al. (1981). Solution properties should be determined, as far as possible, at the same concentrations as those employed for preparing samples for electron microscopy.

The optical properties, such as circular dichroism, of glycoproteins and proteoglycans arise from both protein and carbohydrate components. Amino acid side-chains and the peptide backbone chromophores can make a strong contribution and the optical properties of the carbohydrate component are usually more informative when little protein is present, as for example in hyaluronate (Franks and Morris, 1975). The optical properties of the carbohydrate components are influenced by primary structure and conformation (Wright, 1976; Morris, 1981).

The single most powerful method for determination of the secondary and tertiary structure of polysaccharides is X-ray diffraction from fibres (Winter, 1982). This approach has been applied extensively to glycosaminoglycans (Atkins and Isaac, 1973; Atkins and Sheehan, 1973; Guss et al., 1975; Winter et al., 1975; Arnott et al., 1973, 1974). Samples are usually orientated by preparing partially dried films of glycosaminoglycan which are then stretched under defined conditions of relative humidity. X-ray diffraction patterns are analysed by fitting the repeating distances along and between the molecular axes to structural models derived from the polysaccharide primary sequence and known restrictions on the conformation of the carbohydrate chains. It has been found that a particular glycosaminoglycan, such as hyaluronate, can exist in structurally distinct forms, the transition between different structures being critically dependent on the pH, counter-ions present and the relative humidity (Atkins and Mackie, 1972). This type of analysis gives information about the secondary structures and interactions between glycosaminoglycan chains which occur in partially ordered arrays of molecules. However, in solutions of proteoglycans these interactions are not necessarily stable, or may only exist transiently. It is therefore necessary to correlate structural analysis by X-ray diffraction from fibres with results obtained by techniques such as NMR which give information about the conformation and mobility of proteoglycans in solution.

X-ray diffraction from single crystals of glycoproteins can give a detailed three-dimensional picture of conformation and interactions of both the carbohydrate and the peptide moieties. Primary sequence analysis is a prerequisite for interpretation of structure. To date the number of glycoproteins studied in this way has been quite limited (Deisenhofer, 1981; Sutton and Phillips, 1983; Wilson et al., 1981,

1983); this is due in part to difficulties in crystallising glycoproteins associated with carbohydrate heterogeneity. As with fibre diffraction experiments it is highly desirable to relate the structure found in the crystalline state to information derived from studying the conformation and flexibility of carbohydrate units in solution.

The conformations and molecular motions of glycosaminoglycans, proteoglycans and glycoproteins can be examined in solution by NMR techniques. Both ^{1}H and ^{13}C resonances have been studied. Signal line widths of ^{13}C-NMR spectra are relatively much narrower than those of ^{1}H, thus decreasing problems associated with the overlap of signals (Dill and Allerhand, 1979). However, the intensity of ^{13}C-NMR spectra requires considerable enhancement because of the low natural abundance of the isotope. Measurements of line width, relaxation times and nuclear Overhauser enhancement have been used to assess the molecular motion of the oligosaccharide chains of proteoglycans (Darke et al., 1975) and glycoproteins (Egmond et al., 1979; Dill and Allerhand, 1979; Rademacher et al., 1983). Proteoglycans have been examined in aggregates and in samples of connective tissue (Brewer and Kaiser, 1975; Torchiada et al., 1977). Detailed analysis of the preferred conformations of oligosaccharides or glycopeptides isolated from glycoproteins containing *N*-linked carbohydrate units has also been carried out by ^{1}H-NMR (Brisson and Carver, 1983). It is also possible to determine the ring conformation of sugar residues present in glycoprotein or proteoglycan chains from their NMR spectra (Coxton, 1972; Hamer and Perlin, 1976).

CHAPTER 5

Analysis of constituents

5.1. Introduction

Once a homogeneous preparation of glycoprotein or proteoglycan has been obtained analysis of the constituent amino acids, sugars and their substituent groups can be undertaken. The carbohydrate composition will differentiate proteoglycans from glycoproteins and can give a good indication of the types of carbohydrate structures present. It may be possible to establish the class of proteoglycan carbohydrate units directly from the composition. For glycoproteins the sugar composition will often suggest whether the carbohydrate units are of the 'simple' or 'complex' type or whether they are likely to be linked *O*-glycosidically (see Chapter 2).

The sugars commonly found in proteoglycans and glycoproteins are listed in Table 2.1. Some variation can occur in the substituent groups of certain of the monosaccharides present in glycoproteins and proteoglycans. The amino sugars glucosamine and galactosamine are generally *N*-acetylated in glycoproteins. *N*-sulphated hexosamine occurs in heparin and heparan sulphate. *O*-sulphate groups occur in most proteoglycans and in some glycoproteins. *O*-phosphoryl mannose groups have recently been shown to occur in certain glycoproteins. The sialic acids can contain acetyl and glycolyl substituents of the amino or hydroxyl groups of neuraminic acid. Determination of these substituent groups forms part of the analysis of glycoprotein composition (Section 5.11).

Amino acid analysis (Section 5.4) of glycoconjugates can give information about the presence of repeating structure in the peptide chain and may suggest ways in which protein and carbohydrate could be linked. For example the presence of hydroxylysine would suggest the possibility of a collagen-like structure with protein carbohydrate linkages involving this amino acid. Information from amino acid analysis can also be employed to determine the strategy for further structural analysis, as discussed in Chapter 6.

5.2. *Analytical methods*

The best approach to adopt in the determination of the carbohydrate composition of a glycoprotein or glycopeptide depends on the amount of sample and the analytical equipment available. When very small samples (0.5–10 nmoles of sugar) are to be analysed the micromethods described by Takasaki and Kobata (1978) and by Conrad (1977) have the required sensitivity (Section 5.9). For larger quantities ($\geqslant$ 10 nmoles), gas-liquid chromatography following methanolysis (Section 5.8.5.) can give a complete analysis of neutral and amino sugars from a single sample (Clamp et al., 1971). Automated ion-exchange methods for the separation and quantitation of neutral sugars (as borate complexes) are of comparable sensitivity to the GLC methods but require separate analysis of neutral sugars, hexosamines and sialic acid. However, the precision of the ion-exchange method is good, sample preparation is simple, and specific detection systems can be employed (Section 5.8.3.). Hexosamines can be determined with high precision using standard amino acid analyser equipment (Section 5.4.). Gas-liquid chromatography and ion-exchange chromatography have largely superseded paper chromatography for quantitative analysis. Paper chromatography and thin-layer chromatography retain a place as auxiliary methods for the identification of sugars.

Enzyme assays have been developed for the specific assay of D-mannose, D-glucose, D-galactose, L-fucose, D-glucosamine (Finch et al., 1969; Kornfeld et al., 1971) and sialic acid (Schauer et al., 1974) in sugar mixtures. About 25 nmoles of sugar can be estimated spectrophotometrically and fluorimetric assay can increase the sensitivity ten-fold (Schachter, 1975). These methods do not require the prior separation of sugars and they are stereospecific. They are particularly useful for the repetitive assay of a limited number of sugars, for example when following the action of a glycosidase. Some enzyme assays for monosaccharides are available commercially (see Section 5.8).

When applied to purified glycoproteins colourimetric methods will give an estimate of the amount of a particular sugar, or class of sugar, present in the preparation. These methods, which typically require about 100 nmoles of sugar per assay, do not demand specialised equipment. In most cases they can be applied directly to the glycoprotein without prior hydrolysis. The lack of complete specificity of many of these methods is, however, a limitation.

Colourimetric assay of uronic acid (Section 5.6) is a useful preliminary step in the analysis of the carbohydrate units of proteoglycans. The amino sugars can be estimated following release by hydrolysis using colourimetric methods (Section 5.5) or galactosamine and glucosamine can be estimated separately by means of the amino acid analyser (Sections 5.4 and 5.5). Obtaining quantitative release of the sugars present in proteoglycans by hydrolysis can present special difficulties, particularly when *N*-sulphated hexosamine is present (Sections 5.3, 5.5 and 5.6). A general method for the quantitation of sugars in proteoglycans by radiochromatography is discussed in Section 5.6.

Analysis of substituent groups such as acetyl and sulphate (Section 5.11) are a necessary part of the description of the composition of proteoglycans and glycoproteins. Differentiation between the various substituted sialic acids is described in Section 5.7.

The amino acid compositions of glycoproteins and proteoglycans are normally determined by analysis of hydrolysates using an amino acid analyser (Section 5.4). Minor modifications of the procedures of sample preparation and analysis for non-glycosylated proteins are required to prevent destruction of amino acids and to permit determination of hexosamines.

With the exception of colourimetric assays carried out in strongly acidic conditions the methods for analysis of the composition of glycoconjugates require cleavage of glycosidic linkages by one of the procedures discussed in the next section.

5.3. Hydrolysis and methanolysis

Before separation of individual monosaccharides can be achieved it is necessary to cleave the glycosidic linkages of the glycoprotein or proteoglycan. This can be accomplished by acid hydrolysis or methanolysis.

Hydrolysis

The rate of release of a monosaccharide by acid-catalysed hydrolysis of a proteoglycan or glycoprotein depends on the monosaccharide involved and on its linkage to neighbouring sugars (Marshall and Neuberger, 1972; Conchie, 1976). Liberated sugars differ in their stability to acid. Hence the conditions chosen for acid hydrolysis are usually a compromise between obtaining complete release of each sugar and preventing excessive destruction.

Sialic acid and fucose in glycoproteins (and keratan sulphate) generally occur as non-reducing terminal residues and both sugars form glycosides which are rapidly cleaved by acid hydrolysis. Very substantial losses of these sugars may occur in hydrolysis conditions required to release most of the hexose in a glycoprotein.

To obtain good yields of hexosamine in hydrolysates of glycoconjugates conditions must be chosen in which hydrolysis of the glycosidic linkages of hexosamine residues is substantially faster than deacetylation. If deacetylation occurs first the stability of the glycosidic linkage is greatly increased. Hydrolysis conditions severe enough to release all of the hexosamines are liable to cause destruction of neutral sugars.

The quantitative release of hexosamine from proteoglycans requires more vigorous hydrolysis than is usually needed in glycoprotein analysis. Glycosidic linkages formed by glucuronic acid are resistant to hydrolysis (Section 5.6.). It is difficult to obtain quantitative hydrolysis because of concomitant destruction of the released hexosamine. The occurrence of *N*-sulphated hexosamine can also give rise to problems in analysis. The sulphate group is rapidly released in acid conditions and this leads to difficulty in cleaving the glycosidic linkage of the hexosamine (Section 5.6).

Preliminary analysis of the hexosamine present in glycoconjugates can be performed after hydrolysis in a sealed tube in 4 M HCl for 4 h at 100°C (Chandrasekaran and BeMiller, 1980). Conditions for obtaining optimal release of uronic acids and other components of proteoglycans are discussed further in Section 5.6.

To obtain optimal recoveries of the carbohydrate constituents of glycoproteins four separate acid hydrolyses may be required for the analysis of sialic acid, fucose, hexose and hexosamine. For a preliminary analysis of the sugars present in a glycoprotein hydrolysis in a sealed tube in 2 M HCl or 2 M TFA for 3–6 h at 100°C (Sections 5.9 and 5.8.3) can be employed. A separate hydrolysis for analysis of sialic acid, which is completely destroyed under these conditions, must also be carried out (Section 5.7.3).

Destruction of the sugar released from glycoconjugates on acid hydrolysis can be increased by browning reactions in which free reducing groups react with amino groups of amino acids or peptides. These reactions may take place during hydrolysis but they can be of particular importance when the acid is removed and the sample is concentrated because the reactive species are the unprotonated amino groups. For this reason both the conditions of hydrolysis and the way in which the hydrolysate is dried down can affect the recovery of monosaccharides. Such losses can be minimised by carrying out hydrolysis in dilute solution ($\leqslant$ 1 mg/ml) and by drying the hydrolysate rapidly in a rotary evaporator. It is convenient to use a volatile acid (HCl or TFA) for hydrolysis although slightly less destruction of sugars occurs in H_2SO_4 or methanesulphonic acid compared with HCl.

It is possible to make a correction for the destruction of monosaccharides during hydrolysis by the addition of isotopically labelled sugars. Radioisotope dilution techniques which can be applied to the analysis of monosaccharides without the necessity for optimising hydrolysis conditions are discussed in Section 5.10.

Methanolysis

Cleavage of glycoproteins in anhydrous methanolic HCl produces a mixture of α- and β-methyl glycosides. If carried out under suitable conditions this reaction occurs in high yield and with little destruction of sugars (Chambers and Clamp, 1972; Conchie, 1976). One exception, however, is that the GlcNAc residue involved in GlcNAc-Asn linkage is not converted to its methyl glycosides in significant quantity. Some de-*N*-acetylation of *N*-acetyl hexosamines and sialic acid occurs during methanolysis and a reacetylation step is required. Derivatives of the methyl glycosides can be separated and quantitated by gas-liquid chromatography (Section 5.8.4.2). Each monosaccharide gives rise to multiple peaks on the chromatogram. Methanolysis, followed by the separation of methyl glycosides by GLC, is currently the most widely used method for the analysis of the carbohydrate units of glycoproteins.

In the following section procedures are described for the determination of the amino acid and carbohydrate compositions of glycoproteins and proteoglycans. Valuable accounts are available of the amino acid analysis (Eastoe, 1972) and carbohydrate analysis of glycoproteins (Clamp et al., 1971; Marshall and Neuberger, 1972a) and proteoglycans (Rodén et al., 1972).

5.4. *Amino acid composition*

The glycoprotein (or proteoglycan) preparation should initially be dialysed to remove salts and then lyophilised. If sufficient quantities of material are available it is preferable to equilibrate the glycoprotein with atmospheric moisture and then to weigh samples for determination of moisture, ash, extinction coefficient, carbohydrate and amino acid composition (Glazer et al., 1975). Moisture content is obtained by drying to constant weight at 105°C over P_2O_5 in vacuo (Krupke and Dornier, 1978). It is preferable to use a microbalance with vacuum chamber attachment because some dried glycoproteins take up water rapidly on exposure to air.

If the extinction coefficient of the glycoprotein is known a stock solution of the desired concentration can be obtained by measurement of absorbance at 280 nm. Aliquots of this solution can be analysed for amino acid and carbohydrate composition. Methods for the determination of extinction coefficient are discussed by Allen (1980).

For a modern amino acid analyser such as the LKB 4400 the quantity of amino acids that can be analysed with optimal accuracy (without special modification of the analytical system) is in the range 2–25 nmoles. To obtain a reliable analysis of residues present in low proportion about 3 nmoles of protein is required (i.e. 90 μg for a glycoprotein of molecular weight 30 000). Other analytical systems may require some adjustment of the sample quantity.

Procedure

The sample (90–500 μg) of glycoprotein in a thick-walled Pyrex test tube (125 × 12 mm) is dissolved or suspended in 0.5 ml of distilled water containing an internal standard (e.g. nor-leucine, 10 nmoles) and phenol (0.02% w/v). An equal volume of concentrated HCl (Aristar grade) is added. The end of the tube is drawn out in the flame from a blow torch to give an internal diameter of 1–2 mm. The sample is frozen in a mixture of solid CO_2 and ethanol and evacuated below 0.05 mmHg with a vacuum pump. It is convenient to have a three-way tap inserted in the vacuum line. When the sample has thawed out and evolution of gas has ceased the drawn-out portion of the tube is sealed by heating over a low flame. Hydrolysis is carried out at 110°C in a heating block. After cooling the tubes are opened by scratching all round with a diamond, moistening the scratch slightly with water and then touching the scratched glass with a white-hot Pyrex rod. HCl is removed from the hydrolysates on a rotary evaporator at a temperature of 35°C. Dried samples can be stored (at −20°C) prior to analysis. An aliquot of the sample dissolved in citrate buffer is analysed. (For analysers with separate short and long columns equal aliquots of sample are required for each column.)

When the quantity of sample is limited a single hydrolysis of 24 h should be performed. It is preferable to hydrolyse samples for 24, 48 and 72 h to allow correction for residues which are destroyed during hydrolysis (especially serine, threonine and amino sugars) or which are liberated slowly (e.g. valine, leucine). The internal standard is employed to correct for loss of sample occurring during transfers. The cysteine content should be determined separately on a sample oxidised with performic acid (Hirs, 1967). Tryptophan is destroyed by hydrolysis in HCl and should be determined spectrophotometrically (see Allen, 1980) or after hydrolysis with methanesulphonic acid (Liu and Chang, 1971) or following alkaline hydrolysis (Hugli and Moore, 1972).

Acid hydrolysis of glycoproteins can lead to the formation of yellow or brown colouration of the hydrolysate or, in some cases, to a brown or black precipitate of 'humin'. Any insoluble material should be removed by centrifugation prior to rotary evaporation of the sample. Slow evaporation of acid (e.g. in a vacuum desiccator) should be avoided because of increased losses of amino acids as a result of 'browning reactions'.

5.5. Hexosamines and hexosaminitols

Glucosamine and galactosamine can readily be determined at the same time as the amino acid composition of a glycoprotein or proteoglycan. With the LKB 4400 analyser no modifications of standard operating conditions are required (Fig. 5.1). For other analysers minor changes in buffer systems or operating temperature may be needed. When analyses are carried out with instruments using separate long and short columns conditions should be chosen in which

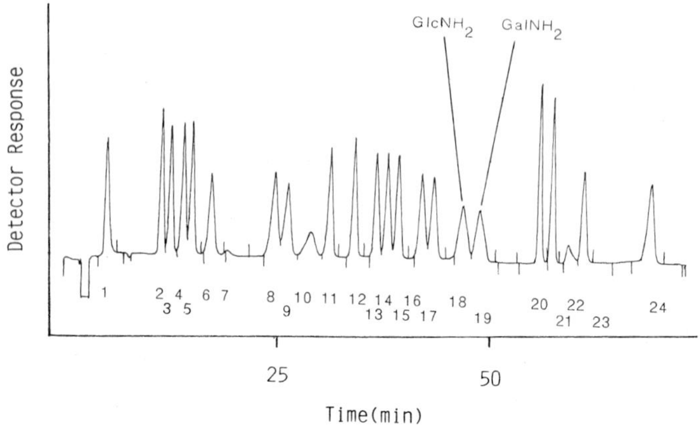

Fig. 5.1. Separation of amino acids and amino sugars. This elution profile was obtained using an LKB 4400 amino acid analyser from a standard amino acid mixture to which glucosamine and galactosamine had been added. The sample contained 5 nmoles of each amino acid (2.5 nmoles Cys) and of the amino sugars and was chromatographed using the normal buffer system and operating conditions for amino acid hydrolysate separation recommended by the manufacturer. Peaks correspond to: 1, cysteic acid; 2, aspartic acid; 3, methionine sulphone; 4, threonine; 5, serine; 6, glutamic acid; 7, (small peak) proline; 8, glycine; 9, alanine; 10, cystine; 11, valine; 12, methionine; 13, isoleucine; 14, leucine; 15, nor-leucine; 16, tyrosine; 17, phenylalanine; 18, glucosamine; 19, galactosamine; 20, histidine; 21, lysine; 22 and 23 ammonia, and 24 arginine.

glucosamine and galactosamine emerge from the long column as discrete peaks following the aromatic amino acids. Suitable modified buffer programmes have been described for Beckman (Fanger and Smyth, 1970; Downs and Pigman, 1976) and Technicon (Spiro, 1972) analysers. Extrapolation of the hexosamine values obtained after hydrolysis in 6 M HCl at 110°C (Section 5.4) for different periods to zero time give satisfactory quantitation. The recovery of glucosamine obtained after 24 h hydrolysis is about 50–75%; the precise value depends both on the glycoprotein or proteoglycan and the efficiency with which oxygen is excluded during hydrolysis. Higher recoveries of glucosamine can be obtained when less severe conditions of hydrolysis are employed. Hydrolysis of glycoproteins with 3–4 M methanesulphonic acid at 100°C for 24 h results in good recoveries of glucosamine (Beeley, 1976). This acid is not volatile and the pH must be adjusted carefully to that of the initial citrate buffer on the amino acid analyser. Alternatively, hydrolysis can be accomplished in shorter times with 4 M HCl at 100°C. Glucosamine is completely released with little destruction from several glycoproteins by hydrolysis at 100°C for 4–6 h in 4 M HCl (Spiro, 1972). Hydrolysis of proteoglycans in 6 M HCl at 100°C for 4 h is recommended (Chandrasekaran and BeMiller, 1980).

Glucosamine and galactosamine produced by hydrolysis of proteoglycans or glycoproteins can be analysed on an amino acid analyser, by a colourimetric method, or by a radioactive micromethod (Section 5.9). Unless very vigorous hydrolysis conditions (i.e. 6 M HCl at 110°C for 24 h or longer) are used it is unlikely that all peptide bonds will be cleaved. Peptides may emerge from the analyser in the same region as the hexosamines and interfere with their quantitation. When mild hydrolysis conditions are employed or the amounts of the amino sugars are low it is advisable to separate the amino sugars from neutral and acidic amino acids and peptides by passing the sample through Dowex-2 (Fanger and Smyth, 1970). This procedure removes artefacts which may be mistaken for amino sugars, particularly galactosamine. When hydrolysates are to be analysed colourimetrically for hexosamines by the Elson-Morgan reaction aminary absorbtion and elution from Dowex-1 is effective in removing most interfering substances (Boas, 1953).

Glucosamine and galactosamine can be reduced by $NaBH_4$ to give high yields respectively of glucosaminitol and galactosaminitol. This reaction forms the basis of a micromethod for estimating these hexosamines (Section 5.9). Hexosaminitols can also be produced during the analysis of carbohydrate structure; for example during alkaline $NaBH_4$ elimination of *O*-glycosidically linked carbohydrate groups (Chapter 6).

Free hexosaminitols react with ninhydrin and the separation and quantitation of glucosamine, galactosamine and their respective alcohols can be achieved by employing an amino acid analyser system with citrate-borate buffers (Donald, 1968; Downs and Pigman, 1976b). An accelerated analytical system allowing complete analysis of hexosamines and their glycitols in less than 4 h on a Beckman analyser was reported by Bella and Kim (1970). Spiro (1972) gives details of the procedure modified for a Technicon analyser.

Procedure for hydrolysis

Glycoprotein (1 mg) is dissolved in 1 ml 4 M HCl (6 M HCl can be employed for proteoglycans) in a thick-walled test tube with a screw-cap lined with Teflon. Oxygen-free nitrogen is bubbled through the sample for 30 s and the tube tightly sealed. After hydrolysis at 100°C for 4 h the sample is cooled and HCl removed completely on a rotary evaporator at 40°C.

Procedures for partial purification of hydrolysates

A. Method of Fanger and Smyth (1970) This method was developed for the removal of substances interfering with the estimation of hexosamines by means of an amino acid analyser.

Thy hydrolysate (1.5 mg glycoprotein) is concentrated to dryness in vacuo, water (2 ml) is added and the solution again dried. *N*-Ethylmorpholine (3 ml of a 2% v/v solution in water) is added and the pH adjusted to 9 with 1 M NaOH. The solution is passed through a column (5 × 0.9 cm) of Dowex-2 × 4 (acetate form) equilibrated and eluted with 20 ml of 1% *N*-ethylmorpholinium acetate, pH 9. Glacial acetic acid (1 ml) is added to the eluate, which is then concentrated to dryness several times with the addition of water. The residue is taken up in the starting buffer for amino acid analysis. Lysine is not removed from the sample by Dowex-2 and the amounts of amino sugars in the sample can be related to the quantity of lysine present.

B. Method of Boas (1953) This purification step is used prior to colourimetric assay of hexosamines by the Elson-Morgan reaction and is a modification of the original method.

The hydrolysate is redissolved in water (1 ml) and applied to a column (1 ml) of Dowex 50 × 4 (H^+ form) and washed on with water (5 ml). These washings are discarded and the hexosamine is eluted with 1 M HCl (5 ml) and dried on a rotary evaporator. After redissolving in water, aliquots of the sample can be assayed for hexosamine colourimetrically by the Elson-Morgan reaction.

Colourimetric assay of hexosamines Total hexosamine can be determined by the Elson-Morgan reaction after hydrolysis. As neutral sugars in the presence of amino acids can give coloured products in this reaction they should be removed as described above by Dowex 50 chromatography. The presence of NaCl depresses colour formation so the HCl used for elution of hexosamines from Dowex 50 must be removed by rotary evaporation.

Many modifications of the original procedure (Rondle and Morgan, 1955) have been proposed. The following method is that described by Davidson (1966).

Reagents. 2,4-Pentanedione (acetylacetone) is redistilled (bp 138–140°C) and stored in the deep-freeze. Other reagents required are *p*-dimethylaminobenzaldehyde (PDMAB), which should be colourless, 0.5 M Na_2CO_3, ethanol (95%) and conc. HCl.

2,4-Pentanedione (2 ml) is dissolved in 98 ml of 0.5 M Na_2CO_3 and PDMAB (677.5 mg) is dissolved in 25 ml of 1:1 ethanol-conc. HCl. These solutions must be prepared on the day of use.

Procedure. Samples containing 2.5–25 μg hexosamine are made up to 1 ml with water in tubes with caps fitted with Teflon liners. (Suitable Teflon-coated septa can be obtained from GLC equipment suppliers.) Tubes are heated in a water bath at 90°C for 40 min and then cooled in water. 95% ethanol (4 ml) is added with mixing followed by 1 ml of PDMAB reagent and the tube contents are mixed again. After standing for 1 h at room temperature the absorbance at 540 nm is measured. Because the chromagen in the reaction is volatile it is essential that tubes are firmly sealed during heating.

An alternative procedure devised by Cessi and Pilego (1960) avoids the need for isolation of hexosamines by Dowex 50 chromatography.

The volatile chromogens produced by reaction of hexosamine with pentanedione are steam-distilled into PDMAB reagent.

Glucosamine and galactosamine give different colour yields in these procedures. When both amino sugars are present the amino acid analyser is the best means of quantitation. However, a colourimetric procedure for determination of glucosamine:galactosamine ratio has been described (Ludoweig and Benman, 1968; Chandrasekaran and BeMiller, 1980). Galactosamine is determined by carrying out the development of chromagen at 0°C.

Colourimetric assay of N-acetylhexosamine *N*-Acetylated derivatives of amino sugars can be released from glycoproteins by enzymatic hydrolysis by *N*-acetylglucosaminidases. Free *N*-acetylhexosamines can be identified and quantitated by gas-liquid chromatography of their trimethylsilyl ethers (Section 5.8.5) or by paper chromatography (Spiro, 1966a). *N*-Acetylhexosamines can also be estimated colourimetrically by the method of Reissig et al. (1955).

Reagents. A borate solution is prepared by suspending boric acid (4.95 g) in about 40 ml water and adding 1 M KOH to bring the pH to 9.1. When the volume is made up to 100 ml the solution is 0.8 M with respect to borate. The other reagent required is prepared as a stock solution containing 10 g *p*-dimethylaminobenzaldehyde (PDMAB) dissolved in 100 ml of glacial acetic acid containing 12.5% (v/v) concentrated HCl. (This solution can be stored refrigerated for one month without deterioration.) It is diluted immediately before use with 9 vols of glacial acetic acid.

Procedure. Samples should contain 5–40 μg *N*-acetylhexosamine and are made up to 0.5 ml with water. Borate solution (0.1 ml) is added and the mixture heated at 100°C in a boiling water bath for 3 min and then cooled in tap water. After adding the diluted PDMAB reagent (3 ml) the samples are mixed and incubated on a water bath at 36–38°C for 20 min. The tubes are cooled in tap water and the absorbance is immediately read at 585 nm. Blanks and standards should be included with every determination.

The absorbance given by *N*-acetylgalactosamine is about one-third that of *N*-acetylglucosamine (approximately 0.6 for 30 μg).

5.6. Uronic acids

The uronic acids which occur in proteoglycans are D-glucuronic acid and L-iduronic acid. One or both of these uronic acids may be present in a single polysaccharide chain. Keratan sulphate and animal glycoproteins (with a few exceptions, cf. Spiro and Bhoyroo, 1971) do not contain uronic acid. Assays for uronic acids can therefore be used for the detection of most proteoglycans as well as for the analysis of their composition.

The simplest procedures for the detection of uronic acids are colourimetric assays based on the carbazole reaction (Dishe, 1947), which can be applied to a proteoglycan without prior hydrolysis. Glucuronic and iduronic acids give different colour yields in this assay. To determine the type of uronic acid(s) present and the relative proportions of the two components when both occur together it is necessary to employ other methods which require hydrolysis, separation and identification of uronic acids or their degradation products.

Quantitative release of uronic acids from proteoglycans presents difficulties. The uronidic linkage is resistant to hydrolysis and significant destruction of released uronic acid is liable to occur before complete cleavage of glycosidic linkages has taken place. A further problem occurs because *N*-acetylated or *N*-sulphated hexosamines are usually glycosidically linked to uronic acid in the carbohydrate chains of glycosaminoglycans. If acid hydrolysis of the *N*-acetyl or *N*-sulphate groups precedes cleavage of the glycosidic linkage the hexosaminidic linkage consequently becomes resistant to hydrolysis, preventing complete release of uronic acid. This problem is particularly acute in the case of heparin and heparan sulphate because the hexosamine *N*-sulphate in these polymers is very rapidly released under acid conditions. Quantitative release and estimation of uronic acid from *N*-sulphated and other proteoglycans can, however, be achieved by the general procedure introduced by Taylor and Conrad (1972) described later in this section.

Carbazole method for uronic acid (Bitter and Muir, 1962)

This colourimetric method is a modification of the original carbazole procedure of Dische (1947). It can be applied directly to uronic acid-containing polymers.

Reagents. Sodium tetraborate (0.025 M) is dissolved in concentrated sulphuric acid (analytical grade). Carbazole (0.125%) in absolute ethanol is prepared and can be stored at 4°C in a brown bottle. Glucuronolactone standard solution, 0.2 μmoles/ml, is stored at 18°C.

Procedure. Tetraborate-acid reagent (5 ml) is placed in tubes cooled to 4°C and 1 ml of the sample or standard (containing 20–200 μmoles uronic acid) is layered on to the surface. Tubes are closed with ground-glass stoppers and the contents mixed with cooling so that the temperature of the mixture does not exceed room temperature. The tubes are then heated in a boiling water bath for 10 min and cooled to room temperature. Carbazole reagent (0.2 ml) is added. The resulting solution is heated for 15 min in a boiling water bath and cooled; the absorbance is read at 530 nm. The absorbance of the blank read against sulphuric acid should be below 0.025.

Dermatan sulphate or iduronic acid derivatives give approximately 80% of the colour obtained from an equivalent amount of glucuronic acid.

Neutral sugars give some colour in the assay when present in large amounts. High concentrations of chloride ion (0.4 M) inhibit colour formation by heparin but not by other uronic acid-containing proteoglycans. Oxidising agents (e.g. nitrites) give a green colour which interferes in the assay.

Determination of uronic acid composition

The method of Taylor and Conrad (1972) is generally applicable to uronic acid-containing polymers including heparin. Quantitative and

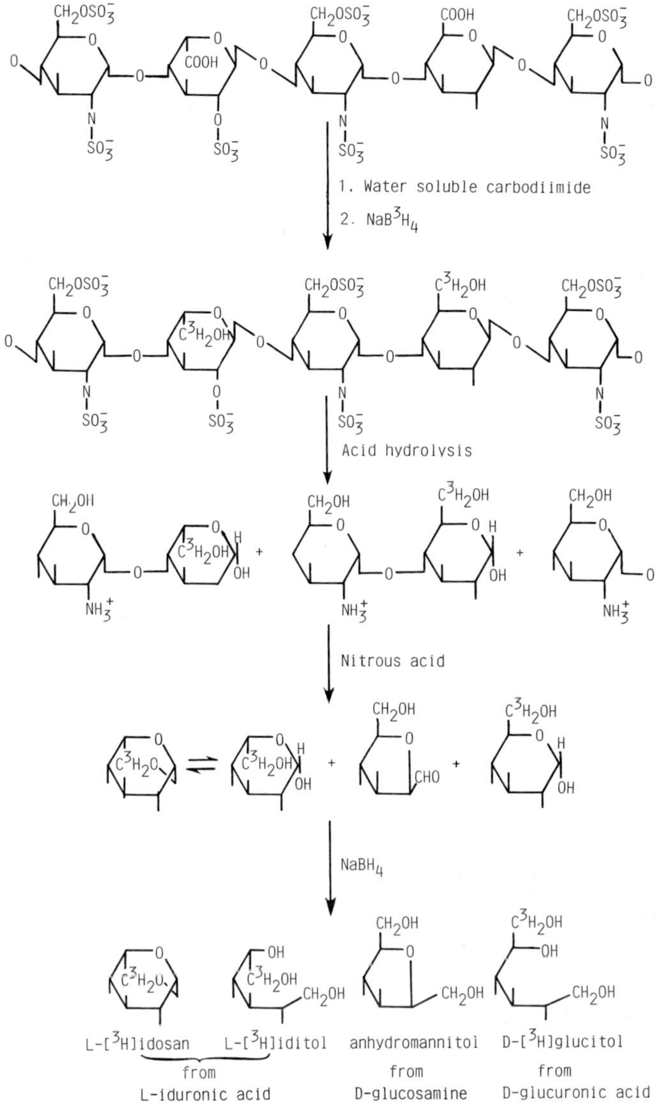

Fig. 5.2. Procedure for degradation of the carbohydrate units of proteoglycans (Taylor et al., 1973).

non-destructive depolymerisation to monosaccharides can be achieved and the products quantitated. The carboxyl groups of the uronic acids are first activated by reaction with a water-soluble carbodiimide to form products (lactones or intramolecular esters) which can then be reduced quantitatively with tritiated sodium borohydride, which converts the uronic acid residues to the corresponding neutral sugars (Fig. 5.2). This reduction incorporates radioactivity stoichiometrically and also facilitates hydrolysis by replacing the acid-resistant uronidic linkages in the polymer with the more acid-labile glycosidic bonds. After acid hydrolysis the remaining acid-resistant amino sugar glycosidic linkages can be cleaved with nitrous acid. The product can then be reduced with unlabelled sodium borohydride to form glycitols which are separated by paper chromatography and quantitated by scintillation counting. For the determination of recoveries it is convenient to add an internal standard of ^{14}C-labelled sugar.

Procedure (Taylor et al., 1973). Proteoglycan sample (7 mg) is allowed to react at room temperature with 19.2 mg of 1-ethyl-3-(dimethylaminopropyl)carbodiimide in 1 ml water maintained at pH 4.75 by automatic additions of 0.1 M HCl in a pH-stat for 1 h. An aliquot (125 μl) of the reaction mixture is added to 250 μl of 3 M NaB^3H_4 (10–20 mCi/mmol) and the mixture is heated at 50°C for 2 h. The sample is cooled and acidified with 3 M H_2SO_4 to decompose excess $NaBH_4$ (fume cupboard). Salts are removed by dialysis and the dialysed sample is transferred to a test tube and evaporated to dryness in a stream of air at 50°C. The [^3H]carboxyl-reduced product is taken up in 50 μl water. 25 μl each of [^{14}C]glucose solution (20 000 cpm/μl, 200 mCi/mmol) and 2 M H_2SO_4 are added and the sample hydrolysed at 100°C for 6 h. An aliquot of the hydrolysate (2 μl) is treated with 5 μl of 5.5 M sodium nitrite. After 10 min at room temperature the sample is neutralised with 2 μl of 1 M Na_2CO_3 and reduced with 5 μl of 0.5 M $NaBH_4$ in 0.1 M NaOH. The vial containing the reaction mixture is closed and heated in a sand bath at 50°C for 40 min. The sample is cooled and excess borohydride destroyed by addition of 10 μl 0.75 M HCl.

Separation of the labelled glucitol, iditol and idose is accomplished by paper chromatography on (1) Whatman No. 1 paper with ethyl acetate-acetic acid-formic acid-water (18:3:1:4), and (2) ethyl acetate - pyridine - 0.005 M H_3BO_3 (3:2:1) on DEAE-cellulose strips sprayed to uniform wetness with 0.5 mM $Na_2B_4O_7$ and then dried. R glucitol values for system (1) are iditol 1.0, idosan 2.5 (Conrad et al., 1973; Shively and Conrad, 1977). In system (2) iditol (R glucitol 0.7) can be resolved from glucitol.

After development the chromatograms are cut into 40 1/2-in segments which are then counted for ^{14}C and ^{3}H in an appropriate scintillant. Counts obtained in the ^{3}H channel must be corrected for those arising from ^{14}C. A plot of ^{3}H cpm against segment number gives a profile of peaks along the chromatogram. The counts under each peak are totalled. ^{3}H counts in glucitol originate from glucuronic acid and those in [^{3}H]iditol and [^{3}H]idosan come from iduronic acid. Hence the molar ratio of iduronic acid to glucuronic acid can be determined directly from the ^{3}H counts. The total hexuronic acid in the original sample can be calculated by use of the ^{14}C counts which correct for sample losses. For the latter calculation it is necessary to determine separately the number of ^{3}H cpm incorporated per uronic acid residue on reduction with $NaB^{3}H_4$. This determination must be carried out for each solvent system (Conrad et al., 1973).

An alternative method developed for the determination of uronic acid composition of heparin-like polymers is described by Höök et al. (1974). In this procedure polymers or oligosaccharides are degraded to monosaccharides by a combination of acid hydrolysis and HNO_2 deamidation. The resulting uronic acids are converted to aldono-1,4-lactones and separated by GLC.

5.7. *Sialic acids*

Sialic acid is a collective term used to describe several derivatives of neuraminic acid. The neuraminic acid occurring in glycoproteins is always substituted with an *N*-acetyl or with an *N*-glycolyl group.

Sialic acid can also be *O*-substituted. For example, a variety of mono-, di- and tri-substituted *O*-acetyl and *O*-glycolyl derivatives of sialic acid have been isolated from equine, bovine or porcine submandibular gland glycoproteins (Schauer et al. 1974; Schauer, 1978); *O*-lactyl substituents have been detected in mammalian sialic acids and *O*-methyl groups have been found in the sialic acids of starfish (Kochetkov et al., 1973).

Sialic acids in mammalian and avian glycoproteins occur as terminal non-reducing residues. For this reason, and because of the lability of its glycosidic linkage, sialic acid can be selectively released by hydrolysis in dilute aqueous acid. For many glycoproteins essentially complete liberation of sialic acid with only limited destruction (about 10%) can be obtained by hydrolysis of glycoprotein (1 mg/ml) in 0.05 M H_2SO_4 at 80°C for 60 min. Under these conditions *O*-acyl substituents are hydrolysed and free *N*-acetyl- or *N*-glycolylneuraminic acid is obtained. In some cases, however, the presence of *O*-acetyl substituents can slow the rate of hydrolysis of the ketosidic linkage of sialic acid (Neuberger and Ratcliffe, 1972, 1973). In these circumstances a longer acid hydrolysis is required for complete release of sialic acid. Alternatively a preliminary mild alkaline hydrolysis (e.g. 0.1 M NaOH for 30 min at 25°C) can be employed to remove acyl groups. The released sialic acid can be identified (Section 5.7.1) and quantitated (Section 5.7.3).

Methanolysis of glycoproteins can be employed to release the methyl glycosides of sialic acid. Quantitation of the total sialic acid present in glycoproteins can be achieved by GLC of these derivatives (Section 5.8.2) or a modified procedure can be employed to determine the amounts of *N*-acetyl- and *N*-glycolylneuraminic acids present as their methyl glycosides.

The type of *O*-acyl sialic acid present in glycoproteins can be determined by releasing the *O*-acyl sialic acid by hydrolysis at pH 2–2.5 at 70°C for 1 h (Schauer et al., 1974) in formic acid or 0.01 M HCl (Schauer, 1978). The sialic acid derivatives can be removed by dialysis and the hydrolysis repeated two or three times to improve the yield. The released derivatives are recovered by lyophilising and can subsequently be purified by ion-exchange chromatography and identified by GLC (Sections 5.7.2. and 5.7.6.).

Several neuraminidases have been isolated which are capable of releasing sialic acid from glycoproteins. Purified enzymes from *Clostridium perfringens* and *Vibrio cholerae* (or *comma*) are available commercially. Release of sialic acids can be carried out (Schauer, 1978) by dissolving the glycoprotein (1–10 mg) in 1 ml of 50 mM acetate buffer, pH 5.5, and adding 5–10 mU of neuraminidase. If the *Vibrio cholerae* enzyme is used 1 mM calcium chloride should be added to the buffer. Incubation is carried out for periods between 30 min and 24 h at 37°C under toluene. The amount of sialic acid released may be measured colourimetrically by the thiobarbituric acid method described later. Both the *Vibio cholerae* and *Clostridium perfringens* enzymes have a wide substrate specificity. However, the glycosidic linkage of *N*-acetyl-4-*O*-acetylneuraminic acid is resistant to hydrolysis. Other *O*-acetylated neuraminic acids are released by these enzymes at slightly reduced rates compared with *N*-acetylneuraminic acid.

5.7.1. Identification of sialic acids

N-Acetyl- and *N*-glycolylneuraminic acids can be separated and identified by paper, thin-layer or gas chromatography after their release (by hydrolysis in 0.05 M H_2SO_4) and purification on Dowex 1. The solvent systems butan-1-ol:propan-1-ol:0.1 M HCl (1:2:1 by vol.) can be employed to separate these compounds on paper (Svennerholm and Svennerholm, 1958) or cellulose thin layers (Schauer et al., 1974; Schauer, 1978) and they can be detected with a silver reagent (Trevelyan et al., 1950) or by more specific stains (Schauer, 1978).

The *N*-glycolyl and *N*-acetyl derivatives of sialic acid can be separated by GLC after release by methanolysis under mild conditions which do not cause de-*N*-acylation (Section 5.7.6.). As with hydrolysis in 0.05 M H_2SO_4 even mild methanolysis results in extensive loss of *O*-acyl groups.

Identification of *O*-acyl sialic acids requires that they are released under very mild conditions (e.g. with formic acid at pH 2.0–2.5). Material released is then purified by treatment with ion-exchange resins and sialic acid derivatives are identified by TLC (Schauer, 1978) or by GLC (Section 5.7.6).

Information about the types of sialic acid derivative present in a glycoprotein can also be obtained as a result of determination of glycolyl groups (Spiro, 1966; Schoop and Faillard, 1967) and *O*-acetyl groups (Downs and Pigman, 1976). Measurement of total acetyl groups (Section 5.10) together with knowledge of the (*N*-acyl) hexosamine content of a glycoprotein can be used to estimate the amount of acetyl groups present in sialic acid.

In addition to the classical chemical and enzymatic methods for identification and quantitation of sialic acids, mass spectrometry (Section 6.9.2) and nuclear magnetic resonance (Section 6.9.3) are now important methods for the identification of sialic acids in glycopeptides or isolated from glycoproteins (Schauer, 1978).

5.7.2. Purification of sialic acid

Sialic acid, which has been released by acid hydrolysis, can be partially purified by ion-exchange chromatography. This procedure removes substances which interfere in analysis by colourimetric methods, GLC or paper chromatography. It is often not necessary to apply this purification step prior to colourimetric assay of purified glycoproteins by the periodate-thiobarbituric acid method but for crude preparations or tissue extracts it is essential. The method described here involves adsorbtion and elution of sialic acid from Dowex 1 (Spiro, 1966a; Codington et al., 1976). Another procedure involving the use of both anion and cation exchangers has been described by Codington et al. (1976).

Procedure. Sialic acid is first released from glycoprotein or glycopeptide by hydrolysis in 0.05 M sulphuric acid at 80°C for 1 h. The hydrolysate is then diluted with four volumes of water and run through a column of Dowex 1- × 8, 50–100 mesh (formate form). For hydrolysis carried out in 1–2 ml of 0.05 M sulphuric acid a column containing 1.0 ml Dowex 1 packed into a glass tube of internal diameter 5 mm is suitable. After the sample has been applied the column is washed with 10 ml of water and the washings are discarded. Sialic acid is eluted with 12 ml of 0.3 M formic acid. Formic acid is removed by freeze-drying. The sialic acids present can be identified and quantitated by GLC or paper chromatography. Recovery of sialic acid standards should be checked on columns run in parallel with those used for hydrolysates of unknowns.

This methodology can be readily applied to sialic acids released by hydrolysis with formic acid or by neuraminidase.

5.7.3. *Choice of sialic acid assay*

Several colourimetric methods have been described which are suitable for measuring the sialic acid content of glycoproteins (for review see Ledeen and Yu, 1976). The resorcinol method (Section 5.7.4.) can be applied directly to an unhydrolysed sample. Colourimetric methods using periodate and thiobarbituric acid are more sensitive and the procedures described by Warren (1959) and Aminoff (1959) have been widely adopted. These methods measure free sialic acid, and prior acid or enzymic hydrolysis is essential.

The results of colourimetric assays for sialic acids must be interpreted with caution when they are applied to samples other than highly purified glycoproteins. In particular the periodate-thiobarbituric acid method is subject to interference from deoxyribose (from DNA) and malonaldehyde (which can be generated by periodate oxidation of unsaturated lipids). It is therefore desirable to carry out an initial purification of the sialic acid as described in the previous section and to compare the results obtained by more than one colourimetric method.

Fluorimetric assays for sialic acid have been described (Hess and Rolde, 1964; Hammond and Papermaster, 1976). The sensitivity of these methods is high but the possibility of non-specific interference requires careful investigation.

A specific enzymic assay for sialic acid has been developed (Section

5.7.5). The fluorimetric version of this method is highly sensitive.

Sialic acid can be quantitated together with neutral and amino sugars by GLC following release by methanolysis (Section 5.8.4.2). Extensive de-*N*-acetylation of sialic acid occurs during methanolysis in 1 M methanolic HCl for 24 h at 80°C and the procedure requires a re-*N*-acetylation step. Hence total sialic acid is quantitated by this method but *N*-acetyl and *N*-glycolyl derivatives are not distinguished.

The proportions of *N*-acetyl- and *N*-glycolylneuraminic acid present in a glycoprotein can be determined if sialic acid glycosides are released under very mild conditions of methanolysis (Yu and Ledeen, 1970). Derivatives of the methyl glycosides of *N*-acetyl- and *N*-glycolylneuraminic acid can be separated and quantitated by GLC (Section 5.7.6). Alternatively, sialic acid can be released by mild acid hydrolysis and, after partial purification on an ion-exchange resin, it can be quantitated by GLC or identified qualitatively by paper chromatography.

For the identification and quantitation of *O*-acyl derivatives of sialic acid, hydrolysis under extremely mild acid conditions is followed by purification on an ion exchanger and TLC or GLC of the products (Section 5.7.6).

5.7.4. Colourimetric assays for sialic acid

Resorcinol method (Svennerholm, 1957)

Reagents. Resorcinol 2% w/v in water. This stock reagent is stable for several months at 4°C.

Resorcinol-HCl reagent is prepared by adding 10 ml of stock resorcinol reagent to a mixture of 80 ml concentrated HCl and 0.25 ml of 0.1 M copper(II) sulphate and making the volume to 100 ml with water. This reagent is stable for 2 weeks at 4°C and should be prepared at least 4 h before use.

Blank reagent is prepared by omitting resorcinol.

Procedure. Samples (2 ml) containing 10–30 μg sialic acid and standards of 0, 15 and 30 μg *N*-acetylneuraminic acid are pipetted into glass centrifuge tubes. Resorcinol reagent (2 ml) is added to the samples and standards. Sample blanks are prepared by adding 2 ml blank reagent to duplicate samples. If the unknown samples also contain other carbohydrates a mixture of these carbohydrates should also be run at two different concentrations. The tubes are heated in a boiling water bath for 15 min. After cooling in running water 5 ml of 3-methyl-1-butanol (amyl alcohol) are added, the tubes are shaken vigourously and then cooled on ice water for 15 min. After centrifuging at 1000 rpm for 1 min the tubes are placed in ice water to prevent fading of the colour until the absorbance of the upper phase has been measured at 450 nm and 580 nm.

The absorbance of samples is corrected by subtraction of the sample blank. A correction for the presence of hexoses can be made on the basis of the absorbance at 450 nm. However, the method is not suitable for samples containing pentoses, glucuronic acid and 2-deoxyhexoses. The effects of interfering substances can be decreased to some extent by extracting the chromophore in a mixture of butylacetate and butanol before measuring the absorbance (Meittinen and Takki-Luukkainen, 1959).

The method gives the total (free and bound) sialic acid content of the sample. *N*-Glycolylneuraminic acid gives about 120% the colour of *N*-acetylneuraminic acid. Because the reactants are heated in concentrated hydrochloric acid *O*-acyl derivatives produce the same colour as the parent *N*-acylneuraminic acid.

Periodate-thiobarbituric acid method (Aminoff, 1959) This method and the similar method described by Warren (1959) measure only free sialic acid. The assay can be applied directly to an acid hydrolysate (in 0.05 M H_2SO_4) or to sialic acid released by neuraminidase digestion.

Reagents. Periodate: 0.025 M periodic acid in 0.0625 M H_2SO_4 (pH 1.2). Sodium arsenite: 2% w/v sodium arsenite in 0.5 M HCl. Thiobarbituric acid: 0.1 M 2-thiobarbituric acid in water adjusted to pH 9.0 with NaOH. This reagent should be made up several hours before use and can be stored for about a month at 4°C in a dark bottle. Acid butanol: butan-1-ol containing 5% (v/v) of 12 M HCl.

Procedure. A solution of sample, blank, or standard, containing 5–40 μg *N*-acetylneuraminic acid in 0.5 ml water (or 0.05 M H_2SO_4) is treated with 0.25 ml of the periodate reagent for 30 min in a water bath at 37°C. Sodium arsenite (1.0 ml) is added to reduce the excess periodate. When the yellow colour of the liberated iodine has disappeared (1–2 min), thiobarbituric acid reagent (2 ml) is added and the samples are covered and heated in a boiling water bath for 7.5 min. The coloured solutions are cooled on ice water and shaken with 5 ml of acid butanol. After a brief centrifugation to separate the two phases the absorbance of the butanol layers is measured at 549 nm. The coulour remains stable for 2 h.

In this assay 10 nmoles of *N*-acetylneuraminic acid will give an absorbance of about 0.35 at 549 nm. *N*-Glycolyl neuraminic acid gives a colour yield of 63% that of the *N*-acetyl derivative (Aminoff, 1959). *O*-Acetylated sialic acids respond differently depending on the position of substitution. The colour produced is decreased slightly by an acetyl substituent on the C_4-hydroxyl, more so with acetyl at C_8 or C_9, and no colour is formed when an *O*-acetyl is at C_7 (Ledeen and Yu, 1976; Schauer, 1978).

There is little interference from other sugars normally present in mammalian glycoproteins in the Aminoff (1959) procedure. However, appreciable depression of the colour yield from *N*-acetylneuraminic acid occurs in the presence of excess fucose using the method of Warren (1959). For this reason the Aminoff version of the method is to be preferred for fucose-containing glycoproteins. When sialic acid assays are carried out on tissue extracts account must be taken of the fact that deoxyribose and malonaldehyde both give high colour yields in both variations of the periodate thiobarbituric acid assay. Warren (1959) has suggested a method for correction for the presence of interfering substances by making measurements at two wavelengths. A better, but more time-consuming, procedure is to purify the sialic acid by adsorption and elution from Dowex 1 as described earlier.

Increased sensitivity can be obtained by scaling down the periodate-thiobarbituric acid procedures so that about 25 ng of sialic acid can be measured (Hahn et al., 1974), or by using a fluorimetric version of the method.

5.7.5. Enzymic assays for sialic acid

Acylneuraminate pyruvate lyase splits sialic acids into pyruvate and *N*-acetylmannosamine. Cleavage is quantitative in the presence of lactate dehydreganase and NADH. The decrease in NADH concentration followed spectrophotometrically (or fluorimetrically) corresponds to the amount of sialic acid in the sample. This method is highly specific for sialic acids and the fluorimetric version of the assay is extremely sensitive; about 50 picomoles of sialic acid can be detected. Both methods are described in detail by Brunetti et al. (1963). A factor which has inhibited the wider use of these methods is the requirement for appreciable quantities of enzyme which should be free of NADH oxidase activity. A method for preparing the enzyme has been described by Brunetti et al. (1963). The aldolase is commercially available (Sigma).

The specificity of the reaction of acylneuraminate pyruvate lyase for the different *N*- and-*O*-acyl sialic acids has been investigated by Schauer et al. (1971).

5.7.6. Analysis of sialic acid by GLC

Gas chromotography can be employed to determine (1) the total sialic acid content of a sample, (2) the proportions of *N*-glycolyl- and *N*-acetylneuraminic acids, (3) to identify and quantitate the *O*-acetyl derivatives of sialic acid.

(1) Total sialic acid The glycoprotein sample is dried and subjected to methanolysis in 1 M methanolic HCl as described in Section 5.8.5.2. Total sialic acid (*N*-acetyl and *N*-glycolyl) is determined as the TMS-ether of *N*-acetylneuraminic acid.

(2) N-Acetyl- and N-glycolylneuraminic acid The ratio of these two derivatives can be determined following methanolysis under mild conditions which releases the sialic acids as their *O*-methyl glycosides without extensive loss of *N*-acyl substituent (Codington et al., 1976). *O*-Acetyl substituents are removed during methanolysis (Yu and Ledeen, 1970).

Glycoprotein or glycopeptide containing 5–100 μg of bound sialic acid is placed in a reaction vial, which has a cap with a Teflon liner, together with *myo*inositol (10 μg) as internal standard. After thorough drying over P_2O_5 in vacuo 0.5 ml of 0.5 M dry HCl in anhydrous methanol (see Section 5.8.4.2) is added and the tube is capped tightly and heated at 65°C for 1 h. Solvent is removed in a stream of dry N_2 and several small aliquots of methylene chloride are added and evaporated from the sample. Trimethylsilylation is carried out by adding 100 μl of a mixture of hexamethyldisilazane:trimethylchlorosilane:pyridine (5:1:1 by vol.). After 60 min at room temperature the sample is dried under a gentle stream of N_2 at 25°C. The sample is then redissolved in hexane (50 μl) and an aliquot is injected onto a column (3 mm × 1.5 m) containing 0.1% OV-17 on GLC-110, 120/140 mesh (Applied Science Labs). The column is temperature-programmed from 80 to 250°C at 10°/min. Unsubstituted sialic acid has a relative retention time of 0.87 and *N*-glycolylneuraminic acid 1.13 compared with the retention time of *N*-acetylneuraminic acid. Quantitation is obtained by measuring peak areas. Correction factors for the peak areas, relative to the internal standard, are calculated from authentic standards of *N*-acetyl- and *N*-glycolylneuraminic acid. Deacylation of *N*-glycolylneuraminic acid occurs at a slightly faster rate than for the *N*-acetyl derivative. A correction factor can be obtained by plotting results for methanolysis at different times (Codington et al., 1976). Separation of these derivatives can also be obtained on 3% OV-1 or OV-225 (Ledeen and Yu, 1976) or by the method of Zanetta et al. (1972) described in Section 5.8.5.2.

(3) Identification and quantitation of O-acyl sialic acids *O*-Acyl sialic acids released by hydrolysis of glycoproteins with formic

acid (Section 5.7) and purified by ion-exchange chromatography (Section 5.7.2) can be determined by GLC of their trimethylsilyl ethers (Schauer et al., 1974; Schauer, 1978).

For trimethylsilylation, sialic acid (10–100 μg) is mixed with *N*-trimethylsilylimidazole (25 μl) for 5 min at 60°C under N_2. Derivatives are separated on a column (1.6 m × 2 mm i.d.) of OV-17 (3%) on Gas-Chrom Q (80–100 mesh) run isothermally at 206°C. Trehalose can be employed as internal standard. The retention indices for several derivatives of sialic acid are listed by Schauer (1978). For the quantitative analysis of sample mixtures it is advantageous to employ shorter columns (0.8 m × 2 mm i.d.) which give improved recoveries of sialic acids. Linear calibration curves can be obtained for *N*-acetylneuraminic acid, *N*-glycolylneuraminic acid and *N*-acetyl-9-*O*-acetylneuraminic acid in the range 0.5–6 μg. These curves, related to the internal standard trehalose, do not pass through the origin.

Identification of sialic acid derivatives is greatly facilitated by combined use of GLC and mass spectrometry. This technique is described by Schauer (1978).

5.8. Neutral sugars

The neutral sugars found in glycoproteins include L-fucose, D-mannose, D-galactose, D-glucose and L-arabinose. D-Galactose also occurs in the repeating disaccharide unit of keratan sulphate and both galactose and D-xylose are present in the protein-carbohydrate linkage region of proteoglycans. Colourimetric assays (Section 5.8.1) for neutral sugars are generally carried out in strongly acidic conditions and they can be applied either to free monosaccharides or to unhydrolysed samples of proteoglycan or glycoprotein. These methods are usually simple to perform and are extremely valuable in monitoring the purification of proteoglycans, glycoproteins and glycopeptides as well as for the quantitative analysis of sugar residues in glycoproteins and proteoglycans.

A major limitation of colourimetric methods is their lack of complete specificity for a single sugar. For example the phenol-sulphuric acid colourimetric assay for neutral sugars does not distinguish between fucose, mannose, galactose and glucose (although the colour yields for different monosaccharides vary). Other methods have been developed which permit the separation and quantitation of individual monosaccharides derived from glycoproteins. The first separation methods adopted hydrolysis followed by paper chromatography. Individual sugars were detected and quantitated either by determination of reducing groups (Section 5.12) or other colourimetric assay. These chromatographic methods have been detailed by Spiro (1966); they have the virtue of simplicity but more sensitive and accurate methods have superseded them.

The most sensitive methods for the analysis of neutral (and amino) sugars involve liberation of monosaccharides by hydrolysis followed by reduction with NaB^3H_4 to produce labelled glycitols. These derivatives are separated either by paper electrophoresis in borate buffers (Section 5.9) or by paper chromatography. The glycitols are identified by their mobilities or R_F values and are quantitated by counting the ^3H incorporated. These radiochromatographic or radioelectrophoretic methods can be applied to very small quantities of sugars (100 ng or possibly less) and when amounts of sample are limiting they are the methods of choice. They do, however, suffer from the problems of incomplete liberation or destruction of sugars inherent in acid hydrolysis (Section 5.3). When larger quantities of sample are available other methods of analysis can give better precision.

Enzyme assays for L-fucose, D-galactose and D-mannose (as well as D-glucosamine) have been described by Schachter (1975). These methods are highly specific and fluorimetric versions of the assays can be extremely sensitive. They are particularly suited to the repetitive determination of a single sugar. Commercial kits for the spectrophotometric assay of galactose are available (Boehringer).

Analysis of neutral sugars (and uronic acids) released by acid hydrolysis of glycoconjugates can be carried out with good precision by ion-exchange column chromatography in borate buffers coupled with continuous colourimetric monitoring of the column effluent (Section 5.8.3). This procedure is analagous to amino acid analysis and the instrumentation can indeed be adapted from an amino acid analyser. Monosaccharides released by acid hydrolysis can also be identified and quantitated by GLC methods. Because the derivatives of α and β anomers of sugars are separable in GLC systems the sugars are usually reduced to alditols so that each sugar only gives rise to a single peak on the gas chromatograph. A technique for separation of alditol derivatives is described in Section 5.8.4.1. In general gas chromatographic methods require a shorter time for analysis than the ion-exchange method. However, the time required for processing the sample prior to analysis is greater for GLC.

The procedure which has been most widely adopted, in recent years, for the analysis of neutral sugars in glycoproteins is GLC of derivatives of the sugars which have been released by methanolysis. Quantitative analysis of neutral sugars, hexosamines and sialic acid can be obtained in a single analysis with consequent economy of material. Methanolysis liberates the methyl glycosides of monosaccharides in high yield and with little destruction. Each monosaccharide gives rise to two or more peaks on the chromatogram. This aids identification of the sugars present but is a disadvantage in quantitation; the precision of the method is rather lower than that of the borate ion-exchange method. Because of the non-specific nature of the detection system (flame ionisation) spurious peaks can arise from contaminants. Rigorously anhydrous conditions are required for methanolysis and care is required in maintenance of the chromatographic system. Perhaps for these reasons some laboratories (e.g. Kennedy, 1979) have experienced difficulties in setting up this method. Two versions of the method are described in Section 5.8.4.2.

5.8.1. *Colourimetric assays for neutral sugars*

General methods for the determination of neutral sugars in purified glycoprotein or proteoglycans include the phenol-sulphuric acid (Dubois et al., 1956), orcinol (Winzler, 1955) and anthrone (Spiro,

1966) reactions, The methyl pentose L-fucose can be estimated specifically by a cysteine-H_2SO_4 method (Dische and Shettles, 1948). Xylose, a pentose which is present in the linkage region of proteoglycans, can be determined using an anthrone method (Tsigonas and Muir, 1966).

Phenol-H_2SO_4 assay for neutral sugars This method is simple and sensitive and the presence of protein has little effect on colour production. The procedure described here is a scaled-down version of the method of Dubois et al. (1956).

Reagents. Concentrated H_2SO_4 (Sp. gr 1.84). Phenol: a 5% wt./vol. aqueous solution of colourless crystals of phenol; this solution is stable for several weeks at room temperature. Standard D-mannose solution (40 µg/ml) stored at −18°C.

Procedure. The reaction is carried out in Pyrex test tubes 14 mm i.d. To a sample containing about 10 µg neutral sugar dissolved in water (0.5 ml) add phenol reagent (0.3 ml) and mix. Then add concentrated H_2SO_4 (2 ml) rapidly from a fast-flowing pipette and mix immediately (care). The heat generated by mixing acid and water is essential for colour development. After standing at room temperature for 30 min the absorbance is read at 484 nm.

A water blank and standards of 5–20 µg mannose are included in each assay.

The colour yields vary for different sugars. Mannose (10 µg) gives an absorbance of about 0.42; the colour yield from fucose is less than half that of mannose. Ribose and deoxyribose also give a colour in the reaction.

Fucose Fucose can be determined in unhydrolysed glycoproteins by the cysteine-sulphuric acid reaction for methyl pentoses (Dische and Shettles, 1948).

Reagents. Sulphuric acid reagent; a mixture of six volumes of concentrated H_2SO_4 with one volume of water. This solution is cooled to 4°C before use. Cysteine hydrochloride, 5% in water.

Procedure. Samples and standards containing 5–20 µg fucose in

1 ml water are placed in 18 × 50 mm Pyrex test tubes. A water blank is also prepared and all the samples are cooled on ice. Sulphuric acid reagent (4.5 ml) is added and the solutions are mixed. The tubes are warmed to 20–22°C for a few minutes and then heated for exactly 3 min on a vigorously boiling water bath and cooled in tap water. Cysteine hydrochloride solution (0.1 ml) is added to each tube with immediate mixing. The absorbance is determined after 1–2 h at 396 nm and at a second wavelength in the region 425–428 nm.

The absorption maximum for fucose is close to 396 nm. Neutral hexoses, including mannose, galactose and glucose, give a colour in this reaction but a correction can be applied for their contribution to the absorbance at 396 nm by reading at a second wavelength, close to 427 nm, at which the absorbance of hexoses equals their absorbance at 396 nm. The precise wavelength at which readings are made to correct for the presence of hexose is determined by including a standard hexose sample (50 μg of mannose, galactose or glucose – all have similar absorption curves) in each assay and finding the wavelength between 425 and 428 nm at which the absorbance is equal to that at 396 nm. Variation can occur between batches of samples in the optimal wavelength for this correction. Fucose contributes little to the absorbance at 427 nm. The difference between the absorbance at 396 nm and 427 nm is used to determine the fucose content.

A correction for the formation of non-specific colour by substances treated with sulphuric acid can be applied by reading the absorbance of samples treated with acid but to which no cysteine hydrochloride has been added.

There is little interference in the assay from amino sugars or protein. Rhamnose gives a product with a similar absorption spectrum and colour yield to fucose.

Fucose can also be estimated colourimetrically by the procedure described for the assay of xylose. Identification of fucose can best be achieved by gas-liquid chromatography or borate complex chromatography.

Xylose The pentose D-xylose occurs in the linkage region of proteoglycans. It can be estimated in proteoglycans without prior hydrolysis by the method of Tsigonas and Muir (1966). The methyl pentose fucose can also be assayed by this method. An alternative method employs orcinol in HCl containing ferric chloride (Chandrasekaran and BeMiller, 1980).

Reagents. Anthrone 1% in sulphuric acid (Sp. gr. 1.84) is prepared just before use. D-Xylose (or L-fucose) standards (5–50 μg/ml) are dissolved in water and stored at $-18°C$.

Glassware: thin-walled test tubes, 20 × 150 mm, are washed in hot dilute sulphuric acid prior to use because traces of detergents used in routine washing interfere.

Procedure. Samples (3 ml) of anthrone reagent are delivered from a burette into each clean test tube, frozen in ethanol-solid CO_2 mixture and allowed to stand in an ice-water bath. Triplicate 1 ml samples of unknown or standard are layered on top of the frozen reagent. Immediately afterwards and while immersed in the ice water, each tube is first swirled gently while thawing and then shaken vigorously to mix the viscous liquid thoroughly. The temperature of the solutions during mixing should not exceed that of the ice bath. Uncontrolled heating will increase interference from other sugars present.

The tubes are transferred to a constant-temperature (40°C) bath and kept immersed for exactly 10 min for xylose and 30 min for other pentoses or methyl pentoses (e.g. fucose). At the end of this time the tubes are cooled in ice water and then left to stand for 20 min in the dark. The absorbance is read at 615 nm for pentoses and 640 nm for methyl pentoses.

Blanks are set up by replacing the sample or standards by 1 ml of water.

The absorption spectra of the anthrone colours given by xylose and fucose overlap considerably but differential analysis after heating for both 10 and 30 min permits quantitation of each sugar even in mixtures. Hexoses give about 5% of the colour yield of xylose (after 10 min heating). Interference from hexuronic acids, hexosamine, sialic acid and deoxyribose and protein is negligible. Samples containing ribose give a colour in the reaction and fructose has a high colour yield. However, these sugars are not present in purified proteoglycans or glycoproteins.

5.8.2. Hydrolysis procedure for neutral sugars

The optimal conditions for hydrolysis depend on the particular sugars present and their linkages (see Section 5.3.) and these conditions should, whenever possible, be developed for each glycoprotein. Of the neutral sugars occurring in animal glycoproteins fucose is particularly easily released and destroyed. If fucose is present it can be determined after hydrolysis in more dilute acid (e.g. 0.3 M HCl at 100°C for 1–18 h) than is employed for release of hexose or by acid-catalysed hydrolysis using the acid form of a cation-exchange resin (see below).

For the initial analysis of neutral sugars other than fucose the glycoprotein (1–5 mg/ml) should be hydrolysed in 1 ml of 1 M HCl (diluted from glass-distilled constant boiling HCl) in a screw-capped tube fitted with a Teflon liner inside the cap. The tube is flushed with oxygen-free nitrogen before sealing and hydrolysis continued for 4–6 h at 100°C. After cooling, water (1.5 ml) is added and the hydrolysate neutralised by passing it through a column of Dowex 1 × 8 (200–400 mesh, CO_3 form) containing 3-4 mequivalents of resin. The column is then washed with 5 column volumes of water and the eluate collected. This neutralised hydrolysate can be analysed directly or if necessary concentrated using a rotary evaporator at 35°. Dowex 1 (CO_3^{2-}) can be prepared by washing a column containing Dowex 1 (Cl^-) with 1 M Na_2CO_3 until the eluate is free of Cl^-, followed by distilled water until the eluate is neutral.

An alternative hydrolytic procedure which is more lengthy but leads to minimal destruction of neutral sugars, including fucose, has been described by Lehnhardt and Winzler (1968). Glycoprotein samples of 0.1–3 mg containing about 100 nmoles of neutral sugar are dried in 6 × 50 mm culture tubes under reduced pressure. A centrifuge bio-dryer (Virtis) is the most satisfactory equipment for this procedure but an Evapomix, or rotary evaporator, can be employed in its place. The dried samples are dissolved in 50 µl H_2O and 50 µl of a 40% w/v suspension of freshly prepared Dowex 50 × 2 (H^+) 200/400 mesh resin in 0.02 M HCl is added. Tubes are sealed with Teflon septa, mixed and incubated at 100°C for 40 h.

Columns are prepared by inserting a small plug of glass wool into a Pasteur pipette and adding 50 µl of a 20% w/v suspension of Dowex 1 × 8 (HCO_3^- form). The hydrolysates and Dowex 50 (H^+) are transferred onto these columns and the hydrolysis tubes are washed with two 0.3-ml portions of H_2O which are also applied to the columns. Elution is continued with 1 ml of 50% methanol:water. Combined neutralised eluates are dried on a rotary evaporator or a bio-dryer.

The sugars present in these deacidified hydrolysates can be identified and quantitated by GLC, ion-exchange chromatography or paper chromatography. For paper chromatography Spiro (1966) recommends that acid hydrolysates should be treated with both Dowex 50 (H^+ form) and Dowex 1 (formate form) to remove charged sugars, amino acids and peptides.

5.8.3. Ion-exchange chromatography of borate complexes

The complexes formed between neutral sugars and borate are negatively charged at slightly alkaline pH values. Separation of the borate complexes of different neutral monosaccharides (or disaccharides) can be obtained by chromatography on a column of an ion-exchange resin in borate buffer. Sugars are eluted by increasing the concentration of borate (or other salts) and the eluted sugars are detected and quantitated by automated colourimetric or fluorimetric assays (for review see Kennedy, 1974).

The equipment required is essentially similar to that needed for amino acid analysis. However, the carbohydrates are usually detected by the orcinol-H_2SO_4 reaction and therefore the pump and tubing exposed to this reagent must be resistant to acid. Carbohydrate analysers are available commercially from JEOL, Technicon and other manufacturers. It is also possible to construct a highly effective and versatile carbohydrate analyser from amino acid analyser and Technicon autoanalyser components (Lee, 1972). Detailed accounts have been published of the use of the JEOL (Japan Electron Optics Co. Ltd., Tokyo) JLC- 6AH (Kennedy and Fox, 1980) and the Technicon carbohydrate analyser (Spiro, 1972). As each analysis takes 2.5–6 h it is advantageous to have automatic buffer-changing and sample-loading equipment.

The following procedure applies to the JEOL JLC-6AH carbohydrate analyser (Kennedy and Fox, 1980). This instrument employs

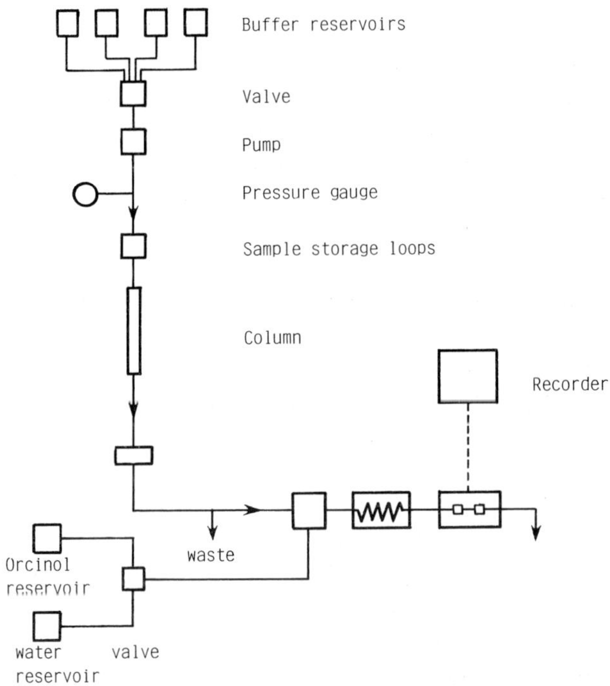

Fig. 5.3. Carbohydrate analysis by ion-exchange chromatography. The diagram illustrates the components of a (single column) carbohydrate analyser. (Modified from Kennedy and Fox, 1980.)

two 12 × 0.8 cm columns of LC-R-3 quaternary ammonium ion-exchange resin (JEOL). The columns are used alternately to prevent delay caused by the time required for column regeneration. Fig. 5.3 shows the general design of the instrument; for simplicity the second column is omitted from the figure.

The sample containing 1–1000 μg free monosaccharides is dissolved in 0.13 M boric acid and is injected into a sample storage loop (capacity 0.8 ml) of the analyser. After initial equilibration of the column with 0.13 M potassium tetraborate buffer, pH 7.5, the sample is applied to the column and elution is performed at 55°C with a buffer programme (Table 5.1). Buffer changes are controlled by a punched-tape programmer. Part of the effluent stream is mixed with an orcinol reagent for detection of carbohydrates. This reagent contains orcinol (1.5 g/l) dissolved in concentrated sulphuric acid. The orcinol reagent is pumped at a rate (24 ml/h) twice that of the column eluate stream (12 ml/h) with which it is mixed in the assay manifold. The reagent and sample are then passed through a heating coil at 95°C for 15 min. The colour which develops is detected by measurement of the absorbance at 425 nm and 510 nm in two flow cells. The data are printed on a point-plot chart recorder. Peak areas can be obtained by manual integration (width at half height × peak height). Alternatively an automatic integrator can be employed. The system is calibrated by applying a standard mixture of monosaccharides in known quantities. Sugars are identified by their time of elution from the column (Table 5.1). The ratio of absorbance at 425 nm and 510 nm can also be of use in confirming the identification of fucose (Kennedy and Fox, 1980). The precision of the method is good (S.D. ± 2.5%).

Faster analyses (2.5 h) can be obtained, with some sacrifice in resolution, by the borate ion-exchange method of Lee, 1972.

Even more rapid resolution of monosaccharides can be achieved by HPLC (McGinnis and Fang, 1980). It seems likely that this technology will in the near future begin to supersede existing liquid chromatography systems, as is the case already with amino acid analysis equipment.

TABLE 5.1

Buffer programme and elution times of some carbohydrates on ion-exchange chromatography in borate buffers (Kennedy and Fox, 1980)

Eluant buffer	Duration of pumping (min)*	Elution time of carbohydrate (h)*	
Carbohydrate separation			
0.13 M potassium tetraborate buffer, pH 7.50	110	2-Deoxy-D-ribose	0.47
		Sucrose	0.55
		Lactose	1.38
		L-Rhamnose	1.62
0.25 M potassium tetraborate buffer, pH 9.08	90	D-Ribose	2.75
		D-Mannose	3.13
0.30 M potassium tetraborate buffer, pH 9.60	190	L-Fucose	4.03
		L-Arabinose	4.12
		D-Fructose	4.33
		D-Galactose	4.46
		D-Xylose	4.75
		D-Glucose	5.38
Regeneration and equilibration			
0.5 M potassium tetraborate buffer, pH 9.60; 0.13 M potassium tetraborate buffer, pH 7.50			

* Some variation in these times may occur depending on the age and state of the column packing but the order of elution remains unchanged.

5.8.4. Gas-liquid chromatography

The speed and sensitivity of GLC methods have led to their widespread adoption for the analysis of carbohydrate in glycoproteins. Analytical techniques have been described for monosaccharides released by hydrolysis of glycoproteins, for glycitols produced by reduction of monosaccharides or for methyl glycosides resulting from methanolysis. Because of the low destruction of sugars occurring on methanolysis recent analytical work has been concentrated largely on methyl glycosides. After methanolysis amino sugars and sialic acid can be determined in the same sample as neutral sugars, whereas at least two hydrolyses and often more are required for the estimation of neutral and amino sugars. Gas chromatography resolves the different anomeric forms of many monosaccharides and methyl glycosides and, in most cases, two or more such forms occur in significant concentration for each sugar. This leads to considerable complexity in the pattern of peaks obtained from a glycoprotein containing several sugars. It is possible to simplify separation and quantitation by reducing monosaccharides to glycitols which give only one peak for each neutral sugar originally present. This procedure is, however, less versatile than the analysis of methyl glycosides.

A number of volatile derivatives of carbohydrates such as acetates, trifluoroacetates and trimethylsilyl derivatives have been examined with a variety of stationary phases being used for their separation. Clamp et al. (1971) give a comprehensive description of earlier work. Newer developments are reviewed annually in the Chemical Society publication Specialist Periodical Reports, Carbohydrate Chemistry.

5.8.4.1. Estimation of neutral sugars as their alditol acetates This method is capable of giving quantitative analysis of the neutral sugars, but not the amino sugars or sialic acid, of glycoproteins or glycopeptides. The glycoprotein is hydrolysed, the monosaccharides are reduced to alditols and the alditol acetates (one peak for each sugar in the glycoprotein) are quantitated. Alditols of neutral sugars produced by Smith degradation and alditols derived from partially methylated sugars obtained during methylation analysis (Section 6.9) can also be quantitated and identified.

The apparatus required is a gas chromatograph with dual-column oven, dual flame ionisation detectors, linear temperature programmer and linear recorder. Nitrogen is used as carrier gas. Two coiled glass columns, one analytical, the other reference (2 mm i.d. × 1.83 m) are packed with 3% w/w ECNSS-M on Gas chrom Q 100–120 mesh.

Procedure (Spiro, 1972). The sample containing 50–300 nmol of each neutral sugar can be hydrolysed and neutralised by either of the methods in Section 5.8.3. If the presence of fucose is expected, hydrolysis with an ion-exchange resin is recommended. Internal standard (150 nmoles of 2-deoxyglycose) is added. After drying in vacuo (Evapomix) the hydrolysate is reduced by addition of freshly prepared 0.25 M $NaBH_4$ (2 ml) and standing overnight at 4°C. Water (3 ml) is added and the solution acidified to pH 4.5 with acetic acid to decompose the borohydride. The sample is passed through a column containing 2 g Dowex 50 × 4 (H^+) resin 200–400 mesh, washed through with water (25 ml) and taken to dryness on a rotary evaporator. Boric acid is removed as the volatile methyl tetraborate by three additions of methanol:HCl (1000:1) with evaporation and the sample is then dried over P_2O_5.

Acetylation is carried out with 0.25 ml pyridine (prepared by refluxing with ninhydrin and stored in a desiccator containing NaOH pellets) and 0.25 ml acetic anhydride in a sealed tube at 100°C for 1 h. Samples are taken to dryness in a stream of dry N_2 at 37°C in an electric heating block. Samples can be stored at − 20° over P_2O_5 until used.

After redissolving in pyridine (20 μl) 1–2-μl samples are injected into the gas chromatograph. Suitable chromatographic conditions are carrier gas flow rate 20 ml/min, initial column temperature 160°C for 4 min and then 1°/min to 205°C.

There is substantial 'bleed' from columns of ECNSS-M and appropriate balance with the reference column is required to maintain a steady baseline.

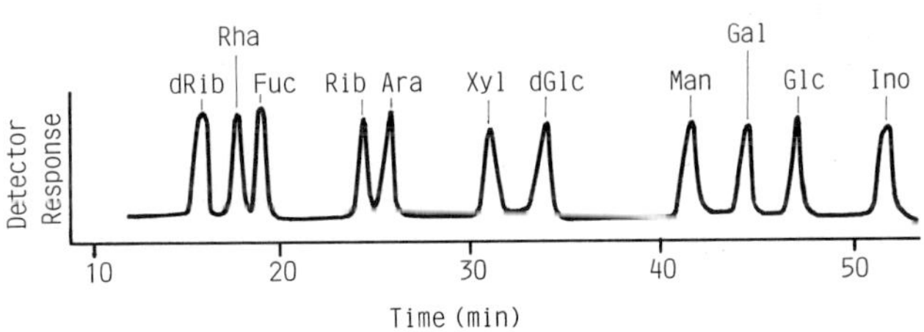

Fig. 5.4. Gas-liquid chromatography of alditol acetates of a standard mixture of neutral monosaccharide on ECNSS-M (Spiro, 1972). The mixture contained 2-deoxyribose, rhamnose, fucose, ribose, arabinose, xylose, 2-deoxyglucose, mannose, galactose, glucose and inositol.

The sugars present can be identified tentatively by their relative retention times (Spiro, 1972) and should be confirmed by use of standard mixtures (Fig. 5.4). Response adjustment factors (the ratio of the area under the 2-deoxy-D-glucitol peak to that of the area under the peak of an equimolar amount of the alditol in question) should be determined for the particular analytical system used. To calculate the amount of each sugar present the following relationship is used

$$\text{nmoles sugar} = \frac{\text{area under sugar peak}}{\text{area under internal standard peak}} \times \text{nmoles internal standard added} \times \text{response adjustment factor}$$

5.8.4.2. Estimation of methyl glycosides Glycoproteins or glycopeptides are first treated with anhydrous methanolic HCl to produce a mixture of methyl glycosides. Volatile derivatives of the glycosides are obtained by formation of trimethylsilyl- (Chambers and Clamp, 1971) or trifluoroacetyl- derivatives (Zanetta et al., 1972). All of the neutral sugars, amino sugars and total sialic acid can be estimated by these methods in a single analysis.

Optimal conditions for the methanolysis of glycoproteins and glycopeptides have been examined by Chambers and Clamp (1971). Yields of methyl glycosides released by methanolysis usually agree well with the yields of monosaccharides obtained by hydrolysis under optimal conditions, except that the GLcNAc residue involved in the GLcNAc-Asn linkage of glycoprotein is not recovered. The importance of strictly anhydrous conditions for methanolysis must be emphasised. Some loss of acetyl groups occurs during methanolysis of glycoproteins. A re-*N*-acetylation step is included in the Chambers and Clamp (1971) procedure. This prevents overlap of peaks from methyl hexosaminides with other peaks in the chromatogram.

Trimethylsilyl derivatives (Chambers and Clamp, 1971). Apparatus required; gas chromatograph with single-column oven, flame ionisation detectors, linear temperature programmer and linear recorder, preferably with disc or other integrator system. Oxygen-free nitrogen is used as carrier and the single coiled glass column (3.2 mm i.d. × 200 cm) is packed with 3.0% SE-30 on Diatoport S (Hewlett-Packard).

The analytical procedure including the determination of response factors has been reported in detail (Clamp et al., 1971: Clamp, 1974). A slightly modified protocol is summarised here.

Procedure. A solution of glycoprotein or standard carbohydrates (preferably 10–100 nmoles of each monosaccharide) is placed in a 2-ml conical Pyrex reaction tube fitted with a PTFE-lined screw cap. After addition of a suitable aliquot (10–50 nmoles) of the internal standard mannitol the mixture is dried thoroughly in a desiccator over P_2O_5. The sample is taken up in dry methanolic 1 M HCl and sealed after bubbling with oxygen-free N_2 for 30 s and incubated at 85°C (Chambers and Clamp, 1971) for 24 h. (Methanol (500 ml) is dried by refluxing with magnesium turnings (2.5 g) and iodine (0.1 g) followed by distillation. Anhydrous methanolic HCl is produced by bubbling dry HCl gas into the dry methanol. Acid strength is determined by titration and the concentration of acid is decreased to 1 M by addition of dry methanol.) When the samples have cooled to room temperature solid Ag_2CO_3 is added to neutralise the acid. Re-*N*-acetylation is accomplished with acetic anhydride (0.05 ml); after thorough mixing the samples are allowed to stand at room temperature for 6–16 h. The precipitate is thoroughly triturated, centrifuged and the supernatant collected. The insoluble residue is extracted three times with portions (0.25 ml) of dry methanol and the combined supernatants are collected and dried on a rotary evaporator at 35°C. The samples are finally dried for 12 h in a vacuum desiccator over P_2O_5.

To the dried material trimethylsilylating agent (0.05 ml) consisting of a mixture of pyridine, trimethylchlorosilane and hexadimethylsilane (5:1:1 by vol.) is added and the tube stoppered and allowed to stand for 30 min at room temperature. After centrifuging, samples of the supernatant (1–5 μl) are injected onto the column. The initial temperature of the column is 140°C and the temperature is programmed to 200°C at 0.5°/min. Alternatively a more recent modification

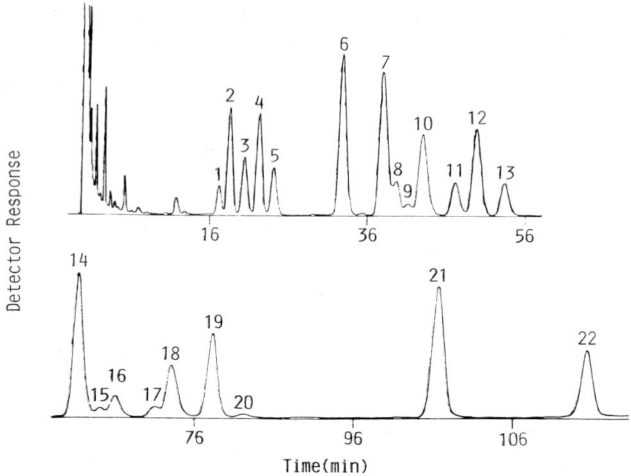

Fig. 5.5. Analysis of monosaccharides by GLC of methyl glycosides (Clamp et al., 1969). The separation of the *O*-trimethylsilyl ethers of the methyl glycosides of a standard mixture of monosaccharides found in glycoproteins (and 3 internal standards) is shown. Peaks are numbered in order of emergence and are identified as follows: fucose 1, 2, 3; xylose 4, 5; (arabinitol 6); mannose 7, 9; galactose 6, 10, 11; glucose 12, 13; (mannitol 14); *N*-acetylglucosamine 15, 17, 19, 20; *N*-acetylgalactosamine 16, 18; (perseitol 21); methyl *N*-acetylneuraminate 22.

of the procedure is to temperature-programme from 120 to 220°C at 1°/min (Clamp, J., personal communication).

Each monosaccharide present in the initial sample gives rise to two or three peaks. The peaks are identified by reference to a standard mixture (Fig. 5.5). Resolution of the major mannose peak and one of the galactose peaks is never complete. Peak areas should be determined and summed for each monosaccharide. Quantitation is then carried out using the expression given in Section 5.8.4.1.

The importance of maintaining anhydrous conditions for the success of this method cannot be overemphasised. Peaks can be lost (particularly those resulting from *N*-acetylhexosamines) and artifactual peaks will appear if water is present. The presence of salts in the original sample can lead to uninterpretable patterns of peaks. Because of the non-specific detection system used artifactual peaks can arise from Tris buffer in the sample or due to contaminating fatty acids in samples originating from membranes. However, extraction of the samples prior to forming the trimethylsilyl derivatives with toluene (3 × 0.5 ml) effectively removes fatty acids.

Methyl glycoside trifluoroacetyl derivatives (Zanetta et al., 1972). Trifluoroacetyl derivatives of methyl glycosides are more volatile than the trimethylsilyl ethers and more polar stationary phases can be used for their separation. *N*-Trifluoroacetyl derivatives can be formed from both free hydroxyl and amino groups (Zanetta et al., 1972). The use of these derivatives therefore provides an alternative to the widely used TMS method for the analysis of glycoproteins and glycopeptides. Complete resolution of all of the peaks arising from monosaccharides in glycoproteins can be achieved.

The equipment required is a dual-column oven, dual flame ionisation detectors, a temperature programmer and linear recorder. Glass columns (2 mm i.d. × 2 m) are packed with 5% (w/w) OV-210 on Varaport 50 (Applied Science Laboratories Inc.)

Procedure. Methanolysis of samples can be performed as described for trimethylsilyl derivatives except that *meso*inositol (40 μg) is added as internal standard and the sample (1–500 μg carbohydrate) is lyophilised and then dried in a vacuum desiccator over P_2O_5 overnight. Zanetta et al. (1972) recommend the use of 250–500 μl of 0.5 M anhydrous methanolic HCl. The tube is closed with a screw-cap lined with PTFE, shaken vigorously, and kept at 80°C for 20 h. Methanolic HCl is removed under a stream of N_2 with the tubes maintained in an electric heating block at 50°C.

The dried samples are dissolved in dichloromethane (100 μl) and trifluoroacetic anhydride (100 μl), stoppered quickly and placed for 5 min on a sand bath at 150°C located behind a shield in a fume cupboard. After cooling to room temperature the sample is again placed in the sand bath for 5 min. After cooling again the samples are ready for the injection of 1–10 μl aliquots onto the column. The injection port temperature is 225°C and the initial column temperature 90°C rising by 1° per minute to 190°C. A flow rate of carrier nitrogen of 7.5 ml/min is employed.

Identification and quantitation of peaks is carried out by the same procedure as for trimethylsilyl derivatives of methyl glycosides. However, re-*N*-acetylation is omitted. The appearance of peaks corresponding to the *N*-TFA derivatives of glucosamine, galactosamine and neuraminic acid in addition to the *N*-acetylives of these sugars is indicative of loss of acetyl groups and the total of all these derivatives should be determined for the purpose of quantitation. The extent to which loss of *N*-acetyl groups occurs during methanolysis may be related to how effectively water is excluded from samples and reagents, as loss of acetyl groups from *O*-methyl derivatives of *N*-acetylhexosamines is not significant (Zanetta et al., 1972). Some loss of acetyl groups from glycoproteins and glycopeptides is always observed.

Zanetta et al. (1972) report that *N*-glycolyl- and *N*-acetylneuraminic acids can be distinguished using their procedure. This is one advantage of omitting re-*N*-acetylation.

This method is more suitable for the quantitation of neutral sugars than amino sugars. The resolution of neutral sugars is good and hexosamines can be determined by some other method such as the use of an amino acid analyser after hydrolysis (Section 5.5).

5.9. Microdetermination of ^{3}H-labelled sugar alcohols

Takasaki and Kobata (1978) have described procedures for the analysis of very small quantities of glycopeptide or glycoprotein. The sample containing 0.1–10 μg sugar is hydrolysed and the monosaccharides reduced to alditols with $NaB^{3}H_{4}$ with the incorporation of label.

$$\mathrm{RCHO} \xrightarrow{\mathrm{NaB^{3}H_{4}}} \mathrm{R\text{-}\overset{\overset{\displaystyle ^{3}H}{|}}{\underset{\underset{\displaystyle H}{|}}{C}}\text{-}OH}$$

After re-*N*-acetylation the *N*-acetylhexosaminitols and other alditols are separated by paper electrophoresis of their borate complexes and quantitated by scintillation counting. *N*-Acetylneuraminic acid can be determined in the same way. The procedure has been employed quite extensively in studies of glycopeptide structure (Yamashita et al., 1978). Alternatively, labelled alditols can be separated by paper chromatography (Conrad, 1977).

Procedure

Samples containing 0.1–10 μg sugar are hydrolysed. For glycoproteins Takasaki and Kobata recommend hydrolysis in 2 M HCl at 100°C for 3 h; for oligosaccharides 1 M HCl at 100°C for 2 h (see also Section 5.3). The sample is brought to dryness in a rotary evaporator and the residue repeatedly redissolved in water and evaporated to remove traces of HCl and then placed in a desiccator containing NaOH pellets in vacuo overnight.

The samples are redissolved in water (0.5 ml) and a known quantity of an internal standard, deoxyribose, is added. After evaporation the sample is redissolved in 20–100 μl of 0.05 M NaOH and labelled with a five-fold molar excess of NaB^3H_4 (dissolved in dimethylformamide, see below) and incubated at 30°C for 4 h. Reduction is stopped by addition of 100 μl 1 M acetic acid (in a fume cupboard to vent gaseous tritium) and evaporated to dryness. The residue is redissolved in a small volume of water and evaporated to dryness again five times.

NaB^3H_4, specific activity 200–500 mCi/mmol, is made up as a 32 mM solution in freshly redistilled anhydrous dimethylformamide. This solution is stored in an airtight screw-capped tube at −10°C and is stable for at least 5 weeks (Takasaki and Kobata, 1978).

After reduction the sample is re-*N*-acetylated by redissolving in 0.2 ml of saturated $NaHCO_3$ and making five additions, each of 10 μl acetic anhydride, with 10-min incubations at room temperature following each addition. The reaction mixture is incubated for 2 min in a boiling water bath and is then passed through a column (internal diameter 5 mm) packed with a layer 4 cm deep of Dowex 1 (OH^-

form) and an upper layer of Dowex 50 (H^+). The labelled sample is eluted with 3 bed volumes of water and the eluate evaporated to dryness.

The labelled, *N*-acetylated sample is applied to Whatman No. 1 paper and electrophoresed in 0.06 M borate buffer, pH 9.0, at 40 volts/cm for 2.5–3 h. A marker of bromophenol blue, which has about the same mobility as the alditol of galactose, can be employed. Radioactive areas can be located with a radiochromatogram scanner and the radioactive regions cut out, eluted with water (1 ml) in a counting vial and radioactivity determined by scintillation counting.

If mannitol and fucitol are not completely separated the region of the electrophoresis paper in which these components might occur is extracted, part of the extract is counted and another aliquot is used to determine the proportions of these two sugars. This is done by first passing the radioactive material through a column of Dowex 50 (H^+) and then evaporating the residue five times with methanol (to remove methyl borate) and then separating the two glycitols by paper chromatography in ethyl acetate-pyridine-water (12:5:4 by vol.).

The precision claimed for this method is lower than that which can be obtained using borate column electrophoresis or gas-liquid chromatography of methyl glycosides. However, the small amounts of sample which can be analysed, which are, in principle, limited only by the specific activity of the NaB^3H_4 employed, are an important feature of the radioelectrophoretic method. Using sodium [^{3}H]borohydride with a specific activity of 200 mCi/mmole it should be possible to determine the sugars present in a sample in quantities as low as 0.1 nmole with an accuracy of about 10% using radiochromatographic analysis (Conrad, 1976).

Because of the variable rates of release and stabilities of the monosaccharides present in glycoproteins (Sections 5.4.1 and 5.4.6.2.) analyses carried out after hydrolysis under the conditions recommended by Takasaki and Kobata (see above) cannot be expected to give a completely accurate sugar composition. Whenever possible the conditions used for hydrolysis should be optimised for each glycoprotein; separate hydrolyses for determination of fucose, hexose and amino sugars may be required.

5.10. Isotope dilution methods

The problem of correcting for destruction of released sugars can largely be overcome by radioisotope dilution methods. Radioactive sugar is added to the glycoprotein and the mixture is hydrolysed. From the hydrolysate the sugar is isolated and its specific activity is determined; the change in specific radioactivity before and after hydrolysis gives a measure of the sugar content.

The procedure has been applied to the determination of galactose, mannose and fucose in several glycoproteins with results (Graham and Neuberger, 1968) which are in agreement with those obtained by other analytical methods. The method has also been applied to the determination of hexosamine in pancreatic deoxyribonuclease (Price et al., 1969).

One advantage of the method is that loss of sugars during the isolation procedure does not affect the final result. Relatively severe conditions of hydrolysis may be used, thus removing the need to establish optimum conditions of hydrolysis. It is necessary to isolate sufficient quantities of each sugar to determine the specific activity; this can be accomplished by separation of sugars by liquid chromatography (Sections 5.8.4. and 5.5) and collecting fractions for the determination of radioactivity by scintillation counting.

Isotope dilution methods deserve wider use in glycoprotein analysis than they have received.

5.11. Acetyl and sulphate

The amino sugars glucosamine and galactosamine occur in glycoproteins and proteoglycans as their *N*-acetylated or occasionally as

N-sulphated derivatives (e.g. heparin and heparan sulphate). The amino groups of sialic acids can have *N*-acetyl or *N*-glycolyl substituents (Section 5.7). *O*-Acetyl and *O*-glycolyl derivatives of sialic acid have been identified in glycoproteins and ester sulphate substituents are found in proteoglycans (e.g. chondroitin 4- and chondroitin 6-sulphates) as well as in some glycoproteins.

Acetyl and sulphate substituents sometimes occur in the peptide chain of glycoproteins as in the *N*-acetylated amino-terminal residue of ovalbumin or the tyrosine *O*-sulphate of fibrinogen.

Methods are described below for determination of the total acetyl and sulphate groups of glycoproteins and proteoglycans. The location of *O*-sulphate groups can often be ascertained by infrared spectroscopy or by enzymatic degradation methods (Section 6.10). *N*-Sulphate can be identified by studies of the hydrolytic conditions required for release of sulphate or by nitrous acid deamination methods (Section 6.10). Methods for the identification of acetyl and glycolyl derivatives of sialic acid have been described in Section 5.7.

Total sulphate Sulphate is completely liberated from proteoglycans or glycoproteins by acid hydrolysis. The inorganic sulphate is then precipitated as barium sulphate, which gives a stable turbidity in the presence of gelatin and can be determined spectrophotometrically. The procedure described below is a minor modification of that of Dodgson and Price (1961).

Reagents. Gelatin (Difco-Bacto), 2 g, is dissolved in water (400 ml) at 60–70°C and then allowed to stand at 4°C overnight.

Barium chloride, 0.5 g, is dissolved in the gelatin solution (100 ml) and the resultant cloudy solution allowed to stand 2–3 h before use. This reagent is stable for at least one week at 4°C.

Trichloroacetic acid 8% in water.

Sulphate standards are prepared by dissolving K_2SO_4, which has been dried at 110°C, in water at a concentration of 100 μg sulphate/ml.

Glassware used in the assay should be cleaned in HNO_3 and rinsed in deionised water.

Procedure. Samples of sulphated polysaccharide (1–3 mg) are hydrolysed in 1 ml of 1 M HCl at 100°C in sealed tubes for at least 6 h. After cooling the tubes are opened and the contents taken to dryness by rotary evaporation at a temperature below 40°C. The residue is redissolved in water (1 ml).

Aliquots (0.1 ml) of the hydrolysate are placed in small test tubes and made up to a volume of 0.4 ml with distilled water. Trichloroacetic acid solution (0.35 ml) and barium chloride gelatin solution (0.25 ml) are added and mixed. Standards containing 10–40 μg sulphate and water blanks are treated similarly. The mixtures are allowed to stand for 15–20 min and their turbidities are measured at 360 nm in a silica semi-micro cell (capacity 1 ml) of 1 cm light path.

Another aliquot of hydrolysate is treated in precisely the same way except that 0.35 ml of gelatin solution is added in place of the gelatin-barium chloride reagent. The absorbance of this solution at 360 nm is subtracted from the value obtained in the presence of barium chloride. This control is included to compensate for any ultraviolet-absorbing materials present in the hydrolysate (Dodgson and Price, 1961).

Acetyl groups The following method can be employed to determine the total *O*- and *N*-acetyl groups present in glycoproteins or proteoglycans. Acetyl groups are liberated as methyl acetate by treatment with anhydrous methanolic HCl. The methyl acetate is isolated by distillation and estimated colourimetrically. This method was described by Ludowieg and Dorfman (1960).

Reagents. Methanolic 2 M HCl is prepared as described in Section 5.8.5.2.

Alkaline hydroxylamine reagent is prepared immediately prior to use by mixing equal volumes of 0.35 M hydroxylamine hydrochloride and 1.5 M NaOH.

Ferric-perchloric acid solution is prepared by dissolving 1.9 g of $FeCl_3 \cdot 6H_2O$ in 5 ml concentrated HCl, adding 5 ml of 70% perchloric acid, taking the solution almost to dryness on a rotary evaporator (care, there is an explosion hazard with dry perchlorates) and then diluting to 100 ml with water. This reagent is stable for up to 1 month at 4°.

Procedure. Sample containing 1–10 μmoles of acetyl groups is placed in a 70 × 7 mm Pyrex tube and dried over P_2O_5 in a vacuum desiccator at 80°C. To the dried residue 0.5 ml of 2 M HCl in anhydrous methanol is added. The tube is cooled in an ethanol-solid CO_2 bath and then sealed with an oxygen torch. Methanolysis is performed at 100°C for 4 h. After cooling the tube is opened and the tube and its contents are carefully placed in chamber A of the distillation apparatus (Fig. 5.6). The chamber is closed with a polyethylene cap.

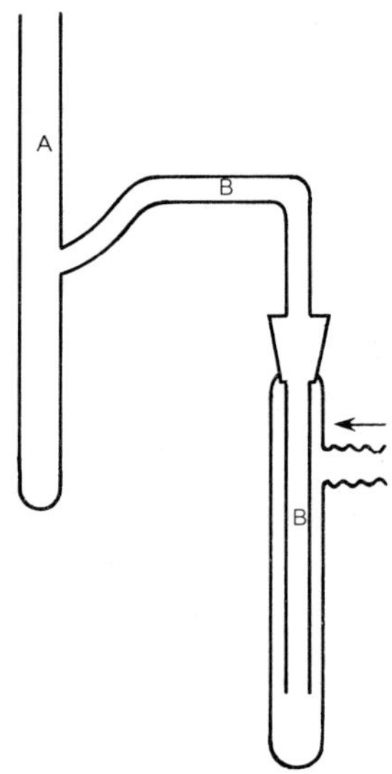

Fig. 5.6. Distillation apparatus for acetyl determination (Ludowieg and Dorfman, 1960). Use of the apparatus is explained in the text.

The distillation apparatus has a sidearm B which extends to about 20 mm above the bottom of suction tube C to which it is attached via a rubber bung. The suction tube is connected to a desiccator tube filled with $CaCl_2$. Between the desiccator tube and the water pump a T-piece is fitted which can be opened to the outside. The suction tube C is cooled to the level of the sidearm with a solid CO_2-ethanol mixture.

To carry out the distillation the apparatus is evacuated to 30–40 mmHg and chamber A is then submerged to a depth within 10 mm of the sidearm in a water bath at 35–45°C. After all of the acid methanol has distilled and collected in tube C the polyethylene cap is removed from A and 0.5 ml of anhydrous methanol is added to the deacetylation tube and the distillation is repeated. Colourimetric assays are carried out on the sample contained in tube C. If several distillations are performed the samples can be stored prior to assay in tube C provided that it is sealed with Parafilm and kept in the freezer.

For assay of methyl acetate tube C is warmed to 20–25°C for 1–2 min in a water bath. Water (1 ml) is added, followed by 2 ml of alkaline hydroxylamine solution. This is prepared just before use by mixing equal volumes of 0.35 M hydroxylamine hydrochloride and 1.5 M NaOH. The contents of tube C are mixed by shaking and allowed to stand for 10 min. Then 2 ml of 0.75 M perchloric acid is added followed by 1 ml of ferric-perchloric acid solution. The colour which develops is read at 520 nm after 5 min. Standard solutions containing 1–10 μmoles methyl acetate in 2 ml methanol:H_2O (1:1) are also assayed and a blank solution of methanol-HCl is employed. Salts of volatile organic acids interfere in the assay.

5.12. Reducing sugars

It is possible to estimate the free reducing groups present in a sample containing carbohydrate by colourimetric reactions based on the reduction of copper (Somogyi, 1952) or iron (Spiro, 1966a). A micromethod suitable for the assay of 2–100 nmoles of reducing sugar has also been reported (Avigard, 1975). The long-established Nelson-Somogyi method is described below.

Reagents. A. Rochelle salt (sodium potassium tartrate), 15 g, and anhydrous Na_2CO_3 are dissolved in 300 ml water. Then add 20 g $NaHCO_3$. A solution of 180 g anhydrous Na_2SO_4 in 500 ml water is heated to expel air and, after cooling, it is combined with the tartrate-carbonate-bicarbonate solution and the mixture made up to 1 l with water.

B. $CuSO \cdot 5H_2O$ (5 g) and anhydrous Na_2SO_4 (45 g) are dissolved in water and made up to 250 ml.

Just prior to use 4 vols. of reagent A are combined with 1 vol. reagent B.

C. Arsenomolybdate reagent is prepared by dissolving 25 g ammonium molybdate in 450 ml water and adding 21 ml concentrated H_2SO_4 with stirring. $Na_2HAsO_4 \cdot 7H_2O$ (3 g) is dissolved in 25 ml water and mixed with the molybdate solution. This solution is incubated at 37°C for 24–48 h and then stored in alass-stoppered brown bottle. It is stable for a year.

Prior to use this reagent is diluted with two volumes of 0.75 M H_2SO_4.

Procedure. The reaction is carried out in small Pyrex tubes (13 × 100 mm) to minimise the surface area for reoxidation. Standards (10–100 μg reducing sugar) and blanks (1 ml) are mixed with the copper reagent (1 ml). Tubes are capped with glass bulbs and heated in a boiling water bath for 30 min. After cooling in water 1 ml of diluted arsenomolybdate reagent is added to each tube and mixed. Samples are then diluted with water (3 ml) and mixed by inversion. The absorbance is measured at 520 nm.

The colour yields from mannose and glucosamine are similar and about twice the value for *N*-acetylglucosamine. Reducing disaccharides also show reducing power. Other reducing agents can interfere in the assay, as can proteins and amino acids (Beeley and Jevons, 1963).

CHAPTER 6

Structural analysis

6.1. Introduction

To describe the primary structure of a glycoprotein or proteoglycan fully the following information is required: (1) the sequence of amino acids in the peptide chain, (2) the number of carbohydrate units, their locations and modes of linkage to protein, (3) the structures of the carbohydrate units and an assessment of the extent of microheterogeneity at each site of carbohydrate attachment. Methods for sequencing proteins, including glycoproteins, have been described in a previous volume of this series (Allen, 1980). In this chapter consideration of protein chemistry will therefore be restricted to the isolation of glycopeptides and the location of carbohydrate units on peptide chains. Methods for the structural analysis of the oligosaccharide chains of glycoproteins and proteoglycans will also be discussed.

Structural features of glycoproteins have been reviewed by Kornfeld and Kornfeld (1976, 1980), Montreuil (1980) and Sharon and Lis (1982). Kennedy (1979) has reviewed proteoglycan structure. Valuable accounts of older methodology can be found in the volumes edited by Gottschalk (1972).

6.2. Strategy for structural analysis

Before becoming involved in the structural analysis of a glycoprotein or proteoglycan it is essential to be sure that the sample contains protein covalently linked to carbohydrate (Section 6.3). Once this has been established the amino acid and monosaccharide units present should be identified and quantitated as described in Chapter 5. The carbohydrate composition provides a strong indication of the type(s) of oligosaccharide structure likely to be present (Table 6.1), while the amino acid composition can, in some cases, give an indication of the ways in which protein and carbohydrate are linked. For example the absence of hydroxylysine excludes the possibility of linkage via this residue.

TABLE 6.1

Monosaccharides of the major types of glycoproteins and glycosaminoglycans occurring in higher animals

	Monosaccharides and substituents
Glycoprotein type	
N-Linked simple	GlcNAc Man
complex	GlcNAc Man Gal (Fuc SA)
O-Linked	GalNAc (Gal GlcNAc Fuc SA O-SO_3^-)
Collagen	Gal (Glc)
Proteoglycan type*	
Chondroitin 4-sulphate	GlcA GalNAc O-SO_3^-
Chondroitin 6-sulphate	GlcA GalNAc O-SO_3^-
Hyaluronate	GlcA GlcNAc
Heparin	IdoA GlcA GlcNAc O-SO_3^- N-SO_3^-
Heparan sulphate	IdoA GlcA GlcNAc O-SO_3^- N-SO_3^-
Dermatan sulphate	IdoA GlcA GalNAc O-SO_3^-
Keratan sulphate I	GlcNAc Gal (Man Fuc SA) O-SO_3^-
Keratan sulphate II	GlcNAc Gal (GalNAc Fuc SA) O-SO_3^-

* With the exception of hyaluronate and the keratan sulphates the protein-carbohydrate linkage regions also contain Gal and Xyl. Most proteoglycans contain carbohydrate chains of different types and also have 'glycoprotein-type' carbohydrate units attached to their peptide chain.

Brackets indicate monosaccharides which may be present in low relative amounts or which may be absent. The abbreviation SA is used for sialic acids, O-SO_3^- for ester sulphate and N-SO_3^- for *N*-sulphated hexosamine.

The nature of the protein-carbohydrate linkages of glycoproteins and proteoglycans can be determined (Section 6.8) by studying their stability in alkali (Section 6.5.1) and by the isolation of glycopeptides (Sections 6.4 and 6.7). Determination of the molecular weights of glycopeptides containing only a few amino acids (Section 6.7), or of oligosaccharides released by cleaving protein carbohydrate linkages with alkali, by hydrazinolysis (Section 6.5) or trifluoroacetolysis (Section 6.5.3) allows an estimate to be made of the number of carbohydrate groups per peptide chain.

Glycoproteins frequently contain more than one carbohydrate moiety and different types of protein carbohydrate bond can exist in the same molecule. It is therefore important to account quantitatively for all of the carbohydrate present in the original molecule. It is also desirable to know at what location on the peptide chain particular carbohydrate groups occur. This can be achieved by selective cleavage using enzymatic or chemical methods (Section 6.4.2) and by isolation of glycopeptides whose amino acid sequences can be placed in the overall sequence of the molecule.

Until recently the structural analysis of the oligosaccharide chains of proteoglycans and glycoproteins seemed to present rather different problems. The structures of the repeating units of most of the glycosaminoglycans were established many years ago by the application of enzymatic and classical chemical methods. Current structural investigations of these molecules commonly involve the identification of the type(s) of repeating unit(s) present and modifications of the repeating units in heteropolymers or mixtures of polymers. This type of information can be obtained by combining compositional analysis with specific enzymatic cleavage by glycosidases, lyases and sulphatases (Section 6.9.4.1) or by use of chemical degradation methods, including periodate oxidation, partial hydrolysis, or deamination (Section 6.9.4). Recently it has been shown that, in addition to their glycosaminoglycan polysaccharide chains, the proteoglycans also contain large numbers of oligosaccharide units resembling those found in glycoproteins (Nilsson et al., 1982). The analysis of the structures of these units presents precisely the same problems and involves the same methods as those which have been developed for structural analysis of glycoproteins.

The carbohydrate units of glycoproteins vary in complexity from simple mono- or disaccharides to large highly branched oligosaccharides. Only in the last few years have definitive structures been obtained for many of the more complex carbohydrate groups. Certain common molecular structural themes seem occur (Chapter 2) but the range of structures present in glycoprotein carbohydrate units is still being actively investigated. Methods for the analysis of these structures continue to develop rapidly. Different combinations of techniques are being applied to the solution of these structures. Some general strategies for the structural analysis of the oligosaccharide units of glycoproteins are given in Section 6.9.

Structural analysis of the carbohydrate moieties of glycoproteins and proteoglycans is normally carried out after the removal of most or all of the peptide moiety. Glycopeptides can be isolated following proteolytic digestion (Section 6.4). Oligosaccharides suitable for structural analysis can be released by chemical cleavage of protein-carbohydrate linkages (Section 6.5) or by splitting off oligosaccharides specifically with endoglycosidases (Section 6.6). Structural heterogeneity occurs commonly even in carbohydrate units attached at a specific site on the peptide chain and careful fractionation of glycopeptides or oligosaccharides is therefore of great importance (Section 6.7). Complete structural analysis of the purified oligosaccharide units requires determination of the nature of the monosaccharide units, their ring form (furanoside or pyranoside), configuration (D or L), anomeric linkage (α or β), the positions of linkages between sugars and their sequence. In addition, the position of any substituent groups (e.g. sulphate) must be established (Section 6.10).

Analysis of the positions of linkages between monosaccharides by methylation (Section 6.9.1) is of fundamental importance in the determination of structures of glycoproteins. This technique can also confirm the ring form of the sugar residues. The separation and identification of methylated derivatives is facilitated by combined use of GLC and mass spectrometry. In addition to the analysis of methylated monosaccharides GLC-MS or MS can be applied directly to methylated oligosaccharides to yield partial (or complete) sequence information (Section 6.9.2).

High-resolution nuclear magnetic resonance has recently become established as an important technique for the structural analysis of oligosaccharides and glycopeptides. Information about composition, sequence and anomeric linkages can be obtained non-destructively (Section 6.9.3).

Specific enzymic cleavage of the carbohydrate units of glycoproteins can be employed to determine monosaccharide sequence, anomeric linkage and ring form (Section 6.9.4.1). In some cases it is also possible to establish the positions of linkages between sugar residues from the specificities of particular enzymes. Enzymatic cleavage can also be applied to identification of the repeating units in glycosaminoglycans (Section 6.9.4.1) and to locate ester sulphate substituents.

The primary sequence of sugar residues in oligosaccharides, glycopeptides or glycosaminoglycans can also be examined by partial degradation using chemical methods (Section 6.9.4) including partial acid hydrolysis, acetolysis, periodate oxidation, Smith degradation, deamination of amino sugars with nitrous acid and chromium trioxide oxidation.

Lectin affinity chromatography can be employed to obtain structural information about oligosaccharide units (Section 6.7 and Chapter 7) and as a fractionation technique. Other separation methods of importance in structural analysis of oligosaccharides and glycopeptides are also discussed in Section 6.7.

6.3. *Is carbohydrate covalently linked to protein?*

Before attempting the analysis of glycoprotein structure it is important to establish that the material under study is not a protein contaminated with non-covalently bound carbohydrate. Tenacious non-covalent interactions can occur between carbohydrates and lectins or enzymes which act on polysaccharide substrates. Carbohydrate contamination can also easily arise during fractionation using polysaccharide supporting media for ion-exchange, gel filtration or affinity chromatography.

In the case of proteoglycans the covalently bound peptide component is always present in small amounts compared to carbohydrate. Proteoglycans can form strong non-covalent interactions with proteins, including the 'link proteins' involved in the assembly of macromolecular aggregates in connective tissues. In addition the high negative charge of proteoglycans can lead to electrostatic interactions with basis residues in proteins. It is therefore of great importance to distinguish between non-covalent binding and the covalent interaction between core protein and polysaccharide chains in a proteoglycan.

If protein and carbohydrate are covalently linked they will co-purify both when the peptide chain is in its native state and under denaturing conditions. Co-purification is indicated by the coincidence of elution of protein and carbohydrate from ion-exchange and gel filtration columns and a constant ratio of protein to carbohydrate during the final stages of purification (Chapter 3). However, it must be born in mind that carbohydrate micro-hetergeneity can sometimes lead to some displacement between the mid-points of carbohydrate and protein elution profiles and can produce some variation in protein carbohydrate ratios.

Attempts should be made to separate protein from carbohydrate under conditions in which the peptide chain is denatured. A simple approach is to carry out SDS-polyacrylamide gel electrophoresis of glycoproteins and stain the gel for protein and carbohydrate (Section 4.10.2) to determine whether these components comigrate. Electrophoretic separation or iso-electric focusing of glycoproteins in high concentrations of urea, or caesium chloride density gradient ultracentrifugation in concentrated urea or guanidinium chloride (Section 3.4.6), can be employed. The latter technique is also applicable to proteoglycans, as is ion-exchange chromatography in the presence of 8 M urea.

Co-precipitation of protein and carbohydrate with protein-denaturants such as trichloroacetic acid or perchloric acid is supporting evidence for a covalent protein-carbohydrate linkage in glycoproteins. Glycosaminoglycans are soluble in these reagents but can be precipitated with anionic detergents instead.

The existence of a covalent protein-carbohydrate linkage can be most convincingly demonstrated by the isolation of glycopeptides containing the linkage region, as described in the following section.

6.4. *Release of carbohydrate units*

The release of the carbohydrate units of glycoproteins or proteoglycans from most or all of their peptide moieties is usually carried out at an early stage in the structural analysis of the carbohydrate component of these molecules. Cleavage of the peptide moieties, or of protein-carbohydrate linkages, is sometimes employed in extraction procedures applied to tissues or membrane preparations (Chapter 3). Carbohydrate units are also released from purified glycoproteins and proteoglycans to allow their fractionation (Section 6.7) and to avoid interference of their peptide moieties in structural investigations. Studies of the conditions required for the release of carbohydrate units by alkali (Section 6.5.1) give valuable information about the nature of the protein-carbohydrate linkage. The isolation of glycopeptides following cleavage of the peptide chains of glycoproteins or proteoglycans is also an important method for determining the nature of protein-carbohydrate linkages and the sites of attachment of carbohydrate units to polypeptide chains.

6.4.1. *The isolation of glycopeptides*

There are two general strategies for the production of glycopeptides from purified glycoproteins (or proteoglycans). One is to degrade the peptide chain very extensively by non-specific proteolysis to give a mixture of glycopeptides containing very short peptide chains and preferably with a high proportion of glycopeptide with a single amino acid attached. This approach produces glycopeptides suitable for studies of the nature of the protein-carbohydrate linkage and for the determination of carbohydrate structure. A second strategy is to cleave the peptide chain in a highly specific manner so that fragments are obtained which yield information about the location of carbohydrate groups on the peptide chain. When there is more than one site of carbohydrate attachment specific cleavage makes possible the separation of carbohydrate units from different sites. Glycopeptides can be freed from contaminating peptides and further fractionated by the methods described in Section 6.7.

6.4.2. Non-specific cleavage

The enzyme which has been most widely used for the non-selective cleavage of the peptide chains of glycoproteins is Pronase, a preparation containing a mixture of proteinases produced by *Streptomyces griseus*. Prolonged Pronase digestion usually results in very extensive degradation of the peptide chains of glycoproteins. Commercial preparations of the enzyme are usually free of exoglycosidase activities but some batches have traces of endoglycosidase activity (Lis and Sharon, 1978). Other enzymes of wide specificity such as papain have also been employed (particularly for digestion of proteoglycans) and use has been made of trypsin, chymotrypsin, collagenase and exopeptidases either sequentially or in combination.

Glycopeptides can be released either from soluble glycoproteins (Spiro, 1972) or from lipid-extracted membrane preparations (Finne and Krusius, 1982) by Pronase digestion. The sample is dissolved, or suspended, at a concentration of 25 mg protein/ml in 0.1 M Tris-HCl buffer, pH 8.0, containing 1 mM $CaCl_2$. A stock solution of Pronase P (Calbiochem-Behring) in the same buffer (10 mg/ml) is preincubated at 60°C for 30 min to inactivate contaminating enzymes which may be present. Digestion of the sample is started by the addition of Pronase (1% by weight) and further additions (0.5% by weight) are made at 24 h and 48 h. The pH should be checked and, if necessary, readjusted to pH 8.0 before each addition of enzyme. Digestion is carried out at 37°C with gentle shaking and toluene is added to suppress bacterial growth. When particulate samples are digested it is essential that they should be well dispersed and not 'clumped' before starting the digestion. Moistening the lipid-extracted residue of membranes with ethanol before addition of buffer aids their dispersion (Finne and Krusius, 1982).

The course of hydrolysis can be followed by analysing aliquots with the ninhydrin reagent (Moore and Stein, 1954) and digestion is continued until no further cleavage of peptide bonds occurs (usually 72–96 h). The digest is lyophilised. Glycopeptides are separated by gel filtration or lectin affinity chromatography (Section 6.7).

For the isolation of glycopeptides from basement membrane and collagens Spiro (1976) recommends an initial digestion with purified collagenase from *Clostridium histolyticum*. The membranes or collagen (25 mg/ml) suspended in 0.15 M Tris-acetate buffer, pH 7.4, containing 5 mM calcium acetate are digested with enzyme initially at 0.7% of the substrate weight. After incubation under toluene at 37°C with shaking further additions of enzyme are made at 24 and 48 h (0.35% and 0.1% of the substrate weight) and after 72 h the mixture is centrifuged to remove a small quantity of undigested material. The supernatant is then adjusted to pH 7.8 and further digested with Pronase (0.5% by wt. added in three portions) for a total of 72 h. After lyophilisation glycopeptides are isolated by gel filtration.

Papain can be employed in place of Pronase for the release of carbohydrate units from glycoproteins or for the solubilisation of protein-bound carbohydrate from cellular samples (Brungraber et al., 1971). Incubations are carried out in 0.1 M sodium acetate buffer, pH 5.5, containing 10 mM cysteine and 10 mM EDTA.

Both Pronase and papain are suitable for the digestion of proteoglycans or the release of the carbohydrate units of these molecules from tissue samples. In the latter case preliminary lipid extraction and effective dispersion of the tissue increases the accessibility to the enzyme. For digestion with papain the substrate (up to 100 mg/ml) in 0.1 M phosphate buffer, pH 6.5, containing 1 M NaCl, 5 mM cysteine and 5 mM EDTA is incubated with activated enzyme (2% by weight) at 65°C for 12 h under toluene. If solubilisation and digestion is not complete more enzyme should be added and digestion continued for a further 12 h. The addition of high concentrations of salt to the buffer prevents inhibition of the enzyme by glycosaminoglycan (Scott, 1960).

Glycosaminoglycans can be recovered in high yield from the medium of cultured cells after Pronase digestion by the method of Merilees et al. (1977). The glycosaminoglycans (and glycopeptides) can be isolated from the digest by gel filtration. Alternatively the digest can be deproteinised by heating in acidified saline and centrifugation. Niebles and Sciffler (1975) found that TCA precipitation can lead to loss of some glycosaminoglycan. Proteoglycans can then be precipitated with ethanol containing sodium acetate before further fractionation or analysis (e.g. by cellulose acetate electrophoresis). The culture medium (6 ml) in a stoppered centrifuge tube is mixed into 1 ml of a solution of Pronase (15 mg/ml) in Tris buffer, pH 7.4. Incubation is carried out at 50°C for 16 h under toluene. After adding 4 ml of 31% NaCl sufficient 3 M acetic acid is added to bring the solution to pH 5, the tube is stoppered and heated at 100°C for 5 min and then cooled on an ice bath before centrifuging at 30 000 *g* for 20 min at 4°C. The supernatant is transferred to a glass centrifuge tube containing 25 ml of ethanol/sodium acetate (0.8 g/l) and left overnight. After centrifugation the supernatant is discarded and the precipitate is taken up in water for further fractionation or analysis.

6.4.3. Specific cleavage

Isolation of glycopeptides after specific cleavage of the peptide chain is an important step in establishing the location of carbohydrate groups in glycoproteins with more than one carbohydrate moiety. The distribution of carbohydrate units on the peptide cores of proteoglycans can be investigated by similar methods. Specific proteolytic cleavage can be carried out with enzymes such as trypsin and chymotrypsin. Chemical methods (e.g. cyanogen bromide cleavage) can also be employed provided that it is established that the carbohydrate groups and the protein-carbohydrate linkages are stable to the chemical treatment.

It is usually desirable to modify any disulphide bonds present in a glycoprotein prior to selective hydrolysis. This can be accomplished by reduction and alkylation with iodoacetamide, iodoacetate or ethyleneimine. The latter reagent converts cysteine residues to *S*-aminoethylcysteine, thus incorporating extra sites on the peptide which can be cleaved by trypsin. Both reduction and alkylation should be carried out in the presence of a denaturant (8 M urea or 5 M guanidinium chloride).

Examples of the isolation of carbohydrate units from specific loci on glycoproteins include work on orosomucoid (Schmid et al., 1977), ovomucoid (Beeley, 1976b) and glycophorin A (Tomita et al., 1978).

In the case of ovomucoid the glycoprotein was first reduced and aminoethylated and then digested at a concentration of 10 mg glycoprotein/ml in 0.1 M ammonium bicarbonate, pH 8.5, with TPCK-trypsin (0.2 mg/ml) for 4 h at 25°C under toluene. After this short period the cleavages which occurred were all adjacent to lysine and arginine residues but some glycopeptides arising from incomplete cleavage were observed. After further digestion for 20 h following a second addition of trypsin small amounts of glycopeptides resulting from cleavages outside the normal range of specificity of trypsin could be detected (Beeley, 1976b). Fractionation of the digest resulted in the isolation of glycopeptides from each of four carbohydrate-attachment sites in the glycoprotein.

Tomita et al. (1978) digested glycophorin A (300 mg, 10 mg/ml) in 40 mM Tris-HCl, pH 8.2, with TPCK-trypsin (9 mg) at 37°C for 24 h. Some of the peptides isolated from digests of the highly sialated native glycoprotein contained more than one lysine and/or arginine residue, suggesting that some peptide linkages involving these residues were resistant to trypsin. Following the removal of sialic acid these resistant linkages became susceptible to further cleavage by trypsin. Thus the carbohydrate moieties of glycopeptides, and particularly sialic acid residues, can inhibit cleavage of peptides by trypsin.

Schmid et al. (1977) removed sialic acid residues from human plasma α_1-acid glycoprotein (by mild acid hydrolysis at pH 1.8, 80°C for 1 h) before digesting the reduced, carboxymethylated glycoprotein with chymotrypsin under conventional conditions (enzyme:substrate ratio 1:50, pH 8.2, 37°C, 3 h). Fractionation of the digest resulted in the isolation of glycopeptides derived from five specific regions of the polypeptide chain. Subsequent Pronase digestion of the chymotryptic glycopeptides from each carbohydrate-attachment site followed by chromatography of the products on Dowex-50 at low ionic strength led to the isolation of a total of 26 homogeneous carbohydrate-carrying peptides. Structural differences between the oligosaccharides attached at different loci on the peptide chain and the microheterogeneity at each attachment site have been examined in detail by the structural analysis of these glycopeptides (Fournet et al., 1978; Schmid et al., 1979).

Specific cleavage of glycoprotein can also be carried out by chemical methods. Consideration must, however, be given to the stability of glycosidic and protein-carbohydrate linkages under the conditions employed. For example, the acid conditions employed in the cleavage of proteins at methionine residues with cyanogen bromide can result in the loss of some sialic acid from glycoproteins (Tomita et al., 1978). However, the *N*-glycosylamine-linked carbohydrate units of ovomucoid remained intact during CNBr cleavage of sialic acid-free ovomucoid (Beeley, 1976a).

The isolation of glycopeptides obtained by specific cleavage of glycoproteins can pose special problems. These are discussed in Section 6.7.

Peptide-chain cleavage by specific proteolysis can also be valuable in examining the organisation of oligosaccharide chains in proteoglycans. For example, Yanagishita and Hascall (1983a) have studied the distribution of carbohydrate chains on a dermatan sulphate proteoglycan of low buoyant density isolated from cultured cells by selective cleavage with trypsin. The proteoglycan, labelled with $^{35}SO_4^{2-}$ and [^{3}H]mannose or [^{3}H]glucosamine was digested with DPC-trypsin, 30 μg/ml in 0.1 M Tris, 0.1 M sodium acetate buffer, pH 7.3, for 3 h and the digests were fractionated by gel filtration on Sepharose 4B or in 0.5 M guanidine hydrochloride on Sephacryl S-200. Three types of fragment were identified in the digest: (a) glycosaminoglycan peptides having an average of two dermatan sulphate chains; (b) clusters of *O*-linked oligosaccharides; (c) *N*-linked glycopeptides.

6.5. *Release of carbohydrate units by cleavage of protein-carbohydrate linkages*

The methods which are available for cleavage of protein-carbohydrate linkages to release oligosaccharides from glycoproteins and proteoglycans are listed in Table 6.2 and are discussed individually in the following sections of this chapter. Of these methods the most widely employed has been the alkaline β-elimination of *O*-glycosyl GalNAc-Ser/Thr linkages, which often produces good yields of oligosaccharides. Hydrazinolysis has been applied quite extensively to release oligosaccharides from linkage to asparagine. This technique can be used to liberate complex carbohydrate units which are not readily cleaved by endoglycosidases (Section 6.6).

TABLE 6.2

Methods for cleaving protein-carbohydrate linkages to release oligosaccharides

Linkage	Cleavage method	Section
O-Glycosyl		
GalNAc-Ser/Thr	alkaline β-elimination	6.5.1.1
Xyl-Ser		
Gal-Ser/Thr		
Man-Ser/Thr		
N-Glycosyl		
GlcNAc-Asn	vigorous alkaline hydrolysis	6.5.1.2
	hydrazinolysis	6.5.2
	trifluoroacetolysis	6.5.3
	enzymic	6.5.4

Cleavage of *O*-glycosyl linkages can also occur during trifluoroacetolysis and hydrazinolysis.

TABLE 6.3
Stability to alkali of protein-carbohydrate linkages

Linkage residues	Glycoside type	Typical cleavage conditions	Examples of molecules containing linkage
Alkali-sensitive			
Xyl-Ser	O	0.5 M NaOH 4° 19 h[1] (β-elimination)	Proteoglycans (except keratan sulphates)
GlcNAc-Ser	O	0.05 M NaOH, 1 M $NaBH_4$ 45° 16 h[2] (β-elimination + reduction)	Glycoproteins (especially mucins)
Gal-Ser	O	0.2 M NaOH, 0.2 M $NaBH_4$ 50° 5 h[3] (β-elimination + reduction)	Plant glycoproteins (potato lectin, extensin)
Gal-Ser -Thr	O	0.1 M NaOH 25° 24 h[4] (β-elimination)	Earthworm cuticle collagen
Man-Ser -Thr	O	0.1 M NaOH 25° 18 h[5] (β-elimination)	Yeast, fungal and *Nerei* cuticle glycoproteins
Hydrolysed under severe conditions			
GlcNAc-Asn	N	1 M NaOH, 4 M $NaBH_4$ 60° 24 h[6] (hydrolysis + reduction)	Glycoproteins
Alkali-resistant			
Gal-Hly	O	Alkali-resistant[7]	Collagen
Ara-Hpr	O	Alkali-resistant[8]	Plant glycoproteins
Gal-Hpr	O	Alkali-resistant[9]	Plant glycoproteins

(1) Anderson, B., Hoffman, P. and Meyer, K. (1965) J. Biol. Chem. *240*, 156–167.
(2) Carlson, D.M. (1968) J. Biol. Chem. *243*, 616–626.
(3) Lamport, D.T.A., Katona, L. and Roerig (1973) Biochem. J. *133*, 125–131.
(4) Muir, L. and Lee, Y.C. (1970) J. Biol. Chem. *245*, 502–509.
(5) Nakajima, T. and Ballou, C.E. (1974) J. Biol. Chem. *249*, 7679–7684.
(6) Zinn, A.B., Marshall, J.B. and Carlson, D.M. (1978) J. Biol. Chem. *253*, 6761–6765.
(7) Spiro, R.G. (1972) Methods Enzymol. *28*, 1–43.
(8) Allen, A.K., Desai, N.N., Neuberger, A. and Creeth, J.M. (1978) Biochem. J. *171*, 665–674.
(9) Fincher, G.B., Sawyer, W.H. and Stone, B.A. (1974) Biochem. J. *139*, 535–545.

6.5.1. *Alkaline cleavage of protein-carbohydrate linkages*

Protein-carbohydrate linkages can be divided into three categories on the basis of their stability to alkali (Table 6.3). The alkali-sensitive *O*-glycosidic linkages (involving Xyl, GlcNAc, Gal or Man and Ser or Thr) are readily split in relatively mild conditions by a β-elimination mechanism resulting in the release of the carbohydrate moiety. Cleavage of the glycosylamine GlcNAc-Asn bond by alkali involves hydrolysis and requires considerably more severe conditions. In the third group of linkages, which involve the side chains of hydroxyproline and hydroxylysine, the protein-carbohydrate bond is more stable to alkali than the glycosidic linkages of the oligosaccharide. The release of carbohydrate units by alkali has been most widely applied to glycoproteins and proteoglycans containing alkali-sensitive *O*-glycosidic linkages, but more severe treatment with alkali has been used in a few cases for the release of oligosaccharides from glycosylamine linkage. Valuable information about the nature of protein-carbohydrate linkages present in a glycoprotein or proteoglycan can be obtained by examination of the effects of alkali on the protein-carbohydrate linkage (Section 6.8). However, it should be recognised that the rate of cleavage of protein carbohydrate linkages of a particular type in alkali can be greatly influenced by the nature of the peptide moiety and by the structure of the oligosaccharide.

Oligosaccharides with a free reducing group can undergo a complex range of alkali-catalysed epimerisations and degradative reactions which include the removal of sugars from the reducing end by 'peeling' reactions. To prevent these side-reactions alkaline cleavage is usually carried out in the presence of a large excess of sodium borohydride, which rapidly reduces the aldehyde group of the reducing terminal sugar to an alcohol.

6.5.1.1. Alkali-sensitive O-linked carbohydrate units Oligosaccharides can be released from glycoproteins or proteoglycans containing this type of linkage (Table 6.3) by treatment with alkaline borohydride. This method has been very widely employed to obtain

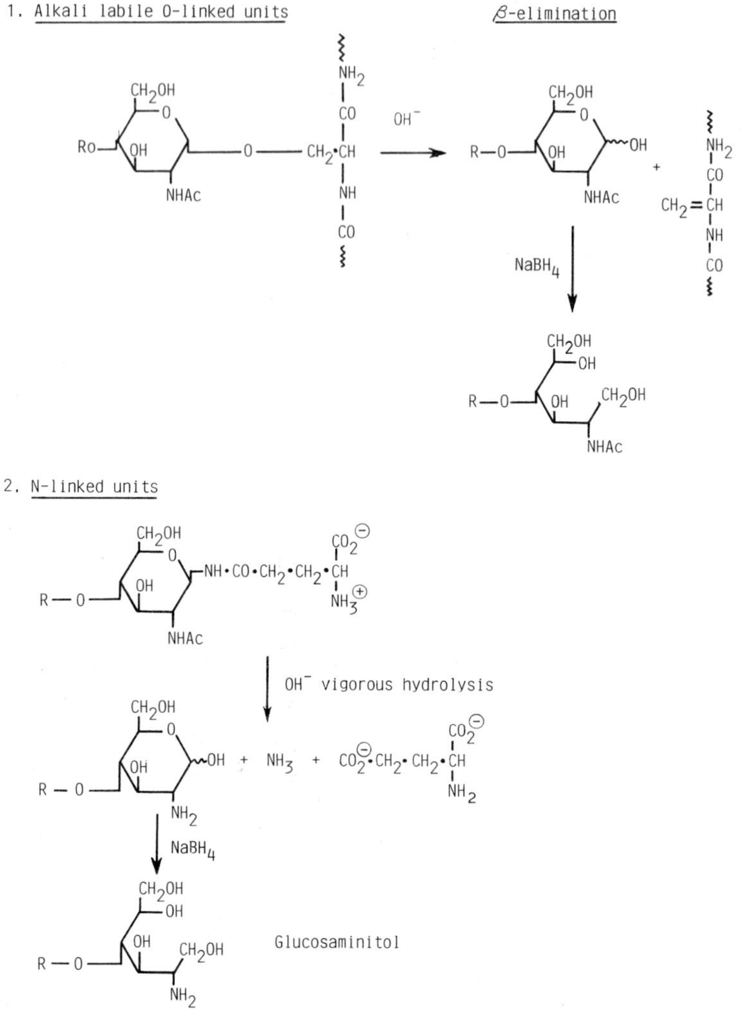

Fig. 6.1. Release of carbohydrate units with alkali.

carbohydrate units from mucins and other glycoproteins containing the GalNAc-Ser(Thr) linkage (Fig. 6.1). The conditions developed by

Carlson (1968) for the release of carbohydrate from pig submaxillary mucin have been successfully applied to other glycoproteins and proteoglycans containing this type of linkage (Rovis et al., 1973; Lohmander et al., 1980). Glycoprotein 25 mg/ml in 0.05 M NaOH containing 1.0 M $NaBH_4$ (freshly prepared) is incubated for 15 h at 45°C (or 50°C; Iyer and Carlson, 1971). Borohydride is destroyed by the addition of 50% acetic acid to bring the solution to pH 5.0. The reaction mixture can be deionised and fractionated by gel filtration. This procedure can produce cleavage of a high proportion (90%) of *O*-glycosidic linkages and high recoveries of sugars can be obtained in the released oligosaccharides (Iyer and Carlson, 1971). However, the conditions giving optimal yields of oligosaccharides should be determined for each glycoprotein.

Spiro and Bhoyroo (1974) have employed alkaline cleavage to selectively release the three *O*-glycosidically linked carbohydrate units of fetuin, which also contains *N*-linked oligosaccharides. The β-elimination reaction was carried out by incubating glycoprotein (5–10 mg/ml) in 0.1 M NaOH containing 0.8 M $NaBH_4$ at 37°C for 68 h.

Treatment of glycoproteins (or proteoglycans) with alkaline borohydride under conditions severe enough to release *O*-linked oligosaccharides in good yields leads to some peptide bond cleavage. Thus the appearance of carbohydrate units of low molecular weight following alkaline cleavage does not necessarily show that the carbohydrate was *O*-glycosidically linked. Glycopeptides arising from *N*-linked carbohydrate units can arise from proteoglycans (Lohmander et al., 1980) as a result of peptide bond cleavage occurring under the conditions described by Carlson (1968) together with oligosaccharides released by β-elimination. The latter can be distinguished by the presence of *N*-acetylgalactosaminitol, formed by borohydride reduction of the reducing terminal sugar.

Alkaline β-elimination can be applied either to intact glycoproteins or to glycopeptides. However, the elimination will not proceed satisfactorily if the linkage amino acid has a free α-amino or α-carboxyl group. For this reason general procedures for the detection of *O*-glycosidic linkages by alkaline elimination of Pronase glycopeptides (Finne and Krusius, 1982) are not entirely satisfactory. Blocking free α-amino groups by *N*-acetylation or maleylation of glycopeptides prior to alkaline elimination may improve the yield of oligosaccharides.

Conditions which have been employed for the release of oligosaccharide units from other types of *O*-glycosidic protein carbohydrate linkage are given in Table 6.3.

6.5.1.2. N-linked carbohydrate units The linkage between *N*-acetylglucosamine and asparagine which occurs in many glycoproteins (and proteoglycans; Nilsson et al., 1982) can be cleaved by vigorous alkaline hydrolysis (Fig. 6.1). Borohydride is added to the reaction mixture to reduce the sugar released from the protein carbohydrate linkage and protect it from alkaline degradation. This type of procedure can be applied to glycopeptides containing one or more amino acids (Lee and Scocca, 1972) but the yield of oligosaccharide may be influenced by the peptide moiety (Zinn et al., 1978). Difficulty may be encountered with the cleavage reaction if the asparagine-linked glucosamine residue is substituted in the 3-position, for example with fucose. If such a substituent is present it may be removed by glycosidase digestion or by Smith degradation prior to alkaline cleavage. This procedure was originally described by Lee and Scocca (1972). A modification of this method, reported to give higher yields of oligosaccharide, is given below (Zinn et al., 1978).

The glycopeptide in a screw-capped 13 × 100 mm test tube is dissolved at a concentration of 1 μmol/ml in 1 M NaOH containing 4 M $NaBH_4$, capped, and heated at 80°C for 24 h. After cooling on ice water the reaction mixture is diluted with water (3 volumes) and adjusted to pH 5.0 with glacial acetic acid. After centrifuging, the while pellet is washed twice with 1-ml portions of water. The supernatants are combined and desalted by chromatography on Sephadex G-25.

As partial de-*N*-acetylation occurs during hydrolysis the product requires re-acetylation. To do this the sample (0.02–5 μmoles) in

0.1–1.0 ml water is treated with 0.1 volume of saturated $NaHCO_3$ and 0.1 ml of 2% v/v acetic anhydride in acetone. After 5 min reaction the mixture is deionised by treatment with Dowex 50 (H^+ form) to remove sodium ions and the solution is taken to dryness on a rotary evaporator (Wheat, 1966).

6.5.2. Cleavage of protein-carbohydrate linkages by hydrazinolysis

Oligosaccharide units can be released from linkage to the side chain of asparagine by treatment of glycoproteins with anhydrous hydrazine. This reagent also brings about cleavage of peptide bonds and the loss of acetyl groups from *N*-acetylhexosamine residues. The free amino groups can be reacetylated prior to the isolation of the oligosaccharides or alternatively degradation can be carried out by deamination (Section 6.9.4.5).

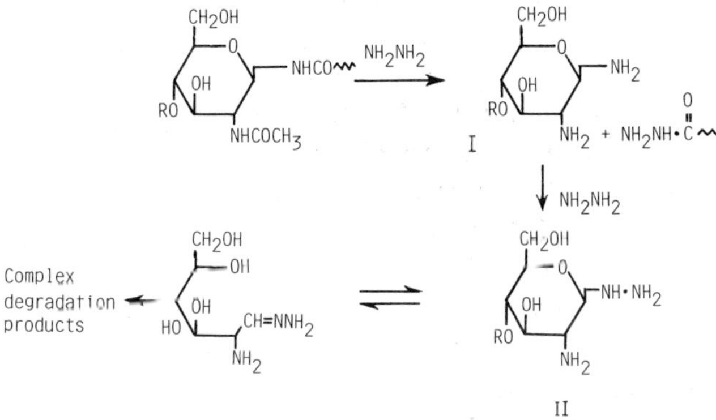

Fig. 6.2. Hydrazinolysis of an asparagine linked oligosaccharide. R represents an oligosaccharide chain.

The reaction of anhydrous hydrazine with asparagine-linked sugar chains probably occurs by the mechanism shown in Fig. 6.2 (Saeed and Williams, 1980; Takasaki and Kobata, 1982). Prolonged exposure to hydrazine can lead to isomerisation and decomposition of the reducing terminal residue. The glycosylamine is readily hydrolysed in neutral or slightly acidic aqueous conditions but the hydrazone II (Fig. 6.2) is relatively stable to hydrolysis. According to Takasaki et al. (1982) *N*-acetylation of II promotes hydrolysis of the glycosylamine linkage, prevents the formation of complex degradation products, and reconverts glucosamine residues in the oligosaccharide back to *N*-acetylglucosamine. Short periods of hydrazinolysis, rapid removal of the reagent and *N*-acetylation of the products promote good yields of oligosaccharide which can be reduced with borohydride to convert the reducing terminal residue to *N*-acetylglucosaminitol (Takasaki et al., 1982).

Hydrazinolysis can be accelerated by the addition of hydrazine sulphate as catalyst (Yosizawa et al., 1966). Both the catalysed (Reading et al., 1978) and the uncatalysed reactions have been applied to glycoproteins. The following method (uncatalysed) is recommended for the release of oligosaccharides from asparagine linkage (Takasaki et al., 1982).

Glycopeptide or glycoprotein (0.2–1 mg) is dried overnight over P_2O_5 and then suspended in 0.5–1 ml of freshly distilled anhydrous hydrazine and heated in a sealed tube at 100°C for 8–12 h. The reaction mixture is dried at room temperature in a rotary evaporator with H_2SO_4 in the trap and traces of hydrazine are removed by repeated evaporation with toluene. After dissolving in saturated $NaHCO_3$ solution the products are acetylated with acetic anhydride (Section 6.5.1.2). The reaction mixture is passed through Dowex 50W (H^+) and washed through with five bed volumes of water. Eluate and washing are combined and dried on a rotary evaporator. The glycopeptides are redissolved in water and applied as a line to Whatman No. 3 MM paper and chromatographed for two days in *n*-butanol/ethanol/water (4:1:1 by vol.). A strip 0–5 cm from the origin containing the glycopeptides is cut out and eluted with water. The glycopeptides are then reduced with sodium borohydride. At this stage a tritium (or deuterium) label can be introduced (from NaB^3H_4)

into the terminal *N*-acetylglucosaminitol. After labelling the glycopeptide mixture is reapplied to paper and the previous chromatographic step is repeated and the glycopeptides are reisolated. This procedure removes radioactive contaminants present in NaB^3H_4. If an internal standard (maltotriose or *N*-acetylneuramin lactose) is added immediately prior to the first paper chromatography step it is possible to calculate the recovery of radioactivity in the standard and the reduced oligosaccharides released from the glycoprotein. This information, together with the molecular weights of the glycoprotein sample and the oligosaccharides, allows the number of sugar chains originating from one mole of sample to be calculated.

It is preferable to determine the optimal time of reaction with hydrazine for individual glycoproteins. Substituents of the *N*-acetylglucosamine residue linked to asparagine can influence the yield of the desired product. Good recoveries are obtained when C-6 of this sugar is substituted with fucose. As well as measuring the radioactivity incorporated into the glycopeptide after tritiated borohydride reduction, the amount of labelled glucosaminitol should be determined (Section 5). If side-reactions have occurred other labelled products are likely to be present. Low yields of terminal *N*-acetylglucosaminitol resulting from side-reactions occurring as a result of hydrazine treatment are a potential source of difficulty in the structural analysis of oligosaccharides released in this way (Saeed and Williams, 1980).

Complications can also arise when *O*-linked carbohydrate units are present in the glycoprotein subjected to hydrazinolysis. Partial release of oligosaccharides from their linkage to serine and threonine occurs together with some further breakdown to monosaccharides and their degradation products (Takasaki et al., 1982). It is therefore advisable to determine whether alkali-labile linkages are present in the glycoprotein (Section 6.8) before attempting hydrazinolysis.

6.5.3. *Trifluoroacetolysis*

This recently introduced method is mainly of interest as a means of releasing oligosaccharides from glycoproteins or glycopeptides containing carbohydrate linked via the amide nitrogen of asparagine (Nilsson and Svensson, 1979; Nilsson et al., 1982). Treatment with mixtures of trifluoroacetic anhydride (TFAA) and trifluoroacetic acid (TFA) at 100°C for 48 h results in cleavage of the glycosylamine protein-carbohydrate linkage. When 1:1 (by volume) TFAA/TFA mixtures are employed for trifluoroacetolysis extensive destruction of both the reducing terminal and the penultimate hexosamine residues occurs. Reducing terminal sugars other than amino sugars are stable to the reagent as are most glycosides (but not those involving sialic acid). Thus the major oligosaccharides isolated after cleavage of desialated *N*-linked carbohydrate units with TFAA/TFA (1:1) have lost the two glucosamine residues closest to the protein-carbohydrate linkage and any fucose attached to the linkage glucosamine (Nilsson and Svensson, 1979; Franzen et al., 1980) but the glycosidic linkages otherwise remain intact. However, it has recently been shown that milder conditions of trifluoroacetoylis (TFAA/TFA 100:1 at 100°C for 48 h) can result in the release of oligosaccharide in which there is no loss of hexosamine (Nilssen et al., 1982).

Trifluoroacetolysis cleaves peptide bonds by transamidation and replaces the *N*-acetyl substituents of amino sugars by *N*-trifluoroacetyl groups. In addition *O*-trifluoroacetylation of free hydroxyl groups occurs. Both *O*- and *N*-trifluoroacetyl groups are subsequently removed, and the free reducing group of the oligosaccharide is simultaneously reduced by treatment with sodium borohydride. The amino sugars can be re-*N*-acetylated subsequently.

Cleavage of the *O*-glycosidic linkage between serine and threonine and *N*-acetylgalactosamine also occurs on trifluoroacetolysis in TFAA/TFA (1:1). Partial degradation of the reducing terminal galactosamine residue occurs. Alkaline β-elimination with NaOH-$NaBH_4$ is therefore to be preferred as a means of releasing this type of carbohydrate unit.

The following method (Franzen et al., 1980) can be applied to glycopeptides or glycoproteins containing the *N*-glycosyl protein-carbohydrate linkage.

Lyophilised desialated glycoprotein or glycopeptide (1–10 mg) is added to 5 ml TFAA/TFA solution (1:1 v/v or 1:100 v/v) and the reaction mixture is sealed in a thick-walled glass vessel which is heated at 100°C for 48 h. (Caution! Corrosive mixture under pressure). The resulting clear dark-coloured solution is cooled and evaporated to dryness. Methanol, 2.5 ml, is added and the solution again evaporated. Acetic acid (50%), 2.5 ml, is added, the mixture is heated for 30 min at 100°C and again evaporated to dryness. The products are partitioned between diethyl ether (5 ml) and water (2.5 ml) and the ether layer is re-extracted twice with water (2.5 ml). The combined water phases are extracted once with diethyl ether (2.5 ml) and are then evaporated to dryness. After redissolving in water (1 ml) sodium borohydride (or borodeuteride), 10 mg, is added and allowed to react for 18 h at room temperature. After destruction of excess reducing agent by acidification with glacial acetic acid the solution is concentrated to dryness. Boric acid is removed by repeated evaporation with methanol and the reduced oligosaccharides are desalted on Sephadex G-25 eluting with water. The product(s) isolated at this stage contain de-*N*-acetylated amino sugars. Selective *N*-acetylation can be accomplished with acetic anhydride in $NaHCO_3$ (Section 6.5.1.2). Alternatively both *O*- and *N*-acetylation can be performed by treatment with acetic anhydride:pyridine (1:1 v/v, 1 ml) at 100°C for 1 h. Following removal of reagents by evaporation, de-*O*-acetylation can be accomplished by treatment with 1 M ammonia in 75% aq. methanol (2.5 ml) at room temperature for 18 h followed by evaporation to dryness. Further fractionation of the oligosaccharide can be carried out as described in Section 6.7.

The recovery of oligosaccharide groups by this procedure is less than quantitative. Franzen et al. (1980) reported the isolation of 4 mg oligosaccharide from 100 mg of antithrombin III containing 10–15% carbohydrate.

6.5.4. *Enzymic cleavage of protein-carbohydrate linkages*

Enzymic cleavage of the protein-carbohydrate linkage in glycoproteins (or proteoglycans) is not usually feasible when the peptide and oligosaccharide chains are intact. One exception to this generalisation is the endo-α-*N*-acetylgalactosaminidase considered in Section 6.6.4. However, enzymes are available which will split protein-carbohydrate linkages after extensive degradation of the peptide chain (GlcNAc-Asn linkage) or of the carbohydrate moiety (GalNAc-Ser/Thr and other linkages, Section 6.8). The GlcNAc-Asn linkage can be cleaved enzymatically in glycopeptides in which both the α-amino and α-carboxyl groups of the linkage amino acid are unsubstituted. Suitable glycopeptides containing a single Asn residue have been isolated from several glycoproteins following Pronase digestion (Section 6.4.1) and purification (Section 6.7). The isolation of aspartylglycosylamine amidohydrolases ('glycosyl asparaginases') from hen oviduct and hog kidney which catalyse this reaction has been described (Tarentino and Maley, 1972; Kohno and Yamashita, 1972). The hen oviduct enzyme has been employed to release the oligosaccharide unit from glycopeptides isolated from ribonuclease B and ovalbumin glycopeptides (Plummer et al., 1968) as outlined below.

Glycopeptide (0.2 μmole) is incubated with aminohydrolase (6 mg of protein, specific activity 18 nmoles per h per mg protein) in 0.165 M potassium phosphate buffer, pH 7.5, at 37°C for 5 h. An aliquot is diluted with citrate buffer and applied to an amino acid analyser. Released aspartic acid and any undegraded glycopeptide (appearing near the breakthrough) can be detected. Plummer et al. (1968) separated the oligosaccharide unit from enzyme by phosphocellulose chromatography and then reduced the product with sodium borohydride. Alternatively the oligosaccharide could be isolated by gel filtration. Because of the labour involved in obtaining and purifying glycopeptides containing only a single asparagine this type of procedure is not normally used in the isolation of oligosaccharides for the analysis of carbohydrate structure. Digestion with glycosylaminohydrolase is valuable as a way of confirming the nature of the protein-carbohydrate linkage. Enzymic release of oligosaccharides for structural analysis can be more readily accomplished by the use of endoglycosidases, as described in the next section.

6.6. Release of carbohydrate units with endoglycosidases

The mildest and most specific means of cleaving oligosaccharide units from glycoproteins or glycopeptides is to employ endoglycosidases. These enzymes specifically hydrolyse internal glycosidic linkages, releasing a part (or in some cases all) of the carbohydrate moiety. The liberated oligosaccharides are eminently suitable for structural analysis (Kobata, 1979). It is also possible to use these enzymes to examine changes in the properties of glycoproteins following the release of much of their carbohydrate.

The substrate specificities of endoglycosidases are directed at oligosaccharide units and are quite exacting. These enzymes can therefore be used to detect particular structural features present in purified glycoconjugates or even in mixtures (e.g. of membrane glycoproteins). The type of carbohydrate units present can often be inferred from the specificity of the enzyme required for its release. Several endoglycosidases which have been purified sufficiently to be applied to the structural investigation of glycoconjugates are listed in Table 6.4. These enzymes fall into three categories; endo-β-*N*-acetylglucosaminidases, endo-α-*N*-acetylgalactosaminidase and endo-β-galactosidases.

The choice of endoglycosidases which may be capable of releasing oligosaccharides from a glycoconjugate can be made on the basis of the evidence (such as sugar composition and stability towards alkali) indicating the types of carbohydrate units which may be present. Table 6.4 gives a general indication of the substrates hydrolysed by different endoglycosidases and more details of their specificities are given later in this section. For some of the endoglycosidases (e.g. endo-D and endo-C_I) it is essential to pretreat substrates with exoglycosidases in order to remove peripheral sugar residues which prevent endoglycosidase action. Purified glycopeptides containing a single amino acid have been used as substrates for the endoglycosidases in several investigations of oligosaccharide structure. Larger glycopeptides and even intact glycoconjugates can, in many cases, also be employed as substrates. For intact glycoconjugates more complete

TABLE 6.4

Endoglycosidases useful in structural analysis of glycoproteins

References to the purification and properties of these enzymes can be found in the text

Enzyme and source		Typical substrate
endo-β-*N*-Acetylglucosaminidases		
D	*Diplococcus pneumoniae*	*N*-linked complex type (peripheral sugars removed)
H	*Streptomyces plicatus* *Streptomyces griseus*	*N*-linked high-mannose and hybrid types
L	*Streptomyces plicatus*	*N*-linked low mol. wt. only
C_I	*Clostridium perfringens*	*N*-linked complex type (peripheral sugars removed)
C_{II}	*Clostridium perfringens*	*N*-linked high-mannose type
F-Gal type	*Sporotricum dimorphosphos-phorum*	*N*-linked complex type (biantennary only, requires terminal Gal)
F	*Flavobacterium meningosepticum*	*N*-linked high-mannose and complex types
endo-α-*N*-Acetylgalactosaminidase		
	Diplococcus pneumoniae	*O*-linked, only Galα1–3GalNAc1–
endo-β-*N*-Galactosidases		
	Diplococcus pneumoniae	Blood group A and B determinants
	Escherichia freundii *Flavobacterium keratolyticus*	Keratan sulphate and oligosaccharides containing sequence R GlcNAcβ1–3Galβ1–4GlcNAc (or Glc)

release of carbohydrate units can sometimes be obtained by prior denaturation of the peptide moiety (e.g. by disulphide-bond modification).

Oligosaccharides released by exoglycosidase digestion can be characterised directly or they can be labelled by reduction with NaB^3H_4

which converts the reducing sugar to the corresponding alditol. The labelled oligosaccharide alditols can then be purified and fractionated by paper chromatography and/or gel filtration (Section 6.7).

6.6.1. *Endo-β-N-acetylglucosaminidases — specificity*

These enzymes act on the asparagine-linked carbohydrate groups of glycopeptides and glycoproteins catalysing the reaction given below:

R-GlcNAc(β1–4)GlcNAc-Asn→R-GlcNAc + GlcNAc-Asn
(R represents a mono- or oligosaccharide)

The reaction products are an oligosaccharide (the glycone) with *N*-acetylglucosamine as the reducing terminal residue and a single non-reducing *N*-acetylglucosamine residue linked to asparagine or an asparagine-containing peptide chain (the aglycone). These enzymes all show marked specificity towards the glycone but vary in the extent to which the aglycone influences hydrolysis (see Kobata, 1979 for review).

Endo-D and C_I Endo-β-*N*-acetylglucosaminidases-D and -C_I, although purified from different micro-organisms, have the same specificity. These enzymes have been used quite extensively to cleave the core of complex-type asparagine-linked glycopeptides. They act on substrates containing the trisaccharide Man(α1–3)Man(β1–4)-GlcNAc within the glycone. The non-reducing terminal α-mannosyl residue is of paramount importance and this residue must be unsubstituted. Substitution of the β-linked mannose (e.g. with Gal(β1–4)-GlcNAc(β1–2)Man(α1–6) does not prevent cleavage. Considerable variation can be tolerated in the aglycone which can consist of GlcNAc, Fuc(α1–6)GlcNAc, GlcNAc-Asn or Fuc(α1–6)GlcNAc-Asn. Hydrolysis, at a greatly decreased rate, also occurs when the aglycone is *N*-acetylglucosaminitol. Because of the glycone requirement for a free terminal Man(α1–3) residue it is essential to pretreat complex-type glycopeptides with exoglycosidases (neuraminidase, fucosidase, β-galactosidase, β-*N*-acetylglucosaminidase) to remove sugars or sugar chains which may substitute this residue. Digestion with Endo-D or Endo-C_I is carried out in citrate-phosphate or phosphate buffers at pH 6.5.

Endo F-βGal This partially purified enzyme has a distinctive specificity in that it cleaves biantennary monosialo- and asialoglycopeptides (containing asparagine as the sole amino acid) of the complex type. The presence of at least one unsubstituted non-reducing terminal galactose is essential for enzyme activity. Thus sialylated biantennary glycopeptides should be pretreated with neuraminidase. Tri- and tetra-antennary carbohydrate units are not cleaved. Fucose-linked α1–6 to the *N*-acetylglucosamine in the aglycone decreases the rate of hydrolysis but does not abolish activity (Bouquelet et al., 1980).

Endo-F This enzyme is of particular interest because it releases oligosaccharides from both the high-mannose and the complex types of carbohydrate units of intact glycoproteins (Elder and Alexander, 1982).

Endo-H This enzyme acts on oligosaccharide units of the high-mannose or hybrid type. The tetrasaccharide glycone structure Man(α1–3)Man(α1–6)Man(β1–4)GlcNAc is required. While the non-reducing terminal mannose residue is important for the specificity of the enzyme substitution at C-2 can be tolerated. Sugar chains with GlcNAc, GlcNAcol or GlcNAc-Asn as aglycone are hydrolysed but Fuc(α1–6)GlcNAc and Fuc(α1–6)GlcNAc-Asn aglycones prevent cleavage. The enzyme has a pH optimum of 5.0–6.0 and full activity is retained in the presence of 0.2% SDS (Tarentino et al., 1978).

Endo-L This endoglycosidase activity was present in early preparations of Endo-H but separation can be obtained by gel filtration (Trimble et al., 1982). Endo-L cleaves the *N*-acetylchitobiose unit of low molecular weight substrates such as Man(β1–4)GlcNAc-(β1–4)GlcNAc-Asn. Increasing the oligosaccharide chain length by the addition of one mannose residue decreases the rate of hydrolysis by a factor of 10^4 and no cleavage occurs when three mannose residues are present. The enzyme appears to be a chitobiase with a very limited range of specificity (Trimble et al., 1982). It has been used for the isolation of the disaccharide Man(β1–4)GlcNAc but, because of its restricted specificity, it has not found wider applications.

Endo-C_{II} Oligosaccharide units of the ‘high-mannose’ but not of the hybrid type are cleaved by Endo-C_{II}. Substrates for this enzyme contain the branched pentasaccharide aglycone

```
Manα1–3Manα1
            \
            6
             Manβ1–4GlcNAc
            3
            /
       Manα1
```

The two terminal mannose residue must either be unsubstituted or substituted only at C-2. Aglycone specificity is the same as for Endo-H. Digestion of substrates with Endo-C_{II} and Endo-H can be used to differentiate between high-mannose and hybrid-type carbohydrate units. High-mannose units are cleaved by both enzymes but hybrid units are hydrolysed only by Endo-H.

The optimal pH for digestion with Endo-C_{II} is pH 7.0.

6.6.2. Endo-β-N-acetylhexosaminidases — assays

The assay of endoglycosidases is carried out with purified glycopeptides as substrates. For example a purified ovalbumin glycopeptide $Man_5GlcNAc_2Asn$ is a suitable substrate for Endo-D and Endo-H. To improve the sensitivity of the assay the glycopeptide can be [^{14}C]-*N*-acetylated (Koide and Muramatsu, 1974) or alternatively glycopeptides can be labelled by reaction with [^{3}H]dansyl chloride (Tarentino et al., 1978).

Assay mixtures contain 10 μl each of substrate (0.15 nmoles of [^{14}C]acetylated glycopeptide, 4 000 cpm), 0.15 M citrate-phosphate buffer (pH 6.5 for Endo-D, pH 5.0 for Endo-H) and enzyme solution diluted, if necessary, with 0.1 M NaCl containing 0.1% bovine serum albumin. After incubation at 37°C for 15 min the enzyme can be inactivated by the addition of ethanol (0.1 ml) and the products separated by paper electrophoresis (Koide and Muramatsu, 1974). Alternatively the reaction mixture is diluted with 0.5 ml of ice-cold water and applied immediately to a column of ConA-Sepharose (0.5 × 1.5 cm) equilibrated with 0.01 M Tris-HCl buffer, pH 7.5, containing 0.15 M NaCl. The column is washed with 2 ml of buffer and the radioactivity in the eluate is determined by liquid scintillation counting. ConA-Sepharose binds the substrate but not the products of the reaction (Muramatsu et al., 1978).

Results of the assay are expressed as moles of substrate released per minute per mg enzyme protein. It should be noted that some authors and commercial suppliers 'correct' the values obtained for the sub-optimal concentration of substrate employed in this assay by multiplying by a factor (51 for Endo-D and 90 for Endo-H) designed to indicate the activity at saturating substrate concentration.

As a rough guide to the amount of enzyme to be used 1–20 milliunits of enzyme are usually required to produce complete hydrolysis of 1 μmole of glycopeptide if incubated at 37°C for 24 h. However, it is desirable to follow the time course of hydrolysis in every case.

6.6.3. Endo-β-N-acetylhexosaminidases — applications

Release of oligosaccharide units for carbohydrate structural analysis (Yamashita et al., 1978) Purified ovalbumin glycopeptides, containing asparagine as the sole amino acid were labelled by *N*-acetylation with [^{14}C]acetic anhydride. Glycopeptides (5.8 mg) were incubated with 1.5 milliunits of Endo-H in 50 μl of 0.15 M citrate-phosphate buffer, pH 5.0, containing 0.1 M 2-acetamido-2-deoxygluconolactone (to inhibit any exo-β-*N*-acetylglucosaminidase contaminating the enzyme preparation) at 37°C for 24 h. The reaction was stopped by the addition of 0.1 ml ethanol. After centrifugation the supernatant was removed and the pellet washed once with 40% ethanol. The supernatants were combined and evaporated. After reduction with NaB^3H_4 the reaction products were separated by paper electrophoresis. Release of [^{14}C]acetylasparaginyl-*N*-acetylglucosamine occurred in 93% yield. The tritium-labelled reduced oligosaccharide unit was subjected to structural analysis.

Characterisation of radioactive glycopeptides (Muramatsu et al., 1976) Glycoproteins of human diploid fibroblasts were labelled by the incorporation of [2-^{3}H]mannose. Glycopeptides were released by Pronase digestion of the cells. The glycopeptides were then hydrolysed with Endo-H and Endo-D to distinguish between carbohydrate units of the high-mannose and the complex types. Digestion with Endo-D (5.2 milliunits) was carried out in the presence of neuraminidase (1.8 milliunits), β-galactosidase (7.1 milliunits) and β-*N*-acetylglucosaminidase (23 milliunits), all from *Diplococcus pneumoniae*, to remove peripheral sugars. Incubation was carried out in 0.1 ml of 0.05 M citrate-phosphate buffer, pH 6.0, for 15 h under toluene. The labelled oligosaccharides were separated by paper chromatography and characterised by Sephadex G-25 column chromatography, by affinity chromatography on ConA-Sepharose and by successive digestion with α-mannosidase and β-mannosidase.

Release of oligosaccharides from glycoproteins (Tarentino et al., 1974) The release of intact oligosaccharides from glycoproteins using Endo-H was examined by Tarentino et al. (1974). Glycoprotein (10–20 mg) dissolved in 1.0 ml of 0.1 M sodium citrate buffer was incubated with 0.5 units of enzyme at 37°C under toluene. Aliquots of the reaction mixture were precipitated with TCA or ethanol. For glycoproteins which cannot be completely precipitated the digest can be fractionated by gel filtration on Sephadex G-100. Under these conditions of digestion 75% of the carbohydrate of deoxyribonuclease was released in 30 min. While carbohydrate was released completely from some other native glycoproteins, denaturation of ovalbumin (by sulphitolysis) was necessary for the removal of carbohydrate. Cleavage, with this enzyme, is also restricted to the high-mannose type of carbohydrate unit.

6.6.4. Endo-α-N-acetylgalactosaminidase

The purification of this enzyme, together with endo-β-galactosidase, endo-β-*N*-acetylglucosaminidase D and the exoglycosidases β-galactosidase and β-*N*-acetylglucosaminidase, from culture filtrates of *Diplococcus pneumoniae* has been described by Kobata and Takasaki (1978). The endo-α-*N*-acetylgalactosaminidase cleaves the *O*-glycosidic linkage between the disaccharide Gal(β1–3)GalNAc and serine or threonine. Both glycopeptides and glycoproteins are substrates but the oligosaccharide specificity is very high in that no other carbohydrate chain apart from Gal(β1–3)GalNAc has been found to be liberated (Kobata, 1979). Because of this restricted specificity this enzyme is far from being a general hydrolytic agent for GalNAc-Ser/Thr linkages.

The enzyme has a pH optimum of 6.0 and some inhibition occurs in the presence of EDTA, Mg^{2+}, Zn^{2+} or *p*-chloromercuribenzenesulphonate. It has found application in structural studies of the carbohydrate units of bovine kininogen (Endo et al., 1977).

6.6.5. Endo-β-galactosidases

Two types of endo-β-galactosidases with quite distinct specificities have been employed in the structural analysis of proteoglycans, glycoproteins and glycopeptides. The endo-β-galactosidase isolated from *Diplococcus pneumoniae* culture filtrate (Kobata and Takasaki, 1978) specifically releases the blood group A and B determinants from type II chains. The H antigen and blood group A determinants on type I chains are not cleaved. The enzyme has been applied to studies of the blood group determinants on cell membranes and to the release of these carbohydrate units from the membrane glycoprotein glycophorin (Takasaki and Kobata, 1976). The pH optimum of this enzyme is about 6.0.

A quite different type of endo-β-galactosidase has been purified from two strains of *Escherichia freundii*. The endo-β-galactosidases from the two strains have the same specificities (Li et al., 1982) but the strains differ in the quantities of other glycosidases which they produce. Substrates for this type of endo-β-galactosidase include the carbohydrate chains of keratosulphate, sulphated porcine mucin (Fukuda and Matsumura, 1976) and other glycoconjugates containing *N*-acetyllactosamine (or lactosamine) units. Examination of the susceptibility of a series of glycolipids to hydrolysis by this enzyme indicated that the common structural element in substrates is

R-GlcNAc(β1–3)Gal(β1–4)GlcNAc (or Glc)

when R can be either hydrogen or a sugar or sulphate substituent. When R is a sialic acid residue the cleavage of the β-glycosidic linkage is enhanced (Kobata, 1979). The endoglycosidase does not attack heparin, hyaluronate, dermatan sulphate, chondroitin 4-sulphate or chondroitin 6-sulphate. The enzyme purified by Li et al. (1982), with nasal cartilage keratan sulphate as substrate has optimal activity between pH 5.5 and 5.8 in 0.05 M sodium acetate buffer. Low concentrations of the enzyme (11 μg/ml) are stable when stored at 4°C (at pH 5.0) but are unstable at 37°C. Addition of bovine serum albumin increases stability of the enzyme at 37°C.

The production of endo-β-galactosidase is induced by the growth of *E. freundii* in a medium containing keratan sulphate. *Flavobacterium keratolyticus* produces an endo-β-galactosidase of very similar specificity without the necessity for induction. Enzyme from this source has been purified and characterised by Kitamikado et al. (1982). The optimal activity of this enzyme in 0.05 M sodium acetate buffer is about pH 6.0.

The hydrolysis of keratan sulphates (or other glycoconjugates) with endo-β-galactosidase from *E. freundii* (or *F. keratolyticus*) can be examined under the following conditions (Li et al., 1982). Keratan sulphate (300 μg) is incubated with 50 units of enzyme at 37°C in

100 μl of 0.05 M sodium acetate buffer, pH 5.5, for 17 h. The incubation mixture is evaporated to dryness, redissolved in the minimum amount of water and oligosaccharides are separated by paper chromatography (with ethyl acetate-pyridine-water 12:5:4, using three developments). Preparations of keratan sulphate from different sources produce distinct patterns of oligosaccharides due to structural variations (in sulphation, chain length etc.) but 6-*O*-sulpho-GlcNAc(β1–3)Gal is the major product.

6.6.6. Problems encountered in the use of endoglycosidases

Whilst the endoglycosidases are very valuable tools for the selective release of carbohydrate units from glycoproteins, proteoglycans or glycopeptides, problems can arise in their use.

Endoglycosidase preparations used for structural studies must be free of all exoglycosidase contamination. Some exoglycosidases which can cleave oligosaccharide substrates cannot be detected with the nitrophenyl glycosides of monosaccharides which are usually employed to assay for these enzymes. It is important to check during the course of digestion with endoglycosidases for the release of monosaccharides, which indicates the presence of exoglycosidases. In some cases it may be possible to selectively inhibit exoglycosidases without inhibition of the endoglycosidase (e.g. by addition of *N*-acetylglucosamine 1,5-lactone to inhibit exo-*N*-acetylglucosaminidase).

Another difficulty which has been encountered is that some batches of commercial enzyme preparations may be contaminated with traces of protease. While this does not greatly hinder studies on released carbohydrate units it can create difficulties in attempts to study deglycosylated proteins.

In some cases, especially when the substrate is membrane-associated or particulate, difficulties can arise because of lack of accessibility of substrate. It may be possible to overcome this problem by use of non-ionic detergents or low concentration of ionic detergents. When glycoprotein substrates are employed prior unfolding of the peptide may improve digestion with endoglycosidase.

6.7. *Fractionation of glycopeptides and oligosaccharides*

A variety of methods have been developed which can be applied to the isolation of homogeneous glycopeptides or oligosaccharides derived from glycoproteins or proteoglycans. The most appropriate fractionation strategy depends on the nature of the starting material, the quantity available and the goal of the investigation. Procedures are described in this section for the isolation of glycopeptides produced by selective cleavage (e.g. with trypsin or chymotrypsin) or non-selective cleavage (e.g. with Pronase) of purified glycoproteins and for the isolation of glycopeptides and glycosaminoglycans from enzymic digests of cell or tissue extracts. Methodology is also described for the purification of oligosaccharides released from glycoproteins or proteoglycans by treatment with endoglycosidases, alkali, hydrazine or by trifluoroacetolysis.

It is very important to ensure that glycopeptides or oligosaccharides are homogeneous before attempting structural analysis. The analysis of mixtures can often produce misleading results. There are several instances in the literature of incorrect structures, particularly those of the *N*-linked carbohydrate units, which have been proposed on the basis of studies carried out on incompletely fractionated glycopeptides. In general it is more difficult to fractionate glycopeptides than the oligosaccharide units they contain because glycopeptides are usually heterogeneous in both carbohydrate and peptide moieties.

6.7.1. *Glycopeptides obtained by non-selective cleavage of glycoproteins and proteoglycans*

Glycopeptides released from purified glycoproteins by exhaustive digestion with enzymes of wide specificity such as Pronase or papain (Section 6.4.2) usually contain short but heterogeneous peptide chains. The glycopeptides can be separated from most of the peptides in the digest by gel filtration. Dialysis is not a suitable procedure for this separation as glycopeptides are often small enough to pass through dialysis tubing. The larger carbohydrate units of glycosaminoglycans are, however, non-dialysable. Gel filtration of glycopeptides may also show whether different molecular weight classes of glycopeptide are present. Glycopeptides can be further fractionated by ion-exchange chromatography (or paper electrophoresis) on the basis of differences in their charged groups, which may include sialic acids, ester sulphate or phosphate groups, and the ionising groups of their peptide moieties. The isolation of glycopeptides containing only a single linkage amino acid residue and some fractionation based on carbohydrate structure can usually be achieved by ion-exchange chromatography. Fractionation on the basis of features of the carbohydrate structure can be obtained by lectin affinity chromatography (Section 6.7.3.1.).

Glycosaminoglycans released from proteoglycans by non-specific peptide-chain cleavage can be isolated by gel filtration, by dialysis, by precipitation with ethanol (Section 6.4.2) and/or with cetylpyridinium salts (Sections 6.7.3 and 3.4.2) or by ion-exchange chromatography (Section 6.7.3).

Procedures Gel filtration of the glycoprotein digest is carried out on a column of Sephadex G-25 (or G-50) or Bio-Gel P4 (or P6). A Sephadex column 2.5 × 110 cm is suitable for initial fractionation of a sample containing 100 mg of glycoprotein digest in 4 ml. The column is equilibrated either with 0.1 M pyridine adjusted to pH 5.5 with acetic acid, with 0.1 M acetic acid or with water. Pyridine acetate should be employed if the glycopeptides contain sialic acid. Peptide can be monitored by the ninhydrin procedure of Moore and Stein (1954). Measurement of the absorbance of the column eluate at 280 nm can be employed as a rapid method for the location of peptides but peptides not absorbing at this wavelength are also likely to be present. All peptides produce quite strong absorbance at 230 nm and monitoring at this wavelength can be employed for detecting peptides, provided that the absorbance of the eluting buffer is not too great. Glycopeptides are located by colourimetric analysis of aliquots of fractions for hexose (Section 5.8.1) or other constituent sugars such as sialic acid. When using Sephadex columns it is important to ensure that the hexose content of column washings has reached a negligible value before application of the sample. The problem of carbohydrate contamination resulting from the gel filtration medium can be circumvented by use of Bio-Gel. After extensive digestion the glycopeptides are generally the largest molecular species present and they are usually well separated from most of the peptides. After location, the fractions containing glycopeptides are combined and lyophilised to remove the volatile buffer. If more than one carbohydrate-containing peak is observed on gel filtration of the digest this may indicate incomplete digestion or heterogeneity in the carbohydrate units present. These possibilities can be distinguished by (1) repeating the digestion and gel filtration and (2) checking whether the same carbohydrate components are present in material from both peaks.

The molecular weight (M_r) of glycopeptides can be estimated by gel filtration on columns of Sephadex (Bhatti and Clamp, 1968). A column of Sephadex G-50 (fine grade) 150 cm × 1.2 cm is packed and equilibrated with 0.15 M NaCl. The column is calibrated with Blue dextran (to obtain V_o, the void volume), glucose (to obtain V_{100}, the total volume available to solutes) and with oligosaccharides and glycopeptides of known molecular weight covering the range of interest. Samples of glycopeptide (0.1 mg) together with glucose (0.1 mg) and Blue dextran (0.1 mg) are applied to the column in a total volume of 0.2 ml and the column is operated at 15 ml/h with samples of 2.5 ml being collected and analysed for hexose by the phenol-H_2SO_4 method. A plot of $(V_e - V_o/V_{100} - V_o) \times 100$ against molecular weight, where V_e is the glycopeptide (or oligosaccharide) elution volume, was found to be linear up to a molecular weight of about 2000 (Bhatti and Clamp, 1968; Zinn et al., 1978; Lis and Sharon, 1978).

Glycopeptides differing in sialic acid content can often be separated by chromatography on DEAE-cellulose or other similar anion-exchangers (Spiro, 1972; Hayes and Castellino, 1979). The pH of the buffers employed should be close to neutrality to avoid loss of sialic acid and the initial buffer concentration often has to be low (e.g. about 0.5–2.0 mM sodium phosphate) or the glycopeptides may pass through the column unretarded. After washing the column with starting buffer a linear salt gradient should be applied to elute glycopeptides.

Sialic acid-free glycopeptides with a low amino acid content can be fractionated on strongly acidic or strongly basic ion-exchange resins with a low level of crosslinking. Chromatography on Dowex 50 × 2 in buffers of low pH and low ionic strength separates ovalbumin glycopeptides containing asparagine as the sole amino acid on the basis of differences in the structure of their neutral carbohydrate units (Cunningham et al., 1965; Huang et al., 1970). Columns (2 × 150 cm) packed with the ion-exchange resin (200–400 mesh) in the OH^- form are equilibrated with sodium acetate buffer, pH 2.6, 1 mM with respect to Na^+, at a constant temperature (25°C). Samples of ovalbumin glycopeptide (obtained by repeated Pronase digestion and Sephadex G-25 filtration) containing up to 275 mg mannose can be applied. When the column is eluted with starting buffer five glycopeptide peaks are obtained; glycopeptides containing amino acids in addition to asparagine are eluted with 0.05 M sodium acetate, pH 6.5. Extensive structural studies have been carried out on the five asparatamidoglycopeptides of ovalbumin (Yamashita et al., 1978). The size of the carbohydrate groups plays a part in the separation; the glycopeptides with the largest carbohydrate groups being eluted first. Volatile buffers can also be employed with this chromatographic system. For example, Lis and Sharon (1978) isolated glycopeptides from soybean agglutinin by Dowex 50 × 2 chromatography using 0.001 M pyridine acetate buffer, pH 2.7, as eluant, while Schmid et al. (1977) separated glycopeptides from α_1-acid glycoprotein in pyridine formate buffers. Glycopeptides emerging as single symmetrical peaks on chromatography on Dowex 50 × 2 may nevertheless contain more than one type of carbohydrate structure (Lis and Sharon, 1978; Kobata, 1979).

Other ion-exchange systems suitable for the fractionation of peptides (e.g. Dowex 1 × 2, Spiro, 1972) can be employed for glycopeptides. However, the pore size of the matrix of the support should be large enough to allow entry of bulky carbohydrate groups. With polystyrene-based ion-exchange resins the crosslinking should not be more than 2%; at high levels of crosslinking (e.g. Dowex 50 × 8) glycopeptides are excluded. This property was made use of in older methods for removal of amino acids and peptides from glycopeptides but has, for most purposes, been superseded by gel filtration, which is more effective and gives better recoveries.

Separation of the negatively charged borate complexes of glycopeptides by high-voltage paper electrophoresis in borate buffers is an alternative method (Narasinhan et al., 1980).

Further fractionation of glycopeptides can be achieved by lectin affinity chromatography (see Section 6.7.3.1 and Chapter 7).

6.7.2. Glycopeptides obtained by selective cleavage of glycoproteins

The glycopeptides produced by selective cleavage methods (e.g. using CNBr, trypsin or chymotrypsin) have larger and more homogeneous peptide chains than glycopeptides resulting from non-selective cleavage. Methods of fractionation are generally similar to those discussed in the previous section except that allowance should be made for the larger size and, in some cases, increased tendency to adsorption of the glycopeptides resulting from selective cleavage. Gel filtration on Sephadex G-50 (or G-25) often gives a useful initial separation of glycopeptides from peptides. Subsequently the glycopeptides can be fractionated by ion-exchange chromatography or reversed-phase HPLC.

Separation of tryptic or chymotryptic glycopeptides on the basis of charge differences can be obtained by chromatography on ion-exchangers with a hydrophilic supporting matrix such as DEAE-cellulose, CM-cellulose, DEAE- or QAE-Sephadex and DEAE- or SP-Sephadex. It is often necessary to apply the sample in a buffer of low ionic strength (I 0.01 to 0.001) for the glycopeptides to bind to the charged groups on the column. For example, the sialic acid-free chymotryptic glycopeptides of α_1-acid glycoprotein (Schmid et al., 1977) were initially applied to DEAE-cellulose at pH 5.0 in buffer I 0.01 and were displaced by gradient elution (I 0.01–0.1; 0.1–0.5).

Subsequently the three glycopeptide fractions obtained were purified further by chromatography on CM-cellulose at pH 5.0 or pH 4.0 (*I* 0.001–2.0). When the glycopeptides do not bind readily to DEAE- or CM-cellulose, ion-exchangers with more strongly acidic or basic groups such as QAE- and SP-Sephadex can be employed with advantage. For example, the tryptic digest of reduced aminoethylated ovomucoid (175 mg) is resolved into four glycopeptide fractions by chromatography on a column (1.5 cm × 30 cm) of QAE-Sephadex A-50 equilibrated with 7.5 mM ammonia adjusted to pH 9.3 with formic acid and eluting with a gradient to 0.5 M ammonium formate (Beeley, 1976b). Homogeneous ovomucoid glycopeptides can be obtained by rechromatography using SP-Sephadex (Beeley, 1976b). The strongly acidic ion-exchange resin Dowex 50 × 2 has also been successfully used to purify chymotryptic glycopeptides (Schmid et al., 1977). However, low recovery of larger glycopeptides is sometimes a problem associated with this type of supporting matrix.

To obtain homogeneous glycopeptides by low-pressure liquid chromatography it is usually necessary to carry out at least two ion-exchange separations on different supporting media in addition to gel filtration.

Reversed-phase HPLC has recently become established as a powerful method for peptide purification. This method can also be applied to the isolation of glycopeptides (Tetaert et al., 1982). Chymotryptic glycopeptides containing the GlcNAc-Asn type of linkage have been resolved by HPLC after initial isolation from a digest by Sephadex G-50 chromatography. A reverse-phase Synchropak RP-P column (0.41 × 25 cm) pre-equilibrated with 0.1% trifluoroacetic acid (as counter-ion) is employed. The glycopeptide sample (50–150 nmoles) is eluted at a flow rate of 0.7 ml/min with a programmed linear gradient of *n*-propanol (0–60%). Peptides are detected by measuring absorbance at 230 nm. Tetaert et al. (1982) found that the separation of multiply glycosylated glycopeptides containing the GalNAc-Ser(Thr) linkage was more difficult but could be achieved by varying the counter-ion (to 0.1% heptafluorobutyric acid) and by rechromatography of glycopeptide fractions.

6.7.3. Glycopeptides and glycosaminoglycans from cells or tissues

Treatment with proteolytic enzymes is often used to release glycopeptides and proteoglycan fragments from cells or tissues. The products can be very complex mixtures arising from many different types of molecule. Two procedures designed to fractionate such mixtures and determine what types of glycopeptides and glycosaminoglycans are present are described below. The method developed by Finne and Krusius (1982) can be used to separate glycopeptides on the basis of differences in lectin affinity and the size of carbohydrate units. Glycosaminoglycans, released by proteolysis from cell surfaces or other cell fractions, can be separated from each other and from glycopeptides by-ion exchange chromatography (Kraemer, 1971).

Fractionation based on affinity chromatography (Finne and Krusius, 1982) Lipid-extracted tissue is digested exhaustively with Pronase (Section 6.4.1). If the sample contains glycosaminoglycans or nucleic acid these are removed by precipitation; otherwise this step is not necessary. The digest (25 mg protein/ml) is diluted with an equal volume of water and 0.1 volume of 0.1 M cetylpyridinium chloride containing 0.1 Na_2SO_4 is added in small portions. After incubating at 37°C for 20 min the sample is centrifuged and the precipitate is retained for subsequent isolation of glycosaminoglycans. After transferring the supernatant to another centrifuge tube the completeness of precipitation is checked by adding a drop of cetylpyridinium solution. If the solution becomes cloudy on standing it is necessary to repeat the precipitation. Excess cetylpyridinium salt is precipitated by the addition of 0.1 vol. of 0.1 M NaSCN and allowing the mixture to stand at 4°C for at least 2 h before centrifuging (20 000 *g* for 30 min).

The glycosaminoglycan fraction can be redissolved in 0.3 M NaCl and sulphated glycosaminoglycans are reprecipitated with cetylpyridinium chloride. Excess cetylpyridinium chloride is removed from the supernatant by precipitation with thiocyanate and hyaluronic acid is recovered after dialysis (Margolis and Margolis, 1973).

The glycopeptides, which are not precipitated with cetylpyridinium chloride (exceptions are some glycopeptides having highly sialylated *O*-glycosidically linked carbohydrate units clustered together on the peptide chain) or thiocyanate, can then be separated from peptides by gel filtration on Sephadex G-25 as described earlier.

Before fractionating glycopeptides by lectin affinity chromatography it is convenient to label them by *N*-acetylation with [^{14}C]- or [^{3}H]acetic anhydride. The glycopeptides (500 nmol) are dissolved in 1 ml of H_2O and 0.2 ml of 1 M $NaHCO_3$ is added to give a final pH of approximately 8. [^{3}H]Acetic anhydride (1 mCi, 500 mCi/mmol, diluted to 0.2 ml with acetone) is added and the reaction mixture is kept at room temperature for 30 min. Unreacted amino groups are acetylated by adding 0.2 ml of 1 M $NaHCO_3$ and 0.2 ml of 2% (unlabelled) acetic anhydride in acetone. After acidification with 0.2 ml of 4 M acetic acid, free radioactivity is removed by evaporation followed by gel filtration on Sephadex G-25 in pyridine acetate buffer (0.1 M). Some [^{3}H]acetylated peptides are not completely removed from the labelled glycopeptides at this stage. It is therefore necessary to distinguish glycopeptides from free peptides at a later stage. Labelling the glycopeptides is advantageous because the glycoside employed as eluant in the affinity chromatography of glycopeptides interferes in carbohydrate assays.

The glycopeptides are next fractionated by affinity chromatography on ConA-Sepharose (Pharmacia). A column containing 25 ml of ConA-Sepharose is prewashed with 0.1 M HCl followed by 0.1 M sodium acetate buffer, pH 5.2, containing 5 mM $CaCl_2$, 5 mM $MnCl_2$ and 5 mM methyl α-D-glucoside and is finally equilibrated with the latter buffer but omitting the methyl glucoside. Columns are operated at 4°C and 1% butanol can be added to all buffers as a bacteriostatic agent. The glycopeptides, dissolved in acetate-calcium-manganese buffer, are applied to the column, which is then washed with 125–250 ml of 0.1 M NaCl in the acetate-calcium-magnesium buffer. The labelled material emerging from the column is designated fraction A. Buffer (125–250 ml) containing 20 mM methyl α-D-glycoside is then passed through the column to elute fraction B glycopeptides. Next 25 ml of buffer containing 200 mM methyl α-D-glucoside is run onto the column, which is then stopped and left overnight. Elution with the same buffer is then resumed and fraction C glycopeptides are collected. Stopping the column sharpens the elution of this fraction. Further glycopeptides which bind strongly to the column may be eluted with 0.1 M HCl. Each glycopeptide fraction is subsequently desalted and separated from methyl glucoside by gel filtration.

Lectin affinity chromatography of glycopeptides on ConA-Sepharose produces separations dependent on carbohydrate structure (Chapter 7). With some exceptions, the fraction C glycopeptides and the glycopeptides eluted with 0.1 M HCl are of the *N*-linked high-mannose type which bind strongly to Con A. Fraction B contains *N*-linked biantennary glycopeptides which bind with low affinity, whereas fraction A glycopeptides are composed mainly of the *N*-linked complex type and *O*-linked glycopeptides which do not bind to Con A. Exceptions to this general pattern of binding can be brought about by the presence in glycopeptides of sugars other than mannose (e.g. α-D-glucose) which influence binding to Con A, by the presence of biantennary glycopeptides deficient in galactose or *N*-acetylglucosamine (which leads to stronger binding and elution in fraction C) and by substitution of the internal mannose residue of the biantennary structure with β1–4-linked *N*-acetylglucosamine (which leads to no binding to Con A and elution in fraction A). The extent of substitution of the ConA-Sepharose can also influence the affinity of binding for different types of carbohydrate unit and it is advisable to check this with glycopeptides of known structure.

The glycopeptides in fraction A consist of complex type *N*-linked units together with *O*-linked units. Because the oligosaccharides units of the *O*-linked carbohydrate groups are generally of smaller size than the complex *N*-linked groups it is often possible to achieve a separation of these two types of carbohydrate group by gel filtration following cleavage of the alkali-labile *O*-glycosidic linkages between peptide and carbohydrate. When larger *O*-linked units are present in significant amounts this separation method is not suitable. This separation can only be accomplished after fractionation on ConA-

Sepharose because fraction B and C glycopeptides obscure the size separation of the *O*-linked and complex-type glycopeptides.

The glycopeptides in fraction A are subjected to β-elimination in 0.05 M NaOH containing 1 M $NaBH_4$ at 45°C for 16 h (Section 6.5.1) and the products are separated by gel filtration on Sephadex G-25 or G-50 as described earlier. Fractions are collected and monitored by colourimetric analysis for carbohydrate. The [^{3}H]acetyl label introduced into the peptide moieties of *O*-linked glycopeptides cannot be employed to locate the oligosaccharides released under alkaline conditions. On gel exclusion chromatography complex-type *N*-linked units should emerge as a carbohydrate-containing peak preceding the small oligosaccharides usually released from *O*-glycosidic linkages to serine or threonine. The conditions employed for alkaline β-elimination may also produce some cleavage of GlcNAc-Asn protein-carbohydrate linkages in complex-type units (Ogata and Lloyd, 1982).

This procedure can give a useful group separation of carbohydrate units. However, each of the fractions isolated is still likely to show considerable heterogeneity of carbohydrate structure and further fractionation may be necessary. There can be some overlap of biantennary structures in fractions A and C for the reasons discussed earlier. In addition, the release of oligosaccharides by β-elimination may be incomplete if the linkage amino acids of *O*-linked glycopeptides have a free carboxyl group. Complete *N*-acetylation of the glycopeptides should enhance the possibility of successful β-elimination of glycopeptides.

Glycopeptides and glycosaminoglycan fractionation based on ion-exchange chromatography (Kraemer, 1971) This procedure is designed to separate metabolically labelled glycopeptides and glycosaminoglycans released from intact cells or cell fractions by proteolysis. A glycopeptide fraction and glycosaminoglycan fractions (hyaluronate, heparan sulphate and chondroitin sulphate) are separated by ion-exchange chromatography.

Glycoconjugates are labelled by growing cells overnight in the presence of [^{3}H]glucosamine and $Na_2^{35}SO_4$. Glycopeptides are released from intact cells by treatment with trypsin and from the 7% TCA-insoluble residue of cells by digestion with activated papain (1 mg/ml for 16 h at 56°C in 0.1 M ammonium acetate buffer, pH 7). This papain digest is cooled and reprecipitated with 7% TCA and the TCA-soluble portion of the digest is retained. After removal of TCA by ether extraction the trypsin and papain glycopeptides are separated from low molecular weight material by gel filtration in 0.1 M ammonium acetate, pH 7, on a Bio-Gel P2 100–200 mesh, 1.2 × 140 cm column operated at a flow rate of 24 ml/h. The macromolecular fraction is pooled and lyophilised to remove the buffer.

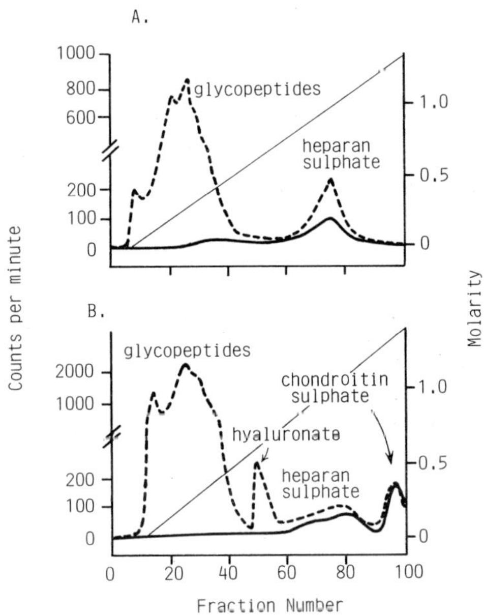

Fig. 6.3. Fractionation of glycopeptides and glycosaminoglycans by DEAE-cellulose chromatography. Cells were labelled with [^{3}H]glucosamine or $^{35}SO_4^{2-}$ and glyconjugates were released from the cell surface with trypsin (A) or by papain digestion of TCA-insoluble cellular material (B). Fractionation was carried out on DEAE-cellulose with a gradient of increasing molarity of sodium acetate. The ^{3}H-counts are indicated by the dotted line and ^{35}S-counts by the continuous line. The figure is redrawn from Kraemer (1971).

Fractionation of glycopeptides and glycosaminoglycans is carried out on a column (0.8 × 40 cm) of DEAE-cellulose (Whatman DE52) equilibrated with 0.01 M ammonium acetate by elution with a linear salt gradient between 0.01 and 2.0 M ammonium acetate in a total volume of 400 ml. Fractions (3 ml) are collected at a flow rate of 24 ml/h. Aliquots are counted for ^{3}H and ^{35}S to locate glycopeptide and glycosaminoglycan peaks.

The results obtained by applying this separation procedure to material released by proteolysis from the cell surface and from the residual TCA-precipitated cellular fraction of cultured CHO cells are shown in Fig. 6.3. The large initial glycopeptide peak is followed by peaks identified as hyaluronic acid, heparan sulphate and chondroitin sulphate (Kraemer, 1971).

6.7.4. *Fractionation of oligosaccharides obtained by chemical or enzymic cleavage*

Oligosaccharides can be released from glycoproteins or glycopeptides by β-elimination, hydrazinolysis, alkaline hydrolysis, trifluoroaceto-

TABLE 6.5
Methods for fractionating oligosaccharides obtained from glycoproteins

Technique	Method described in Section
Paper electrophoresis	6.7.4.1.
Ion-exchange chromatography	6.7.4.1.
Paper electrophoresis of borate complexes	6.7.4.1.
Gel filtration on Sephadex or Bio-Gel	6.7.4.2.
Carbon-Celite chromatography	6.7.4.4.
Paper chromatography	6.7.4.3.
Thin-layer chromatography	6.7.4.3.
High-performance liquid chromatography	
Gel permeation	6.7.4.2.
Ion exchange	6.7.4.1.
Normal phase	6.7.4.5.
Reverse phase	6.7.4.5.
Gas-liquid chromatography	6.7.4.6 and 6.9.1.
Lectin affinity chromatography	6.7.4.7 and 6.7.3.2.

lysis or by endoglycosidases (Sections 6.5 and 6.6). The oligosaccharide units are then separated from products derived from the peptide moiety (peptides, amino acid hydrazides etc.). This is most often done by gel filtration on Sephadex or Bio-Gel but other approaches such as the removal of amino acid hydrazides by paper chromatography (Section 6.5.2) can also be applied. Further fractionation of oligosaccharide units can be achieved by one or a combination of the methods listed in Table 6.5. These separation techniques are also applicable to oligosaccharides occurring in biological fluids such as milk or urine.

The most appropriate method or combination of methods for a particular separation depends on the nature and amounts of the oligosaccharides and on the equipment which is available. Separation methods depending on charge such as paper electrophoresis, ion-exchange chromatography and certain HPLC systems (Baenziger and Natowicz, 1981) are applicable to oligosaccharides containing sialic acid residues or phosphate or sulphate groups. Further fractionation of charged or neutral oligosaccharides can be accomplished by the other methods listed in Table 6.5. In recent years the fractionation techniques most widely used have been gel filtration, paper chromatography and lectin affinity chromatography. Each of these techniques can be applied to trace amounts of labelled material or for the isolation of milligram quantities of oligosaccharides. However, HPLC techniques, although introduced only recently, have advantages in speed and resolving power which seem likely to lead to their increasing adoption for analytical and small-scale preparative work. The analytical-scale separation of oligosaccharides by GLC was, until a few years ago, restricted to molecules containing only a few neutral sugar residues. Advances in derivatization now make this technique applicable to larger oligosaccharides (up to about seven saccharides), including those containing sialic acid (Section 6.7.4.6).

When oligosaccharides are only available in small amounts radioactive label can be introduced to permit their detection. This can be done by metabolic labelling or by introduction of isotope by reduction of the residue at the reducing terminal end with tritiated borohydride or by re-*N*-acetylation (after hydrazinolysis) of amino sugars with labelled acetic anhydride. The labelled oligosaccharides (or oligosaccharide alditols) can then be detected by liquid scintillation counting or by radiochromatogram scanning. Oligosaccharides available in larger amounts can be located and quantitated by the methods described in Chapter 5.

In addition to producing a homogeneous product, fractionation techniques can give information about the nature of the oligosaccharide. Useful estimates of the molecular weight of oligosaccharides can be obtained by gel filtration (Section 6.7.4.2) and information about the number of sialic acid residues per molecule can be obtained from examination of the electrophoretic behaviour of oligosaccharides (Section 6.7.4.1). Combination of separation with characterisation by mass spectrometry of oligosaccharides can give valuable structural information (Section 6.7.4.6).

Some examples of the application of oligosaccharide fractionation procedures are given in the remainder of this section.

6.7.4.1. Oligosaccharide separation — Paper electrophoresis and ion-exchange chromatography Separation methods depending on molecular charge are employed to resolve neutral oligosaccharides from those containing sialic acid, uronic acid, ester sulphate or phosphate substituents. Oligosaccharides of similar size, but containing different numbers of sialic residues, can be separated by paper electrophoresis, DEAE-cellulose chromatography or HPLC utilising an anion-exchanger. It is also possible to separate negatively charged borate complexes of neutral oligosaccharides by paper or gel electrophoresis in borate buffers (Yamashita et al., 1977; see also Narasinhan et al., 1980).

The following procedures for the electrophoretic fractionation of the complex-type sialic acid-containing oligosaccharide units released by hydrazinolysis has been described by Takasaki et al. (1979). After removal of hydrazine the oligosaccharides are re-*N*-acetylated, reduced with NaB^3H_4 and freed from salt and radioactive impurities originating from the NaB^3H_4 (see Section 6.5.2). The labelled oligosaccharides are spotted onto Whatman 3 MM paper alongside markers of Bromophenol blue and ^{3}H-labelled lactitol. After moistening the paper with pyridine:acetic acid buffer, pH 5.4 (pyridine:acetic acid:water, 3:1:387 by volume) electrophoresis is carried out at 73 V/cm for 1.5 h. The Bromophenol blue marker should move about 40 cm from the origin. Radioactive peaks, corresponding to sialylated oligosaccharides, migrate towards the anode (relative to the neutral marker lactitol) and are detected by radiochromatogram scanning or by cutting the paper into strips for scintillation counting. The oligosaccharides can be eluted from the paper, incubated with neuraminidase and re-electrophoresed to establish that their anodic mobility was due to the presence of sialic acid. Partial acid hydrolysis of sialic acid-containing oligosaccharides (in 0.01 M HCl at 100°C for 2.5 min) followed by re-electrophoresis has been used to establish the number of sialic acid residues present. Takasaki et al. (1979) showed that an oligosaccharide A_4 from cold insoluble globulin gave, after partial acid hydrolysis, three additional acidic oligosaccharides together with some unchanged A_4. Neuraminidase treatment of A_4 gave a single electrophoretically neutral product. These results indicated that four sialic acid residues were present in A_4.

Acidic oligosaccharides can be fractionated by ion-exchange chromatography on DEAE-cellulose in similar fashion to sialylated glycopeptides (Section 6.7.1). Oligosaccharides containing one, two or more sialic acid residues can be separated by gradient elution (Zinn et al., 1978).

The separation of anionic oligosaccharides by high-performance liquid chromatography has been described (Baenziger and Natowicz, 1981). Oligosaccharides are labelled by NaB^3H_4 reduction and are detected by liquid scintillation counting. Chromatographic separations utilise a 4 mm × 30 cm column containing the anion-exchanger MicroPak AX-10 (Varian Associates) using 25–500 mM KH_2PO_4 titrated to pH 4.0 with phosphoric acid as mobile phase. Samples are injected in 10–500 μl of 25 mM KH_2PO_4, pH 4.0, and columns are operated at a flow rate of 1 ml/min. After eluting initially with 25 mM phosphate a linear 25–500 mM phosphate gradient is applied.

Separation of oligosaccharides bearing zero, one, two, three or four sialic acid residues can be obtained with this system in less than 45 min. Recoveries of samples are reported to be in the range 90–100% and sample sizes from 500 pmole to 1 μmole can be employed. The resolving power and the recoveries obtained with this technique are superior to DEAE-cellulose chromatography and high-voltage paper electrophoresis. If the necessary HPLC equipment and columns are available this is the method of choice for separation of charged oligosaccharides. Phosphorylated oligosaccharides can also be separated using this system (Baenziger and Natowicz, 1981).

6.7.4.2. Gel filtration (or gel permeation) Oligosaccharides can be separated from lower molecular weight products resulting from peptide degradation by alkali or hydrazine using gel filtration on Sephadex G-50 (or G-25) or Bio-Gel P6. The conditions employed are the same as those described for the isolation of glycopeptides (Section 6.7.1). If major heterogeneity in the size of carbohydrate units exists this may be revealed in the elution profile. For example, the existence of low and high molecular weight oligosaccharides (corresponding to 'poly(glycosyl) peptides' or 'erythroglycan') released by hydrazinolysis of human erythrocyte band 3 glycoprotein was detected by chromatography on Sephadex G-50 (Tsuji et al., 1980).

Effective fractionation of oligosaccharides containing up to about 15 monosaccharide units can be obtained using columns (2 m × 2 cm) of Bio-Gel P4 (under 400 mesh) at 55°C as described by Yamashita et al. (1982). Details of the methodology are given in Section 6.9.4.1. The resolution of oligosaccharides obtainable by this method, while basically depending on molecular size, is enhanced because *N*-acetylhexosamines, hexoses and methylpentoses have different effective sizes. In addition the position to which a particular sugar is linked can modify its effect on the elution volume of an oligosaccharide. Thus oligosaccharides containing the same number of monosaccharide residues but differing in the types of sugar present (or the linkages between sugars) can, in many cases, be separated by gel filtration on Bio-Gel P4. The elution behaviour of a large number of oligosaccharides of defined structure has been tabulated by Yamashita et al. (1982).

An accelerated version of this procedure in which the chromatography time is decreased from 30 h to 3 h by employing narrower columns operated at pressures in the range 55–100 p.s.i. has been described (Natowicz and Baenziger, 1980).

M_r values for oligosaccharides can be estimated from their gel filtration behaviour on Sephadex or Bio-Gel columns provided that suitable markers are available for column calibration (see also Section 6.7.1). The presence of ionising groups (e.g. sialic acid) produces anomalous elution behaviour unless gel filtration is carried out in buffers of high ionic strength.

6.7.4.3. Paper and thin-layer chromatography Paper chromatography provides an effective, simple and inexpensive means of fractionating oligosaccharides and their alditols. This technique has been extensively applied to the separation of oligosaccharides released (by hydrazinolysis or endoglycosidases) from glycoproteins containing the GlcNAc-Asn type of linkage and to complex oligosaccharides of milk and urine. A limitation of the technique is that chromatography times of several days are often required for the separation of larger oligosaccharides, particularly when they contain sialic acid. Thin-layer chromatographic techniques can produce more rapid separations. As well as serving to fractionate oligosaccharides paper chromatography can be used to separate oligosaccharides from products resulting from peptide-chain degradation by hydrazinolysis (Section 6.5.2).

Fractionation of oligosaccharides (or their NaB^3H_4 reduced derivatives) is carried out by dissolving the sample (which must be free of salts) in water and applying it as a spot to a strip of Whatman No. 3 paper 57 cm in length. Appropriate standards are also applied along the origin at 2-cm intervals. The end of the paper furthest from the origin is serrated by cutting with pinking shears. Descending chromatography is carried out in the solvent described by Fisher and Nebel (1955), ethyl acetate:pyridine:acetic acid:water (5:5:1:3 by vol.) in a glass chromatography tank maintained at a constant temperature of 30°C. Chromatography is continued for a period between 14 h and 40 days depending on the R_F of the oligosaccharides. Solvent is allowed to drip from the paper into the bottom of the tank; the serrated edge prevents beads of solvent from building up at the end of the paper. After drying, the chromatogram is developed by staining with alkaline silver or aniline oxalate reagents or radioactive oligosaccharide derivatives can be located by radiochromatogram scanning (or by cutting the paper into strips and scintillation counting). Alkaline silver staining is carried out in the following way.

1. A saturated solution of $AgNO_3$ in water (0.5 ml) is diluted by the addition of 100 ml acetone.
2. NaOH (0.5 g) dissolved in 5 ml water is diluted to 100 ml with ethanol.

The paper is dipped in the silver reagent and acetone is allowed to evaporate. When dry the paper is dipped in the alkaline ethanol and the solvent is blown off with a cold stream of air (hair dryer). Dark brown or black spots develop at room temperature within a few minutes. This stain is sensitive but reacts with substances other than carbohydrates. More specific stains (e.g. analine oxalate) should be applied when oligosaccharides are likely to be contaminated with extraneous substances (e.g. in urine samples).

Paper chromatography can be used to isolate oligosaccharides on a preparative scale. Whatman 3 MM paper is employed and oligosaccharides are applied as a line along the origin with 2 mg of sample per cm. Guide strips at the edges of the chromatogram are stained to locate the oligosaccharides which are eluted chromatographically with water.

Paper chromatography can resolve complex mixtures of oligosaccharides. In general R_F values decrease with increasing molecular weight but other structural features can also be important. Charged oligosaccharides give low R_F values. Chromatograms showing the behaviour of a variety of normal and abnormal human urinary oligosaccharides, many of which are derived from glycoproteins, have been published (Strecker and Montreuil, 1978). When attempting to identify an oligosaccharide by its chromatographic behaviour it is desirable to run an authentic sample of established structure on the same chromatogram.

The long time taken to achieve separations of oligosaccharides of high molecular weight (or those containing sialic acid) is a disadvantage of the method. Much more rapid separations can be obtained by thin-layer chromatography (Bayard et al., 1979; Bouquelet et al., 1980). The sample is applied to 20 × 20 cm silica gel G plates and developed 5 h using butanol:ethanol:water:acetic acid:pyridine (10:100:3:10:30 by vol.). Detection is by use of an orcinol reagent (0.1% orcinol in 20% H_2SO_4 developed by heating plates at 150°C) or radioactively labelled oligosaccharide derivatives can be detected by fluorography. Sialylated oligosaccharides migrate more slowly than their unsialylated counterparts. Oligosaccharides containing bi-, tri- ans tetra-antennary structures can be distinguished.

6.7.4.4. Adsorbtion chromatography on charcoal This is one of the traditional methods of oligosaccharide fractionation which is still occasionally used for the removal of non-carbohydrated contaminants from oligosaccharides or glycopeptides and for preliminary fractionation of complex mixtures. For example, the fractionation of urinary oligosaccharides by carbon-Celite and paper chromatography has been described by Strecker and Montreuil (1978). Adsorbtion chromatography on carbon-Celite can also be employed in large-scale isolation of oligosaccharide standards as in the preparation of di-*N*-acetylglucosamine from partially hydrolysed chitin (Zehavi and Jeanloz, 1971).

After adsorbtion to carbon from aqueous solution oligosaccharides are displaced by elution with increasing concentrations of ethanol. Adsorbtion is influenced by the size and monosaccharide composition of the oligosaccharide. Large size and the presence of sialic acid are factors which favour strong adsorbtion (Spiro, 1966). For glycopeptides with short peptide chains adsorbtion is mainly determined by the carbohydrate moiety unless strongly absorbed aromatic amino acids are present.

To prepare the adsorbent activated charcoal (e.g. Darco G60) is mixed in the dry state with an equal weight of acid-washed Celite 535 (Johns-Mansville) which is added to improve the flow rate. The mixture is prewashed with 70% ethanol and then with water. After thorough mixing the slurry of carbon-Celite is packed into a chromatography column. For oligosaccharide samples at least one ml of column volume should be available per 2 mg of oligosaccharide. A 3×20 cm column is suitable for the initial isolation of oligosaccharides from 1 l of deionised human urine (Strecker and Montreuil, 1979). Elution is carried out initially with water followed by stepwise or gradient elution with ethanol up to a final concentration of 70% by volume. Recoveries of carbohydrate of about 90% can be obtained.

6.7.4.5. High-performance liquid chromatography In the last few years methods have been developed for the separation of oligosaccharides by means of HPLC. These methods have advantages of resolving power and speed which will inevitably lead to their increasing use.

Traditional liquid chromatography is a slow separation technique in which solvent flows through a vertical column under gravity. The increased speed of high-performance liquid chromatography is achieved by pumping sample and solvents through the column at high inlet pressures (about 1000 p.s.i.). Improvements in resolution arise from the nature of column packing materials and from their small diameter and high surface-to-volume ratios.

Several types of HPLC can be applied to the separation of oligosaccharides and their aldehyde-reduced derivatives (Table 6.5). Ion-exchange and gel permeation methods have been discussed alongside conventional liquid chromatography methods in earlier sections. Separation methods based on the affinity of neutral oligosaccharides for the solid packing material of the HPLC column will be considered here. These separations can be carried out with normal or reverse-phase systems. In normal-phase chromatography the column packing (stationary phase) is generally a polar material. Reverse-phase systems employ non-polar packing materials and these can be used for the fractionation of oligosaccharide derivatives in which the polar hydroxyl groups have been modified by peracetylation (or permethylation). Both normal and reverse-phase separations are capable of high resolution.

The essential components of a high-performance liquid chromatography system are a pump, sample injection port, a column containing appropriate packing material and a detection system. Additional devices for the formation of solvent gradients, a column oven and guard column (to prevent contamination of the separating column) may be necessary. The most sensitive method for detection of oligosaccharides is to pre-label them, collect fractions of the column eluate and carry out scintillation counting. Labelling can be carried out by reduction with NaB^3H_4. Oligosaccharides can also be detected, non-specifically, by means of a differential refractometer. This requires 15–30 μmoles of sample (Ng Ying Kin and Wolfe, 1980) and cannot be used with gradient elution. Detection of non-volatile carbon compounds can also be carried out with a moving wire detector.

Procedure for neutral oligosaccharide fractionation by normal-phase HPLC (Mellis and Baenziger, 1981) This method is suitable for the fractionation of oligosaccharide units released from uncharged oligosaccharides of the 'high-mannose' or 'complex' types released by hydrazinolysis or by endoglycosidases. The oligosaccharides are reduced with NaB^3H_4 to facilitate detection by scintillation counting. Separation is carried out on a column (4 mm × 30 cm) of Micro Pak AX-10 (Varian Associates). The column is initially equilibrated with acetonitrile (HPLC grade):H_2O (distilled and deionised), 65:35 by volume. Following injection of the sample dissolved in up to 200 μl of acetonitrile:water elution is carried out at a flow rate of 1 ml/min with a gradient in which the acetonitrile content of the solvent is decreased at 0.5% per min. Fractions are collected at 3-min intervals and radioactivity is determined by scintillation counting.

The separation of di-, tri- and tetra-antennary oligosaccharide units can be obtained and the presence of a single fucose residue on any of these structures results in a distinct peak. High-mannose-type oligosaccharides differing by a single mannose unit can also be resolved. Elution times for these neutral oligosaccharides usually lie within the range 100–150 min. Amounts of oligosaccharide up to 1 μmole can be chromatographed without loss of resolution (Mellis and Baenziger, 1981). The sensitivity of this method to oligosaccharide structure is indicated by the observation that double peaks are obtained when high-mannose-type oligosaccharides are exposed to strongly alkaline conditions (during borohydride reduction in the presence of 0.05 M alkali), which produces partial epimerisation at the C-2 position of the reducing terminal amino sugar.

This method provides a very powerful means of separating neutral oligosaccharides released from glycoproteins but is not suitable for direct application to charged oligosaccharides. Prior removal of sialic acid by treatment with neuraminidase or by mild acid hydrolysis is required for the fractionation of sialylated oligosaccharides. In this way sialylated oligosaccharides separated on the basis of their charge by HPLC (Section 6.7.4.1) can be further resolved.

Procedure for neutral oligosaccharide fractionation by reversed phase HPLC (Wells et al., 1982) This method has been applied to the separation of oligosaccharides released by endoglycosidase H digestion of ovalbumin glycopeptides and to dolichol-derived oligosaccharides. The oligosaccharides are reduced with NaB^3H_4 and *O*-acetylated before chromatography on a reverse-phase column.

The following procedure is recommended to obtain complete *O*-acetylation without di-*N*-acetylation. Acetylation of the dried sample is carried out by the addition of 2 ml formamide, 1 ml acetic anhydride and 0.8 ml of dried pyridine. The sample is placed in an ultrasonic bath to solubilise the oligosaccharide and incubation is continued overnight at room temperature. Water, 5 ml, is added to destroy excess acetic anhydride and the mixture is extracted with two 5-ml portions of chloroform and the combined chloroform extract is washed with three 5-ml portions of water. The chloroform extract is dried on a rotary evaporator and redissolved in acetonitrile prior to column chromatography.

Liquid chromatography is carried out using a water-jacketed column (0.65 × 15 cm) packed with LC-18 (Supelco), a 5μm C-18 support, and maintained at 65°C. Oligosaccharides are eluted for 100 min with a linear gradient of water-acetonitrile from 7:3 to 1:1 followed by isocratic elution for 100 min at a flow rate of 1 ml/min. Fractions (2 ml) are collected and radioactivity is determined by scintillation counting.

This separation technique has approximately similar resolving power to the HPLC separation described in the previous section (Mellis and Baenziger, 1981). The former procedure has the advantage that formation of *O*-acetyl derivatives (and subsequent removal of *O*-acetyl groups) is not required. However, the reverse-phase technique does provide a useful alternative basis for achieving separations. Reverse-phase HPLC would also appear to be well suited to the fractionation of permethylated oligosaccharides derived from glycoproteins. This approach has been succesfully adopted in structural investigation of several polysaccharides not containing amino sugars (McNeil et al., 1982). With methylated oligosaccharides advantage can be taken of mass spectrometry for detection and analysis.

6.7.4.6. Gas-liquid chromatography Suitably derivatised oligosaccharides can be separated on an analytical scale by GLC. It is possible to identify oligosaccharides by comparing their retention times to standards and the analysis of peaks emerging from the GLC column by mass spectrometry can give valuable structural information. Both GLC and GLC-MS of oligosaccharides are discussed in Section 6.9.2. Recent developments in the derivatisation of oligosaccharides have extended the applicability of GLC to oligosaccharides containing seven, or sometimes more, monosaccharide units. This technique is valuable analytically but not preparatively.

6.7.4.7. Lectin affinity chromatography This is a simple but valuable method for the fractionation of preparative or analytical quantities of oligosaccharides (Ogata et al., 1975 and see Chapter 7). The methodology employed is the same as that described for glycopeptide fractionation (Section 6.7.3.1) except that oligosaccharides are usually labelled by reduction with NaB^3H_4 to facilitate detection.

6.8. Determination of the nature of protein-carbohydrate linkages in glycoproteins and proteoglycans

The major types of protein-carbohydrate linkages which have so far been elucidated are listed in Table 6.3. Linkages differ in the amino acid and sugar residues involved and in their relative stability to cleavage by chemical reagents such as alkali.

In practice the type (or types) of linkage most likely to be present in a particular glycoconjugated preparation is often suggested by the origin of the material and its carbohydrate composition (Section 6.2). For example, the occurrence of several of the linkage types listed in Table 6.3 appears to be restricted to a limited group of organisms. One of the linkages (Gal-hydroxylysine) is found exclusively in collagen-like proteins of animals, and others such as the Ara-Hpr linkage have been detected only in plants. It should of course be borne in mind that as more glycoproteins are investigated the range of organisms in which a particular linkage is found may be extended. If information is available regarding the stability of the protein-carbohydrate linkage towards alkali the likely possibilities may be further decreased. This section considers methods for obtaining proof of the nature of the linkage.

Two general strategies can be employed for the identification of protein-carbohydrate linkages. One approach is to digest the peptide chain of the glycoprotein or proteoglycan with proteolytic enzymes and then isolate a glycopeptide containing only the linkage amino acid. The carbohydrate moiety of the glycopeptide is then degraded by partial hydrolysis (with acid or enzymes) to obtain the protein-carbohydrate linkage unit consisting of one sugar and one amino acid residue. This linkage unit can be purified and identified by comparison with synthetic model compounds of known structure. A second type of strategy, applicable to protein-carbohydrate linkages which are cleaved by alkaline β-elimination, is to identify the amino acid and carbohydrate residues which have been modified by treatment with alkali.

More than one type of protein-carbohydrate linkage may occur in a single glycoprotein or proteoglycan molecule. It is therefore essential to examine all of the glycopeptides which can be isolated to ascertain whether they have the same protein-carbohydrate linkage. All of the carbohydrate present in the original molecule should be accounted for as quantitatively as possible to ensure that a group of glycopeptides has not been lost during purification. It is also important in carrying out alkaline borohydride cleavage to analyse the *N*-acetylhexosaminitol products carefully (Section 6.8.2).

6.8.1. Glycosylamine linkage to asparagine

The glycosylamine linkage involving *N*-acetylglucosamine and the side chain of asparagine has been found in glycoproteins and proteoglycans of animal origin and in plant and yeast glycoproteins. Preliminary evidence for the occurrence of this linkage can be obtained from studies of alkali stability but proof requires the isolation of the linkage compound GlcNAc-Asn (2-acetamido-1-*N*-(4-L-aspartyl)-2 deoxy-β-D-glucopyranosylamine) and comparison of its properties with the synthetic compound. Where quantities of material are restricted fluorescent or radioactively labelled derivations of the GlcNAc-Asn can be employed.

The stability to alkaline cleavage of the GlcNAc-Asn linkage is greater than that of most *O*-glycosidic protein-carbohydrate linkages other than the Gal-Hly linkage (Section 6.8.3). Hence if no oligosaccharides are released from a glycoprotein (not containing hydroxylysine) treated with alkali using conditions favouring β-elimination of *O*-linked carbohydrate units (e.g. 0.1 M NaOH at 25°C for 24 h) this constitutes preliminary evidence for the presence of the glycosylamine linkage. However, the rates of alkaline elimination of *O*-glycosidically linked oligosaccharides cover a wide range and the conditions required for complete removal of *O*-linked carbohydrate may cause some hydrolysis of GlcNAc-Asn-linked carbohydrate units. Indeed the conditions of alkaline cleavage widely employed for release and reduction of oligosaccharides *O*-glycosidially linked via *N*-acetylgalactosamine to serine or threonine (0.05 M NaOH, 1 M $NaBH_4$ at 45°C for 24 h) may also produce detectable cleavage of GlcNAc-Asn linkages (Ogata and Lloyd, 1982) with the formation of oligosaccharides containing *N*-acetylglucosaminitol.

For proof of the presence of the glycosylamine linkage by identification of the linkage compound GlcNAc-Asn the first step is the isolation of a glycopeptide containing asparagine (or aspartic acid after hydrolysis) as the sole amino acid. This is accomplished by digestion with a non-specific protease such as Pronase as described in Section 6.4.2. Glycopeptides are separated from peptides by gel filtration (Section 6.7.1) and further purified and fractionated by ion-exchange chromatography (e.g. on Dowex 50 × 2, Section 6.7.1). The purified glycopeptide should be analysed after hydrolysis. It should contain one residue of aspartic acid per mole and must be free of other amino acids. The yield of such glycopeptides is seldom quantitative due to the low rate of enzymic hydrolysis of short peptides. Marshall and Neuberger (1976) improved the yield of oligosaccharide-Asn glycopeptides obtained from ovalbumin by first blocking the *N*-terminal residue of the Pronase glycopeptide and then digesting with carboxypeptidase A (which will not cleave a dipeptide if the *N*-terminal group is free) and subsequently removing the blocking group. Amino acids on the *N*-terminal side of the carbohydrate-linked Asn may be removed by Edman degradation. Glycopeptides containing Asn as the sole amino acid can be detected with ninhydrin and characterised on an amino acid analyser (as can the Asn-GlcNAc linkage compound), where they emerge before aspartic acid.

The Asn-GlcNAc compound can be released from glycopeptides by mild acid hydrolysis (Spiro, 1972; Marshall and Neuberger, 1976).

Glycopeptide (3–100 μmoles) at 10 μmoles/ml is treated with 2 M HCl at 100°C for 10–20 min and the hydrolysate is cooled and lyophilised. Asn-GlcNAc is obtained in 25% yield from ovalbumin glycopeptide after hydrolysis for 20 min. The linkage compound can be identified by comparison with the authentic substance using an amino acid analyser; GlcNAc-Asn has an elution time of 0.27 relative to aspartic acid using a Technicon Analyser system (Spiro, 1976). GlcNAc-Asn has an R_F value in butanol:acetic acid:water (4:1:5 by vol.) relative to Asp of 0.54; on descending chromatography on Whatman No. 1 paper in phenol:water (4:1 w/v) the R_{asp} is 1.10 and using the same solvent on cellulose thin layers the R_{asp} is 2.10. The compound gives a brown stain with ninhydrin in acetone containing 2% pyridine when heated at 80°C. On acid hydrolysis the compound yields one mole each of aspartic acid, ammonia and glucosamine (Marshall and Neuberger, 1976).

When large quantities of glycopeptide are not available the fluorescent dansyl derivatives of the linkage compound can be identified. The following method is that applied to ribonuclease glycopeptide by Plummer et al. (1968).

Glycopeptide (0.25 μmole) is dissolved in water (0.5 ml). Dimethylaminonaphthalenesulphonyl chloride (0.75 mg) in acetone (0.45 ml) and 0.2 ml of 0.1 M $NaHCO_3$ are added. After reaction for 15 h at room temperature the sample is diluted to 5 ml with water and passed through a column (1 × 4 cm) of Dowex 1 (formate). The column is then washed with water (10 ml) followed by 0.1 M formic acid, which elutes the fluorescently labelled glycopeptide. (Fluorescently labelled ovalbumin glycopeptide is eluted from this column with water.) Fractions containing the DNS-glycopeptide are pooled and dried on a rotary evaporator.

Acid hydrolysis of the DNS-glycopeptide is carried out in 1 M HCl at 100°C for periods of 10, 20 and 30 min. Samples are cooled, neutralised and passed through a 4 × 1 cm column of Dowex 1 (formate) and eluted with 30 ml water followed by 20 ml of 2 M formic acid. After concentrating to dryness in a rotary evaporator the residue is redissolved in 0.1 ml water and identified by comparison with authentic samples of DNS-Asp, DNS-Asn and DNS-(GlcNAc)Asn by paper electrophoresis in 0.05 M ammonium acetate buffer, pH 4.5.

Further increase in the sensitivity of detection of DNS-(GlcNAc)Asn can be achieved by the use of ^{14}C-labelled DNS-Cl and detection of products by fluorography after separation by thin-layer electrophoresis (Farwell and Dion, 1981). The GlcNAc-Asn linkage compound obtained from 10–20 μg of glycopeptide can be characterised. In all of these procedures the main limitation is the ability to obtain glycopeptides with only one amino acid residue.

Other methods available for characterising the Asn-GlcNAc linkage include the use of the enzyme glycopeptide aminohydrolase, which releases aspartic acid from glycopeptides containing this amino acid unsubstituted in either α-amino or α-carboxyl groups (Plummer et al., 1968) and the use of vigorous alkaline hydrolysis and borohydride reduction to cleave the protein-carbohydrate link (Section 6.5.1.2). The nature of the protein-carbohydrate linkage can also be determined by ^{1}H-NMR (Section 6.9.3).

6.8.2. O-Glycosidic linkages to serine or threonine

Five sugars are known to form *O*-glycosidic linkages with the side chains of serine or threonine (Table 6.3). There appear to be some limitations on the distribution of linkages involving different sugars. The GalNAc-Ser/Thr linkage, for example, occurs widely in animal glycoproteins but has not been found in plants, whereas the Ara-Hpr linkage is apparently confined to plants. However, linkages between xylose and hydroxyamino acids occur both in the connective tissue proteoglycans of animals and in protein-polysaccharides of algae and higher plants. In studying all members of this class of protein-carbohydrate linkage it is possible to take advantage of their susceptibility to alkaline cleavage.

Alkaline β-elimination of *O*-glycosides of serine or threonine residues follows the route shown for the GalNAc-Ser linkage in Fig. 6.1. The oligosaccharide unit is released with the linkage sugar as the reducing terminal residue and an unsaturated amino acid derivative is formed. If the alkaline elimination is carried out in the presence of $NaBH_4$ the linkage sugar is reduced to the corresponding alcohol. This minimises alkaline degradation of the oligosaccharide by 'peeling reactions' and allows identification of the linkage sugar. Changes in amino acid side-chains resulting from the β-elimination reaction give evidence about the sites of attachment of carbohydrate on the peptide chain. The unsaturated amino acids produced by alkaline β-elimination of *O*-glycosidic protein carbohydrate linkages involving Ser and Thr are unstable when hydrolysed in acid. These residues can be modified either by the formation of bisulphite derivatives or by reduction with sodium borohydride and palladium chloride to form other products (alanine and α-aminobutyric acid) which survive acid hydrolysis and which can be quantitated. Reduction by borohydride alone will convert α-aminoacrylic acid (derived from serine) to alanine but the efficient conversion of α-aminocrotonic acid (from threonine) to α-aminobutyric acid requires the presence of palladium chloride.

Alkaline β-elimination will occur readily only when both α-amino and α-carboxyl groups of the linkage amino acid are blocked by the formation of peptide bonds or have other similar substituents. Cleavage occurs slowly or not at all when carbohydrate is attached to a free *N*- or *C*-terminal residue in a glycoprotein or glycopeptide. This must be borne in mind particularly when alkaline elimination is applied to glycopeptides containing *O*-glycosidic protein-carbohydrate linkages. Despite this potential difficulty the isolation of glycopeptides containing the protein-carbohydrate linkage should be carried out to confirm the nature of the linkage.

The *O*-glycosidic linkage between serine or threonine and *N*-acetyl-D-galactosamine which occurs widely in mucins has been extensively studied. Methods developed for the identification of this linkage, which are discussed in the following section, are applicable with minor modification to the other alkali-labile *O*-glycosidic linkages.

6.8.2.1. N-Acetylgalactosaminyl linkage to serine or threonine The conditions which have been employed for the alkaline elimination of carbohydrate from glycoproteins include NaOH concentrations of 0.05–0.5 M, temperatures between 0 and 45°C and incubation times of 1–216 h (Spiro, 1972). As there can be considerable variation in rates of elimination between serine and threonine residues located in different sequences it is desirable to establish optimal conditions for each glycoprotein.

A simple procedure to obtain preliminary evidence for the presence of *O*-glycosidic linkages is to treat samples of glycoprotein with 0.2 M NaOH at 45°C for 2, 5 and 10 h followed by neutralising the sample with HCl and hydrolysing in sealed tubes in 6 M HCl at 110°C for 22 h (Downs and Pigman, 1976a). The hydrolysates are analysed for amino acids (Section 5.4); loss of serine or threonine compared with a control not incubated with alkali suggests the presence of *O*-glycosidic linkages.

The time course of formation of unsaturated amino acids arising from the β-elimination reaction can be followed by measurement of the increase in absorbance at 240 nm. On acid hydrolysis these olefinic amino acids (α-aminoacrylic and α-aminocrotonic acid from serine and threonine respectively) are converted respectively to pyruvic acid and α-ketobutyric acid. It is possible to assay these hydrolysis products (Planter and Carlson, 1972). Alternatively the unsaturated amino acids can be modified prior to hydrolysis either by reaction with bisulphite (Harbon et al., 1968) or by reduction (Downs et al., 1973; Downs and Pigman, 1976a). The modified amino acids are then determined, after hydrolysis, using an amino acid analyser. Of these procedures the one most generally found to give satisfactory results is the bisulphite method described below.

Procedure for making sulphite derivations (Spiro, 1972) If the sample contains cystine or cysteine this should initially be oxidised to cysteic acid by treatment with performic acid (Hirs, 1967). Optimal conditions for alkaline β-elimination should be established (e.g. 48–72 h in 0.1 M NaOH at 37°C, Spiro, 1972). The glycoprotein, glycopeptide or proteoglycan sample (8 mg/ml) is then incubated with shaking in 0.1 M NaOH containing 0.5 M sodium sulphite (freshly prepared) for 48–72 h at 37°C. The reaction is stopped by acidifying with HCl and the mixture is dried in a rotary evaporator at 40°C. Hydrolysis is carried out in 6 M HCl under nitrogen in a sealed tube for 24 h. A control sample, acidified prior to the addition of alkaline sulphite, is hydrolysed in the same way. Portions of the hydrolysates are applied to an amino acid analyser for quantitation of the loss of serine and threonine and for measurement of the cysteic acid peak. This peak contains cysteic acid arising from performic acid-oxidised cysteine as well as cysteic acid arising from *O*-glycosidically linked serine. In addition the α-amino-β-sulphonylbutenoic acid arising from *O*-linked threonine will also elute with the cysteic acid peak. Separation of cysteic acid and α-amino-β-sulphonylbutenoic acid can be accomplished either by chromatography on Dowex 50 × 8 (Spiro, 1972) or by gas-liquid chromatography (Simpson et al., 1972).

The yields which are obtained for the sulphite derivatives of *O*-linked serine and threonine are almost quantitative and about 75% respectively (Spiro, 1972). Other substituents of serine or threonine side-chains which can undergo alkaline elimination, such as phosphate groups, will cause the loss of hydroxyamino acids and the formation of sulphite derivatives.

Procedure for reducing unsaturated amino acids (Downs and Pigman, 1976a) Aliquots (1 ml) of a solution containing 1–5 mg/ml glycoprotein or glycopeptide in 0.1 M NaOH and 0.3 M $NaBH_4$ (freshly prepared) are placed in 150 × 20 mm culture tubes with Teflon-lined screw caps and are incubated for 0, 5, 10, 15 and 20 h at 45°C and then cooled to 20–25°C. A Teflon-covered magnetic stirring bar and one drop of 1-octanol, to prevent foaming, are added to each tube. The tube contents are stirred vigorously and neutralised by the addition of 1 ml of 0.4 M HCl immediately followed by 0.1 ml of 0.08 M aqueous palladium chloride solution prepared as described by Tanaka and Pigman (1965). $NaBH_4$ (2 ml of a 0.66 M solution in 0.1 M NaOH) and palladium chloride (2 ml of a 0.16 M solution in 0.8 M HCl) are then added *simultaneously*, dropwise from separate pipettes. Each tube is acidified by the addition of 6.1 ml of conc. HCl.

Tubes are firmly capped, heated at 110°C for 22 h, cooled and dried on a rotary evaporator. The residue is dissolved in citrate amino acid analyser buffer, filtered and the amino acid composition is determined. *O*-glycosidic linkages involving serine or threonine result in a decrease (relative to the zero-time control) in the linkage amino acid with an increase in alanine (Ser linkage) or the appearance of α-aminobutyric acid (Thr linkage).

This procedure has the advantage that the determination of alanine and α-aminobutyric acids are readily performed. It is claimed that this procedure gives better quantitation of linkages involving threonine than sulphite procedures although other protocols for the palladium chloride-sodium borohydride procedure are reported to be less effective (Spiro, 1972). For the reductive method to proceed satisfactorily it is essential that the palladium should be maintained in a colloidal state, otherwise the less efficient catalyst palladium black is formed (Downs et al., 1973).

When only small quantities of sample are available the linkage amino acid may be identified qualitatively by a method described by Farwell and Dion (1981). The glycoconjugate is treated with alkaline NaB^3H_4, hydrolysed, and the amino acids are dansylated and fractionated by thin-layer electrophoresis and two-dimensional thin-layer chromatography. The radioactively labelled amino acids alanine and α-aminobutyric acid can be located by fluorography. Both of these products were detected when the procedure was applied to bovine submaxillary mucin. Because the reduction of α-aminocrotonic acid by $NaBH_4$ is reportedly far from quantitative this procedure would be expected to be relatively insensitive for the detection of glycosidic linkages involving threonine.

Procedures for identification of the linkage sugar. The glycoprotein, glycopeptide or proteoglycan is treated with alkali in the presence of $NaBH_4$ (0.15–1.0 M), the oligosaccharides which are released are separated from peptide material and the sugar alcohol corresponding to the linkage sugar is identified. Alkaline borohydride treatment and the purification of products is carried out by the method of Carlson (1968) described in Section 6.5.1.1. Analysis of the sugar alcohol(s) produced is carried out after hydrolysis for 4–6 h in 4 M HCl. This results in de-*N*-acetylation of *N*-acetylgalactosaminitol and *N*-acetylglucosaminitol. The amino sugar alditols can be separated, identified and quantitated directly by means of an amino acid analyser (Spiro, 1972). Alternatively if β-elimination is performed using NaB^3H_4 the labelled amino sugar alditols can be re-*N*-acetylated after hydrolysis and identified and quantitated by high-voltage paper electrophoresis in borate buffer as described in Section 5.9. The qualitative identification of the tritium-labelled hexosaminitols arising from alkaline borohydride cleavage of small quantities of glycoprotein (e.g. 15 μg bovine submaxillary mucin) can also be achieved by separation of dansylated aminohexosaminitols using thin-layer electrophoresis in borate buffer (Farwell and Dion, 1981). It is essential to distinguish between the formation of glucosaminitol and galactosaminitol rather than measuring total hexosaminitol because some glucosaminitol may arise from the alkaline borohydride treatment when glycosylamine linkages are present as well as galactosaminitol from *O*-glycosidic linkages (Ogata and Lloyd, 1982).

In glycoproteins where it has been established that only GalNAc-Ser/Thr is present the cleavage of these linkages can be followed by the incorporation of ^3H from NaB^3H_4 by methodology described by Aminoff et al. (1980).

Anomeric configuration of the protein-carbohydrate linkage. A simple method for showing that the glycosidic linkage between D-GalNAc and Ser or Thr has the α-configuration is to demonstrate cleavage of the linkage with exo-α-*N*-acetyl-D-galactosaminidase (Section 6.9.4.1). Prior removal of external sugar residues with exoglycosidases or by Smith degradation is required.

6.8.2.2. Galactosyl-serine linkage D-Galactopyranoside *O*-glycosidically linked to serine has been identified in the plant cell wall glycoprotein extensin (Lamport, 1973) and in potato lectin (Allen et al., 1978). The linkage is alkali-labile. Lamport et al. (1973) isolated, from an acid-treated cell wall preparation, a tryptic glycopeptide containing two galactosylserine units. The sequence of the peptide was:

```
Ser-Hyp-Hyp-Hyp-Hyp-Ser-Hyp-Lys
 |                   |
Gal                 Gal
```

Treatment of the glycopeptide with 0.2 M NaOH and 0.2 M $NaBH_4$ for 5 h at 50°C resulted in β-elimination at the internal serine residue. However, β-elimination at the *N*-terminal serine occurs only after blocking the free amino group (e.g. by maleylation). Allen et al. (1978) have demonstrated that the galactose linked to serine in potato lectin can be released by treatment with α-galactosidase (from coffee beans) but not β-galactosidase and the configuration of the linkage is therefore α.

In extensin and potato lectin which have not been acid-treated the hydroxyproline residues are substituted with oligosaccharides of arabinose. The proximity of these carbohydrate groups can modify the properties of the galactosyl-serine linkage. Removal of galactose by α-galactosidase from potato lectin or its glycopeptides is hindered if the arabinose is not first removed and alkaline β-elimination is also inhibited by the arabinosides, presumably because of the ionisation of hydroxyl groups on arabinofuranoside at high pH (Allen and Neuberger, 1978).

The presence of an *O*-glycosidic linkage between galactose and both serine and threonine has been demonstrated in the collagen of earthworm cuticle (Muir and Lee, 1970). Carbohydrate units were released on incubation in 0.1 M NaOH at room temperature for 24 h.

6.8.2.3. Mannosyl-threonine or -serine linkages *O*-glycosidic linkages have been found in soluble or cell envelope glycoproteins of yeast and fungi (Nakajima and Ballou, 1974; Raizada et al., 1975) and in the cuticle collagen of the worm *Nereis* (Spiro and Bhoyroo, 1971).

Occurrence of the linkage has been shown by alkaline β-elimination. For example, treatment of yeast mannan with 0.1 M NaOH at 25°C for 18 h results in cleavage of *O*-glycosidic linkages (Nakajima and Ballou, 1974). *N*-Glycosidically linked carbohydrate groups present in mannan are not cleaved under these conditions but their existence can be demonstrated by alkaline hydrolysis under more vigorous conditions (1 M NaOH, 0.5 M $NaBH_4$ at 100°C for 5 h).

6.8.2.4. Xylosyl-serine or -threonine linkages *O*-Glycosidic linkages of xylose to serine have been identified in many proteoglycans of animal connective tissues, including the chondroitin sulphates, dermatan sulphate, heparin and heparan (Rodén, 1980). Included in the linkage region of these proteoglycans are two galactose and one glucuronic acid residue, the latter being bound to the first regular repeating disaccharide of the polysaccharide chain in the sequence:

Polysaccharide-GlcA-Gal-Gal-Xyl-Ser(Protein core)

Alkali-labile linkages between xylose and both serine and threonine have been identified in a protein polysaccharide from red algae (Heaney-Kieras et al., 1977) and a Xyl-Ser linkage occurs in the root tip slime glycoprotein of maize (Green and Northcote, 1978).

Initial identification of the linkage amino acid in proteoglycans was established by proteolytic digestion. Muir (1958) isolated chondroitin sulphate proteoglycan (under conditions which avoided treatment with alkali or protease), and digested exhaustively with papain. After purification of the digested chondroitin sulphate amino analysis showed that serine was the only amino acid present in quantity sufficient to account for the protein-carbohydrate linkage. The predominant amino acid in commercial samples of heparin isolated by proteolytic digestion was also found to be serine (Lindahl and Rodén, 1965).

A variation on this approach is to first digest the polysaccharide moiety of proteoglycan enzymatically and then obtain glycopeptides by further proteolytic cleavage. Gregory et al. (1964) digested chondroitin 4-sulphate proteoglycan first with hyaluronidase and isolated the protein-containing fraction by gel filtration on Sephadex G-75. This material was then digested with papain, and a glycopeptide fraction separated from keratosulphate by gel filtration (on Sephadex G-25). Following further digestion with Pronase, carboxypeptidase and leucine amino-peptidase, glycopeptides were obtained which had serine as sole amino acid. The smallest of these glycopeptides contained Ser, Xyl, Gal, GalNAc and GlcA.

Identification of the protein-carbohydrate linkage can be achieved by the isolation of the linkage compound Xyl-Ser after partial acid hydrolysis of glycopeptides. Lindahl and Rodén (1966) hydrolysed glycopeptides from chondroitin sulphate at pH 1.5 for 4 h at 100°C. The linkage compounds, Xyl-Ser and Gal-Xyl-Ser were isolated by high-voltage paper electrophoresis or by chromatography on Dowex 50. Xyl-Ser is cleaved into xylose and serine by acid hydrolysis at pH 1 at 100°C at a rate consistent with xylose occurring as the pyranoside rather than the furanoside. Both Xyl-Ser and Gal-Xyl-Ser can also be isolated from hydrolysates of heparin containing serine as the sole amino acid.

The serine-xylose linkage can also be identified by alkaline β-elimination. Anderson et al. (1963) treated samples of chondroitin 4- and 6-sulphates (isolated by a procedure involving protease treatment) containing covalently attached peptides with alkali and examined the effect on the amino acid composition. Treatment with 0.5 M NaOH at 4°C for 19 h resulted in loss of most of the serine. Identification of the unsaturated amino acid resulting from β-elimination can also be accomplished by sulphitolysis (Simpson et al., 1972; Section 6.8.2.1). Alkaline cleavage in the presence of borohydride (or borotritide) leads to the formation of xylitol and to reduced derivatives of the unsaturated amino acids resulting from β-elimination (Section 6.8.2.1).

6.8.3. Galactosyl linkages to hydroxylysine

Occurrence of this type of linkage is restricted to the relatively small group of proteins containing hydroxylysine. Carbohydrate units consisting of galactose or glucosylgalactose in β-*O*-glycosidic linkage with the side-chain oxygen of hydroxylysine have been identified in collagens, including that of basement membrane.

This protein-carbohydrate linkage is resistant to alkaline hydrolysis and its presence can be demonstrated by isolation of Gal-Hyl and Glc-Gal-Hyl from alkaline hydrolysates of glycoproteins and glycopeptides. These compounds can be separated and identified using an amino acid analyser.

Glycopeptides can also be isolated by gel filtration following proteolytic digestion as described in Section 6.4.3.

Procedure for alkaline hydrolysis to release Gal-Hyl and Glc-Gal-Hyl (Spiro, 1972) Glycoprotein (15–20 mg/ml) or glycopeptides (5 μmoles/ml) are hydrolysed in 2 M NaOH at 105°C for 24 h in tightly capped polypropylene tubes. After neutralising with HCl and making up to a known volume an aliquot is analysed directly on an amino acid analyser for Gal-Hyl, Glc-Gal-Hyl and free Hyl. A modified pH 5 elution system is described in detail (Spiro, 1972) for optimising the resolution of the protein-carbohydrate linkage compounds. Recovery of these compounds and hydroxylysine are all about 85% under these conditions of alkaline hydrolysis.

Alternatively Gal-Hyl and Glc-Gal-Hyl can be identified by paper chromatography or electrophoresis after desalting. Glc-Gal-Hyl can be converted to Gal-Hyl by hydrolysis in 0.05 M H_2SO_4 at 100°C for 28 h (Spiro, 1967). The linkage between galactose and hydroxylysine is stabilised towards acid by the proximity of the positively charged ε-amino group of lysine.

6.8.4. Arabinosyl and galactosyl linkages to hydroxyproline

Hydroxyproline residues in plant glycoproteins can be linked to β-arabinofuranosides (Allen et al., 1978), α-arabinofuranosides (Yamagishi et al., 1976) or galactopyranosides (Fincher et al., 1974; Miller et al., 1972).

Hydroxyproline *O*-glycosidically linked to arabinose occurs widely in plants but has not been found in animal tissues (Lamport and Miller, 1971). However, the anomeric configuration (α or β) has been ascertained in few cases (Allen et al., 1978). The linkage is stable to alkali and it can be characterised by the isolation of glycopeptides containing hydroxyproline as the sole amino acid. Because of its furanoside structure arabinose can be readily released by mild acid hydrolysis. Lamport et al. (1973) found that arabinose could be released from the extensin of plant cell walls by hydrolysis at pH 1 for 1 h at 100°C; Allen et al. (1978) removed arabinose from potato lectin by heating with 12.5 mM oxalic acid at 100°C for 5 h.

Procedure for isolation of hydroxyproline arabinosides (Lamport and Miller, 1971) Samples of cell wall or soluble glycoprotein are hydrolysed by refluxing for 6 h in 0.22 M $Ba(OH)_2$. The hydrolysate is neutralised with H_2SO_4, centrifuged and rotary evaporated to a small volume at 40°C. A sample of the hydrolysate containing at least 200 μg hydroxyproline in 0.5 ml water (at pH 3.7) is chromatographed on a column (750 × 6 mm) of Technicon Chromobeads B resin (H^+ form) equilibrated with water. Elution is carried out with a gradient produced in a two-chambered vessel. Initially the mixing chamber contains 100 ml water and the reservoir 100 ml of 0.5 M HCl. After pumping at 60 ml/h for 3 h the column is pumped with 0.5 M HCl for a further 2 h. Samples of the eluate are monitored for hydroxyproline and pentose. Lamport and Miller (1971) describe procedures for continuous analysis of both components using a Technicon autoanalyser system.

During alkaline hydrolysis racemisation of the hydroxyproline glycosides occurs; these isomers of the glycosides and of free hydroxyproline are resolved. Arabinosides having the compositions Hyp_1 Ara_1, Hyp_1 Ara_2, Hyp_1 Ara_3 and Hyp_1 Ara_4 are separated in this system.

The anomeric configuration (β) of the Ara-Hyp *O*-glycosidic linkage has been deduced from the optical rotation of the potato lectin glycoprotein (and its glycopeptide) before and after release of arabinose by acid hydrolysis (Allen et al., 1978).

Glycopeptides have been obtained from plant primary cell wall (Lamport et al., 1973) and from potato lectin by proteolytic cleavage. Arabinose oligosaccharides linked to hydroxyproline occur adjacent in the peptide chain to serine residues linked *O*-glycosidically to galactose. Arabinose residues adjacent to galactose inhibit the removal of the *O*-glycosidically linked galactose by alkaline β-elimination (Section 6.8.2.2).

The existence of an *O*-glycosidic linkage between galactose and hydroxyproline in an arabinogalactan peptide from wheat endosperm has been reported by Fincher et al. (1974). Arabinose can be stripped from the molecule by mild acid hydrolysis, leaving a core of galactose linked to peptide. The protein-carbohydrate linkage is resistant to alkaline digestion in 5 M NaOH at 100°C for 24 h. A glycopeptide containing only galactose and with Hpr as the major amino acid can be isolated by gel filtration of the NaOH digest of arabinose-stripped material.

6.9. *Strategies for the structural analysis of carbohydrate moieties*

The general methods available for structural analysis of glycoproteins and proteoglycans have been outlined in Section 6.2. Structural analysis of the carbohydrate units of these molecules requires selection of a combination of methods which have to be varied according to the type of sample, the amounts available and the facilities to which the investigator has access. Some examples of the combinations of these methods which are currently in use are given below. Details of these methods are given elsewhere in this chapter.

Exoglycosidase method for the analysis of glycoprotein oligosaccharide units

This approach has been developed particularly by Kobata (1979) and is applicable to very small quantities of sample.

(1) Carbohydrate units are released as oligosaccharides by endoglycosidases, hydrazinolysis (*N*-linked units) or alkaline elimination (*O*-linked units).

(2) The oligosaccharide or glycopeptide is radioactively labelled by reduction with NaB^3H_4.
(3) The labelled oligosaccharide alditol is purified.
(4) The monosaccharide composition is determined.
(5) Methylation analysis of the glycosidic linkages is performed.
(6) The carbohydrate unit is digested sequentially with exoglycosidases.
(7) In some cases additional information obtained by other methods (e.g. Smith degradation or direct inlet MS) may be required to elucidate the complete structure.

This strategy has been most extensively applied to *N*-linked carbohydrate units released by hydrazinolysis. The site of attachment of carbohydrate units to the peptide chain cannot be ascertained but the average number of carbohydrate units per polypeptide chain can be determined.

Structural analysis of glycoprotein oligosaccharide units by ^{1}H-NMR and methylation

This approach is associated particularly with the laboratories of Montreuil and Vliegenthart (Vliegenthart et al., 1983).
(1) Carbohydrate units are released as glycopeptides by non-specific or specific proteolysis.
(2) The glycopeptides are purified.
(3) The amino acid and monosaccharide composition is determined.
(4) Methylation analysis of glycosidic linkages is performed.
(5) The ^{1}H-NMR spectrum of the glycopeptide is determined at high resolution (360 MHz or 500 MHz).

This approach has been applied most extensively to *N*-linked carbohydrate units but it can also be used to examine *O*-linked oligosaccharides. The sensitivity of this method is limited by the requirement for ^{1}H-NMR and larger quantities of sample are required than for the exoglycosidase method. Access to high-resolution NMR equipment is obviously essential. If reference spectra of closely related structures are available it is possible to determine the structure definitively. Simple mixtures of glycopeptides can be analysed. When glycopeptides are released by specific proteolysis their amino acid sequences can indicate the location of a particular carbohydrate unit on the polypeptide chain.

Analysis and characterization of glycosaminoglycans by enzymatic digestion

This method has been described by Murata (1980).
(1) Acidic glycosaminoglycans are obtained from proteoglycans by proteolytic digestion and fractionation.
(2) The monosaccharide and sulphate composition is determined.
(3) The glycosaminoglycan is digested with specific enzymes to yield unsaturated disaccharides.
(4) The disaccharides are separated and identified by paper chromatography.
(5) Confirmation of the location of sulphate substituents is obtained by paper electrophoresis and digestion with specific sulphatases.

This procedure is applicable to about 1–10 mg of proteoglycan and can be employed to analyse the chondroitin sulphates, dermatan sulphate and hyaluronic acid. Other methods such as electrophoresis on cellulose acetate (Section 4.3.2) are required to identify glycosaminoglycans which are not digested by the chondroitinases employed.

Analysis of labelled glycosaminoglycans by specific depolymerisation

This method, developed by Hart (1976, 1978) can be employed to determine the amounts of different biosynthetically labelled glycosaminoglycans present in a mixture.
(1) Biosynthetically labelled glycosaminoglycans (GAG) are released by Pronase digestion and isolated as radioactive material appearing in the excluded volume after gel filtration on Sephadex G-50.
(2) The glycosaminoglycans are treated with HNO_2 to degrade heparan sulphate and undegraded GAGs are isolated in the excluded volume after gel filtrations on Sephadex G 50.

(3) The HNO_2-resistant GAGs are treated with hyaluronidase and undegraded GAGs are isolated by gel filtration on Sephadex G-50.
(4) The HNO_2- and hyaluronidase-resistant GAGs are treated with chondroitinase ABC and undegraded GAGs are isolated by gel filtration on Sephadex G-50.
(5) The HNO_2- hyaluronidase- and chondroitinase-resistant GAGs are digested with keratan sulphate endo-β-galactosidase and again subjected to gel filtration.

The proportions of heparan sulphate, hyaluronate, chondroitin sulphates plus dermatan sulphate and keratan sulphate present in a mixture can be analysed. Very small amounts of sample are required.

Structural analysis of carbohydrate moieties

6.9.1. Methylation

Methylation techniques have been very extensively employed to determine the positions of the glycosidic linkages between monosaccharide units in polysaccharides, oligosaccharides, glycopeptides and glycolipids. The initial step in the method is to convert all of the free hydroxyl groups of the carbohydrate polymers to their methyl ethers. The permethylated carbohydrate polymer is then cleaved by acetolysis/hydrolysis, or methanolysis and the partially methylated products are separated, identified and quantitated. The location of free hydroxyl groups in these products reveals the position of glycosidic linkages in the original polymer (Fig. 6.4).

This type of linkage analysis does not by itself give sufficient information to define the complete structure of an oligosaccharide. Other techniques are required to establish the anomeric linkages and the sequence of more complex oligosaccharides. For example a combination of linkage analysis with ^{1}H-NMR spectroscopy (Section 6.9.3) has been employed to establish the complete primary structures of several *N*-linked glycopeptides of the complex type (Fournet et al., 1978). Methylation can also be effectively employed in conjunction with degradative techniques such as partial hydrolysis, acetolysis,

Smith degradation or digestion with endo- or exoglycosidases to identify the positions of glycosidic linkages between fragments of the oligosaccharide or glycopeptide (Rauvala et al., 1981).

Methylated oligosaccharides and glycopeptides can also be analys-

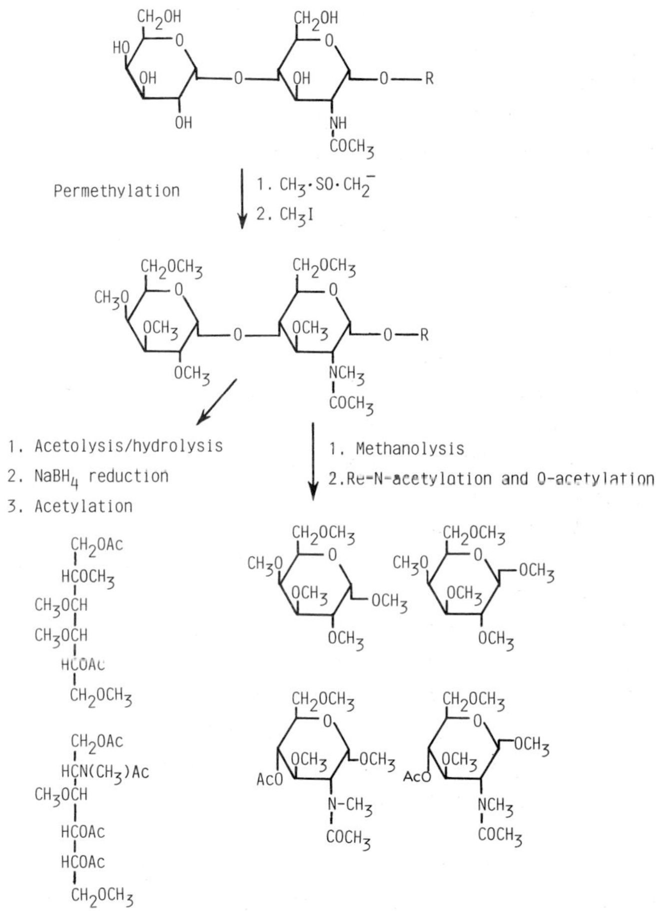

Fig. 6.4. Permethylation of an oligosaccharide followed by acetolysis/hydrolysis or methanolysis.

ed directly by mass spectrometry. This technique, which is discussed in Section 6.9.2, can give sequence information but does not reveal the anomeric configuration of glycosidic linkages.

The partially methylated monosaccharides obtained in the linkage analysis of permethylated oligosaccharides are generally separated and quantitated by gas-liquid chromatography. Because of the large number of structurally similar partially methylated sugars it is technically difficult to separate all of the possible positional isomers. These compounds are therefore usually identified both by their retention times and by mass spectrometry of the material separated by GLC. The techniques of GLC and mass spectrometry give complementary information and allow unambiguous identification of partially methylated monosaccharides. To carry out this type of analysis it is highly desirable to have access to suitable GLC-MS equipment. However, it is possible to identify methylated derivatives by GLC alone, particularly when advantage is taken of column packings giving high resolution (Akrem et al., 1979).

Reviews of different aspects of methylation techniques are available, relating to glycopeptides and glycolipids (Rauvala et al., 1981; Marshall and Neuberger, 1972), to oligosaccharides containing 2-deoxy-2-acetamido hexoses (Lindberg and Lönngren, 1978) and to other polysaccharides (Lindberg, 1972; Bjorndahl et al., 1970).

Preparation of permethylated derivatives. The technique of Hakomori (1964) has largely superseded other methylation methods such as those of Purdie, Haworth and Kuhn. In the Hakomori procedure the oligosaccharide or glycopeptide dissolved in dimethyl sulphoxide is reacted with the strongly basic methylsulphinylcarbanion. The sugar alkoxides produced are readily methylated with methyl iodide. Both solubilisation and reaction of oligosaccharides are promoted by carrying out the permethylation in an ultrasonic bath (Bjorndahl et al., 1970).

The procedure can be applied to glycopeptides or oligosaccharides released by hydrazinolysis or alkaline β-elimination. Oligosaccharides with a free reducing group should be reduced to the alditol with borohydride (or borodeuteride) before methylation. This avoids 'peeling' reactions in the strongly basic conditions used for methylation. Intact glycoproteins or polysaccharides which do not swell in DMSO are unsuitable for methylation by this method.

O-Acetyl substituents (e.g. of neuraminic acids) are lost as a result of exposure to the strongly basic methylsulphinylcarbanion but *N*-acetyl groups are stable. *N*-Methylation of *N*-acetylhexosamine residues occurs during the Hakomori methylation procedure. Sulphate or phosphate substituents of oligosaccharides or glycopeptides should preferably be removed before methylation. Uronic acid-containing oligosaccharides can usually be permethylated in good yield (Lindberg, 1972). However, it is generally preferable to carry out reduction to the corresponding alcohol before methylation. This can best be accomplished by the method described in Section 5.6.

Methylsulphinylcarbanion reacts rapidly with water (Corey and Chaykovsky, 1965). The oligosaccharide should therefore be thoroughly dried and preparation of the reagent and the methylation reaction must be carried out under anhydrous conditions.

The presence of water or the absence of a sufficient excess of carbanion can result in lack of complete methylation. A fully methylated oligosaccharide should not exhibit OH absorbance in the 3400–3600 cm^{-1} region of the infrared. If the amount of sample available is not sufficient for infrared analysis it is useful to test for the presence of an excess of methylsulphinylcarbanion immediately prior to addition of methyl iodide by reaction with triphenylmethane (Rauvala, 1979).

The amount of sample employed for permethylation depends on availability and on the sensitivity of detection methods employed for the analysis of partially methylated monosaccharides. Quantities in the range 0.2–1 mg have been sufficient for the linkage analysis of glycopeptides containing complex carbohydrate units.

It is advisable to use redistilled reagents stored in bottles with Teflon-lined screw caps for the following procedures because plasticisers present in the caps of bottles in which some solvents are commercially supplied may interfere in the analysis of products.

Preparation of methylsulphinylcarbanion (MSC). This reagent is prepared by the addition of sodium hydride to dimethyl sulphoxide in strictly anhydrous conditions.

$$NaH + CH_3SOCH_3 \rightarrow CH_3SOCH_2^- Na^+ + \tfrac{1}{2}[H_2]$$

Sodium hydride, 2.5 g, as a 50% suspension in mineral oil, is placed in a 100-ml round-bottomed flask. The mineral oil is removed by washing by decantation with four 25-ml portions of hexane. Hexane remaining after the final decantation is removed by repeatedly evacuating the flask and flushing with nitrogen. Dry dimethyl sulphoxide, 30 ml, is then added and the reaction mixture is incubated under nitrogen at 55°C on an ultrasonic bath (40 kc/s) until evolution of hydrogen ceases. Aliquots of the MSC solution are stored under nitrogen in screw-capped tubes with Teflon liners in a desiccator at 4°C. The reagent is stable for several weeks and should be thawed at room temperature and centrifuged before use. The concentration of the base can be determined by titration with formanilide using triphenylmethane as indicator.

Permethylation. The sample of glycopeptide or oligosaccharide (0.2–1 mg) is dried over P_2O_5 in a Pyrex tube to which a screw cap with Teflon septum can be fitted. Anhydrous dimethyl sulphoxide (0.5 ml) is added to the tube, which is flushed with nitrogen and closed with the septum. When the sample has dissolved completely, MSC (0.5 ml) is added by injection and the reaction mixture is placed on an ultrasonic bath for 1 h at room temperature. A sample can be removed at this stage and tested for the presence of excess MSC. Methyl iodide (0.5 ml) is added cautiously with external cooling on ice water and the tube containing the reaction mixture is returned to the ultrasonic bath for a further 2 h. Water (0.5 ml) is added and the permethylated products are extracted with three 5-ml portions of chloroform. The organic phase is extracted three times with water (5 ml) and concentrated on a rotary evaporator. Complete removal of DMSO from the permethylated product is achieved by gel filtration of the sample on a column (1 × 20 cm) of Sephadex LH-20 in chloroform:ethanol (1:1). Fractions (1 ml) of the eluate are collected. Permethylated oligosaccharides are detected by the phenol-sulphuric acid procedure (Section 5.8).

An alternative procedure for purifying the permethylated derivative by adsorption on silica gel followed by elution with methanol has been described by Endo et al. (1979).

To check for the presence of an excess of unreacted carbanion prior to addition of methyl iodide a small aliquot of this solution is added to dry powdered triphenylmethane. An instantaneous red colour, fading rapidly, indicates the presence of carbanion. The completeness of methylation of the end product can be examined by infra-red spectroscopy. If complete methylation has not been achieved the whole permethylation procedure should be repeated using the partially methylated oligosaccharide. If uronic acids are present, however, it may be preferable to repeat the methylation with fresh sample under more vigorous conditions (Lindberg, 1972).

Depolymerisation of permethylated oligosaccharides. Cleavage of the permethylated glycopeptide or oligosaccharide to its constituent units can be achieved either by methanolysis or by a combination of acetolysis and hydrolysis. Because *N*-acetylhexosamines are almost completely *N*-methylated in the Hakomori procedure the positively charged nitrogen in these derivatives makes glycosidic linkages involving these sugars resistant to hydrolysis. Stellner et al. (1973) showed that complete cleavage of *N*-acetylhexosamine-containing glycolipids could be obtained by acetolysis followed by hydrolysis. This procedure has been applied successfully to other glycoconjugates.

Before analysis of the partially methylated monosaccharides produced by acetolysis/hydrolysis by GLC-MS the monosaccharides are reduced to alditols with borohydride (or borodeuteride) and free hydroxyl groups are acetylated. Each methylated monosaccharide gives rise to a single peak on GLC. This analytical approach has been employed very extensively (Rauvala et al., 1981; Lindberg and Lönngren, 1978; Lindberg, 1972) and a compilation of GLC and MS data on methylated and acetylated monosaccharides is available (Jansson et al., 1976). This procedure is not suitable for analysis of sialic acid-containing oligosaccharides, which are destroyed. Oligosaccharides containing aminohexitol residues, such as those produced by alkaline borohydride cleavage of *O*-linked glycosidic chains of glycoproteins or by reduction of oligosaccharides released from *N*-linked oligosaccharides by hydrazinolysis or endoglycosidase treatment, also can give difficulties when analysed by this method (Rauvala et al., 1981). Both sialic acid- and aminohexitol-containing samples can be analysed after cleavage by methanolysis.

Methanolysis of permethylated oligosaccharides results in the formation of a mixture of the α- and β-methyl derivatives of the various partially methylated monosaccharides which are present. Before analysis by GLC the free hydroxyl groups are derivatised, usually by acetylation or trimethylsilylation. The mixture of products obtained as a result of methanolysis is more difficult to resolve completely by GLC than the products of acetolysis/hydrolysis. However, the formation of α- and β-methyl glycoside peaks can aid the identification of the methylated monosaccharides. Methodology for the analysis of all of the methylated sugars obtained from complex-type glycopeptides has been reported (Fournet et al., 1981) and has recently been successfully applied to many glycoproteins containing *N*-linked carbohydrate units. The *N*-acetylglucosamine residue involved in linkage to asparagine in this type of glycoprotein is, however, not recovered quantitatively after methanolysis; hydrolysis is required for the release of this residue.

Both the acetolysis/hydrolysis and methylation procedures have features which recommend them for particular types of sample. One general procedure is to split the permethylated sample and to subject part to acetolysis/hydrolysis. The remainder can be cleaved by methanolysis and employed for the analysis of sialic acids and aminohexitols (Rauvala et al., 1981). For small quantities of *N*-linked glycopeptides it may, however, be advantageous to methanolyse most of the sample and analyse by the method of Fournet et al. (1981).

Procedure for acetolysis/hydrolysis and peracetylation. To the permethylated sample, dried in a Pyrex tube with a Teflon-lined screw cap, add 0.3 ml of 0.25 M H_2SO_4 in 95% acetic acid (prepared by mixing 5 ml of 5 M H_2SO_4 and 95 ml glacial acetic acid) and heat at 80°C for 18 h. Add 0.3 ml water and heat at 80°C for an additional 5 h (Stellner et al., 1973).

The hydrolysate is applied to a column (0.5 × 3 cm) of AG3 × 4A (Bio-Rad; acetate form) and is washed through with methanol. Eluate and washings are pooled and dried under a stream of nitrogen. The residue is dissolved in water, 0.2 ml, and $NaBH_4$ (10 mg) is added. After 2 h at room temperature excess borohydride is decomposed by the addition of a drop of glacial acetic acid and the reaction mixture is brought to dryness in a rotary evaporator. The residue is redissolved and evaporated three times with 3 ml of methanol to remove methyl borate. Acetylation is carried out in acetic anhydride-pyridine (1:1), 2 ml, for 1 h at 100°C. Toluene (5 ml), which forms an azeotrope with acetic anhydride, is added and the mixture evaporated to dryness. The residue is redissolved in acetone and samples are analysed by GLC or GLC-MS.

Procedure for methanolysis and peracetylation. The dry permethylated oligosaccharide or glycopeptide is treated with 1 ml 0.5 M anhydrous methanolic HCl at 80°C for 24 h (see Section 5.8.5.2). After cooling, the methyl glycosides are dried in a stream of nitrogen. Peracetylation is carried out in pyridine:acetic anhydride (1:1), 500 μl, at 100°C for 30 min.

Identification of the methyl derivatives of glucosamine residues involved in GlcNAc-Asn linkages is carried out after (1) hydrolysis of the fully methylated glycopeptide in 4 M HCl for 4 h at 100°C, (2) peracetylation in pyridine-acetic anhydride (1:1) at 100°C for 30 min, (3) treatment with 0.5 M methanolic HCl at 80°C for 4 h and (4) a further peracetylation (Fournet et al., 1981) as in step (2).

Separation and identification of partially methylated monosaccharide derivatives. The partially methylated methyl glycosides or alditols arising from cleavage of permethylated oligosaccharides can be separated by several methods (Marshall and Neuberger, 1972) but the speed, sensitivity and excellent quantitation of GLC make this the method of choice. Conventional packed columns can be employed but more effective separations are obtained with capillary columns in which the walls are coated with the liquid phase (i.e. wall coated open tubular (WCOT) columns).

A preliminary identification of derivatives separated by GLC can be made on the basis of retention times related to an internal standard. Extensive compilations of the retention times of partially methylated monosaccharides are available from studies of compounds of established structure.

Methods for the preparation of standard partially methylated derivatives which are not generally available commercially can be found in articles by Rauvala et al. (1981) and Fournet et al. (1981). Gas-liquid chromatography with a flame ionisation detector allows quantitation of partially methylated derivatives of neutral sugars. Molar ratios of sugars are generally obtained from measurements of relative peak areas. The molar response of the different methylated sugars (of one particular type of derivative) is generally assumed to be the same, or alternatively effective carbon response can be used (Sweet et al., 1975). Quantitation of methylated derivatives can be affected by demethylation during solvolysis or by the loss of more volatile components during sample concentration. Response factors for the derivatives of amino sugars differ from those of neutral sugars and quantitation of amino sugars is more satisfactory if based on the original composition of the oligosaccharide. Data obtained from mass spectrometry can also aid in quantitation (Rauvala et al., 1981).

Analysis by electron-impact mass spectrometry of the methylated monosaccharide derivatives as they are separated by GLC provides an extremely valuable method for establishing the pattern of methylation. The identities of methylated derivatives are ascertained either by direct comparison with the mass spectra of known compounds or by consideration of general rules which have been established by studying the ionised fragments formed from methylated carbohydrates of known structure. Mass spectra of alditol acetate derivatives (Jansson et al., 1976; Wong et al., 1980) and methyl glycosides (Fournet et al., 1981) have been tabulated and rules governing the interpretation of the mass spectra of these derivatives have been reported (see Bjorndahl et al., 1970; Lindberg, 1972; Rauvala et al.,

1981, for alditol acetates and Fournet et al., 1981, for methyl glycosides). Detailed discussions of the interpretation of mass spectra of methylated carbohydrate derivates are outside the scope of this monograph and only a few general points will be made here.

Peaks in the mass spectrum arise from positively charged ions resulting from fragmentation of the molecule introduced into the spectrometer. Primary fragments result from fission between carbon atoms in the main chain. Secondary fragments can be formed from the primary by loss of acetic acid (*m/e* 60), methanol (*m/e* 32), ketene (*m/e* 42) and formaldehyde (*m/e* 30) singly or consecutively. Generally the intensities of fragments decrease with increasing molecular weight.

The mass spectra of isomeric alditols with the same methylation pattern but different configurations (e.g methylated derivatives of mannose and galactose) are almost indistinguishable. Their differentiation is based on their GLC retention times; hence the information obtained by the two techniques is complementary.

The incorporation of deuterium into methylated carbohydrates by reduction with borodeuteride can be employed to facilitate the identification of derivatives by mass spectrometry. For instance the symmetrical 3,4-di-*O*-methyl derivative of mannose which occurs commonly in glycoproteins gives two primary cleavage fragments at 189 *m/e* which could be confused with the cleavage fragment of *m/e* 189 given by C-1 to C-3 of the 3,6-di-*O*-methyl derivative of mannose. When reduction is carried out with deuterohydride only half the intensity at *m/e* 189 is shifted to *m/e* 190, whereas in the latter case all of the intensity is moved to *m/e* 190. The introduction of deuterium by reduction is also valuable in the identification of uronic acid residues.

Identification of the methyl substitution pattern of hexosamines and hexosaminitol derivatives from their mass spectra is more difficult than for hexoses because of a high tendency for cleavage to occur between C-2 and C-3. Other primary fragments tend to produce signals of low intensity but reliable identification of the position of methyl substitution can be obtained using a magnetically scanning mass spectrometer (Rauvala et al., 1981).

Separation of partially methylated, acetylated alditols. Effective separation of these derivatives can be obtained on packed or wall coated glass capillary columns of ECNSS-M, OV-225 and OV-210 (Jansson et al., 1976; Lindberg and Lönngren, 1978). SE-30 and OV-101 are preferable for separation of amino sugar derivatives (Rauvala et al., 1981).

A practical difficulty with ECNSS-M columns is that the liquid phase bleeds at a significant rate at normal operating temperatures. This results in decreased retention times and decreases the detector sensitivity which can be employed.

Akrem et al. (1979) have reported that a stationary phase consisting of 0.3–0.4% OV-225 on Chromosorb modified with high molecular weight polyethylene glycol gives better resolution and greatly decreased bleed compared with packed columns of OV-225.

Separation of partially methylated, acetylated methyl glycosides. Separation of partially methylated *O*-acetylated derivatives of mannose, galactose, glucose and *N*-acetylglucosamine are obtained on wall coated capillary columns (60 m × 0.35 mm) of silicone OV-101 (110–240°C temperature programmed at 3°/min) and carbowax 20M (140–225°C at 2°/min). The ratio of areas of α and β derivatives was reproducible. Retention times and the mass spectra obtained are given by Fournet et al. (1981).

Mass spectrometry. Several different types of instrument have been applied to methylation analysis (Lindberg, 1972; Stellner et al., 1973; Wong et al., 1980; Fournet et al., 1981). Stellner et al. (1973) have reported that poor recoveries of partially methylated hexosaminitol acetates can result from the design of certain GLC-MS instruments. Contact of these derivatives with metal surfaces should be minimised. Rauvala et al. (1981) found magnetically scanning mass spectrometers superior to quadrupole instruments for identification of amino sugar derivatives.

Mass spectra are usually recorded in the range 40–600 *m/e* or 40–300 *m/e* using an electron energy of 70 eV and an ionising current of 0.2–0.3 mA. For the analysis of complex mixtures the output from the mass spectrometer can with advantage be analysed with the aid of a computer (Wong 1980; Fournet et al., 1981).

Mass spectra should be taken at the maxima of small peaks and at the beginning, maximum and end of large peaks. It is preferable that mass spectra of standard methylated carbohydrate derivatives should be obtained with the instrument to be employed for methylation analysis as some differences in intensities of peaks between instruments have been found to occur. This problem is accentuated for derivatives of amino sugars (Rauvala et al., 1981).

6.9.2. Mass spectrometry

This section deals with the structural analysis of oligosaccharides and glycopeptides by mass spectrometry and by combined GLC-mass spectrometry. Other uses of mass spectrometry considered elsewhere include the analysis of methylated monosaccharide derivatives (Section 6.9.1) and the identification of substituted sialic acids (Section 5.7.1).

Analysis of disaccharides, trisaccharides and some higher oligosaccharides can be performed by GLC-MS of suitable volatile derivatives. The structural information obtained can include molecular weight, composition (i.e. the number and class of monosaccharide units), positions of linkages between some or all of the sugars and sugar sequence. Mass spectrometry will not generally differentiate between sugar isomers (e.g. mannose, galactose and glucose), although classes of monosaccharides such as hexosamine, hexose and methylpentose are readily distinguished. The nature of anomeric linkages is not obtained from the mass spectrum. GLC-MS is frequently employed to characterise fragments produced by degradative methods such as partial hydrolysis, acetolysis or deamination (Section 6.9.4).

Less volatile oligosaccharides and glycopeptides can be analysed by directly introducing their permethylated derivatives into the mass spectrometer (i.e. the 'direct inlet' or 'direct probe' method). Sequence information can be obtained from the primary fragments which are produced by cleavage at glycosidic bonds and from secondary fragments resulting from elimination of methanol, ketene, etc.

Analysis of glycopeptides, and oligosaccharides, by mass spectrometry has the advantage that only small quantities of material are required. Derivatisation of 100 μg or less of an oligosaccharide or glycopeptide can readily be accomplished and a mass spectrum can be obtained with quantities of the order of 10 μg. This method is potentially applicable to the small quantities of sample which are often all that can be obtained from membrane preparations. The rapidity with which mass spectra can be obtained also permits the handling of a considerable number of samples. Mixtures of oligosaccharides can be separated by GLC immediately prior to mass spectrometry. It may also, in the future, be possible to analyse mixtures of less volatile glycopeptides on the basis of molecular weight differences using the direct inlet method by fractional evaporation of the sample in the ionisation chamber.

Because unsubstituted oligosaccharides are not volatile and decompose when heated it is essential to prepare derivatives for mass spectrometry (except for fast atom bombardment mass spectra). Both methyl and TMS-ethers have been employed. The methyl ethers have the advantage of lower molecular weight and they give rise to less complex mass spectra. Methylation is therefore the preferred form of derivatisation for larger oligosaccharides and glycopeptides. TMS-ethers have been employed successfully in the analysis of disaccharides containing sialic acid (Kamerling et al., 1974). Before oligosaccharides are methylated it is advantageous to reduce the free aldehyde or ketone group with sodium borodeuteride. The oligosaccharide alditols produced in this way are more amenable to structural analysis by GLC-MS than are oligosaccharide methyl glycosides (Kärkkäinen, 1971) both because of simplified separation on GLC and because their mass spectra are generally more straightforward. Incorporation of two atoms of deuterium in the reduction ensures that the alditol is asymmetrical, so allowing analysis of its linkage to the adjacent sugar by MS. For oligosaccharides having 2-deoxy-*N*-acetylhexosamine as their reducing terminal residue reduction with borohydride can be employed because of the asymmetry resulting from the *N*-acetyl group (Kärkkäinen, 1970).

The volatility of oligosaccharides containing *N*-acetylglucosamine residues is increased when the *N*-acetyl substituent is replaced by an *N*-trifluoroacetyl group (Nilsson and Zopf, 1982). *N*-trifluoroacetylated oligosaccharides can be released from glycoproteins as described in Section 6.5.3. After reduction of the *N*-trifluoroacetylated oligosaccharides to alditols and permethylation, the derivatives can be analysed by GLC-MS. The GLC separation of *N*-trifluoroacetylated oligosaccharides can be followed selectively by an electron capture detector. As well as improving volatility the *N*-trifluoroacetyl group stabilises ions resulting from primary cleavage at hexosaminic linkages produced by electron impact. This can facilitate the analysis of sugar sequences. Methodology for the preparation and separation of these oligosaccharide derivatives is described in detail by Nilsson and Zopf (1982).

Another type of modification of the *N*-acetyl substituents of amino sugars has been developed by Karlsson et al. (1978). This methodology was originally used for the analysis of glycolipids by mass spectrometry but can also be applied to glycopeptides. The permethylated glycopeptide is treated with $LiAlH_4$, which reduces the amide groups of amino sugars to amines and the (esterified) carboxyl group of sialic acid and that of the *C*-terminal amino acid are converted to alcohols. This procedure increases volatility and stabilises molecular ions by removal of oxygens, so enhancing molecular fragmentation. It may also be valuable to apply this reductive procedure to oligosaccharides containing sialic acid before analysis by MS (Rauvala et al., 1981). The free hydroxyl groups of the alcohols produced by reduction of carboxyl groups can be converted to trimethylsilyl ethers.

Most studies of the structure of the oligosaccharide chains of glycoproteins have been carried out using electron impact or chemical ionisation mass spectrometry (Rauvala et al., 1981). Fast atom bombardment mass spectrometry has recently been added to the repertoire of techniques used to obtain structural information about these molecules (Dell et al., 1983; Kamerling et al., 1983). This method can be applied to very small samples (about 5 μg) of underivatised, or permethylated or acetylated oligosaccharides. Structural information about relatively large oligosaccharides such as tetradecasaccharides (Dell et al., 1983) or even larger molecules (Spooner et al., 1984) can be obtained. For underivatised samples it is possible to determine the molecular weight up to at least the tetrasaccharide level from the mass of the pseudomolecular ion ($M + Na^+$). Permethylated oligosaccharides give not only the pseudomolecular ion ($M + H^+$) of high intensity but also fragment ions of high diagnostic value for sequence analysis (Dell et al., 1983). Analysis of fragments obtained from peracetylated samples can also give valuable structural information (Spooner et al., 1984).

Preparation of derivatives

Oligosaccharides. These should be reduced to the corresponding alditol with $NaBD_4$ unless it is known that the reducing sugar is *N*-acetylhexosamine, when $NaBH_4$ is used as reducing agent. Small oligosaccharides released by alkaline borohydride elimination of *O*-glycosidically linked carbohydrate groups can also be employed. Permethylation is carried out as described in Section 6.9.1. Suitable stationary phases for GLC include SE-30, OV-225 and QF-1.

Glycopeptides. These are first *N*-acetylated and then permethylated (Section 6.9.1) and then reduced with $LiAlH_4$ (Karlsson, 1974). The permethylated glycopeptide (1.5 mg) in diethyl ether (0.15 ml) is reduced with 1.5 mg $LiAlH_4$ at room temperature with continuous shaking. Free hydroxyl groups resulting from the reduction of the carboxyl groups of sialic acid or amino acid residues can be silylated in hexamethyldisilazane-trimethylchlorosilane-pyridine 2:1:10 (by volume) for 1 h at room temperature. The sample is introduced into the electron impact mass spectrometer by the direct inlet method. The probe should be fitted with a separate heater.

Direct probe MS analysis of the oligosaccharide of human transferrin Karlsson et al. (1978) purified a glycopeptide from human transferrin by fractionation of the Pronase digest of the glycoprotein by gel filtration on Sephadex G-25 and by ConA-Sepharose chromatography. Part of the glycopeptide was treated with neuraminidase and the desialylated glycopeptide was reisolated after chromatography on Sephadex G-50. Both the sialic acid-free and the sialic acid-containing glycopeptides were *N*-acetylated (Section 6.5.2) and then

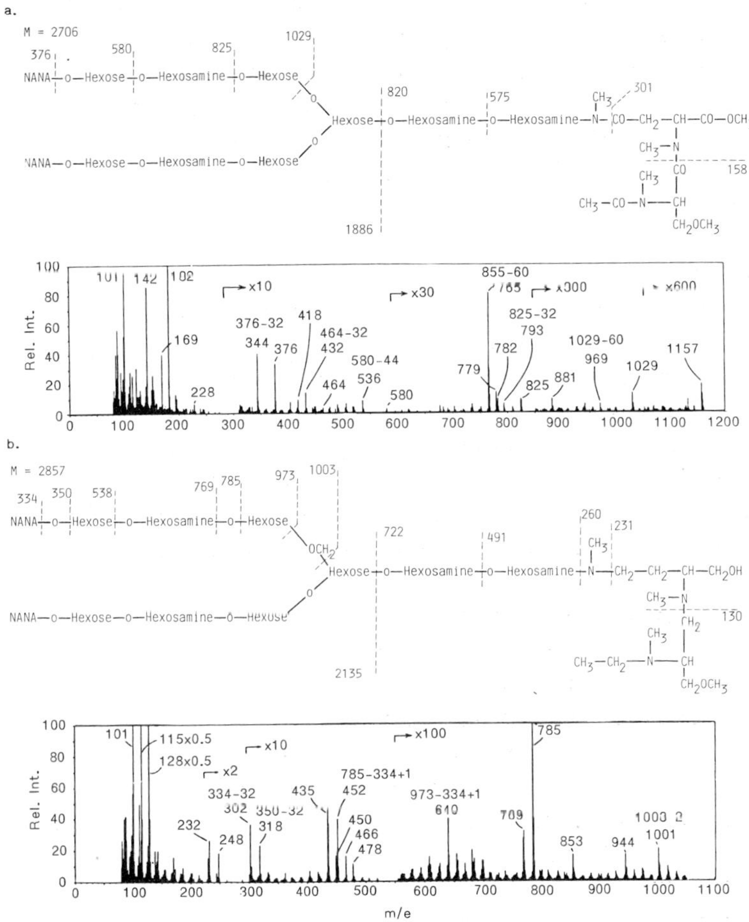

Fig. 6.5. Mass spectra of derivatives of transferrin glycopeptides (Karlsson et al., 1978). (a) the structure, expected primary fragments and mass spectrum of an *N*-acetylated, permethylated, biantennary glycopeptide. (b) the structure, primary fragments and mass spectrum of the same glycopeptide after *N*-acetylation, permethylation and reduction with lithium aluminium hydride.

reduced with $LiAlH_4$. Mass spectra were recorded for the *N*-acetylated, permethylated derivatives before and after reduction and with and without silylation.

The mass spectrum obtained from the *N*-acetylated permethylated transferrin glycopeptide is shown in Fig. 6.5a together with the structure of the glycopeptide (independently confirmed by NMR spectroscopy) on which are marked the *m/e* values of the primary cleavage products. The presence of terminal sialic acid is indicated by *m/e* 376 and 344 (376 – methanol), the terminal disaccharide is shown by fragments at *m/e* 580 and 536, the trisaccharide by *m/e* 825, 793 and the tetrasaccharide branch by *m/e* 1029 (Fig. 6.5a). Further information obtained from the mass spectrum of the reduced derivative (Fig. 6.5b) and from the *N*-acetylated permethylated desialylated glycopeptide (not shown) allowed the monosaccharide sequence to be deduced. It was also possible to infer from the mass spectra that the sialic acid was the *N*-acetyl (rather than *N*-glycolyl) derivative and that 1–4-linked *N*-acetylglucosamine was present in the structure. No evidence as to the nature of the peptide moiety was obtained.

It may be possible to develop this technique to allow fractionation of mixtures of glycopeptides by temperature-programmed evaporation in the ion source. This technique has been applied to glycolipids (Breimer et al., 1979).

Direct-probe MS analysis of a glycopeptide from porcine glycophorin Kawashima et al. (1982) purified chymotryptic glycopeptides from digests of porcine glycophorin. *O*-Linked oligosaccharides were released from the glycopeptides by alkaline borohydride treatment and subfractionated and purified by chromatography on Sephadex G-25, DEAE-cellulose and Bio-Gel P-4. Oligosaccharide CHl-S4-l was shown to be a trisaccharide containing *N*-glycolylneuraminic acid, galactose and *N*-acetylgalactosaminitol. The oligosaccharide was permethylated and subjected to linkage analysis (Section 6.9.1) and to direct-probe mass spectrometry.

The mass spectrum showed the presence of ions at *m/e* 219 and 187 (219 – methanol), indicating that galactose was in a non-reducing terminal position, and ions at *m/e* 406 and 374, indicating a non-reducing *N*-glycolylneuraminic acid.

6.9.3. *Nuclear magnetic resonance*

Glycoproteins The structural analysis of the oligosaccharide groups of glycoproteins by proton nuclear magnetic resonance has developed dramatically in recent years as a result of the application of instruments producing very high magnetic fields. Using these powerful magnetic forces it is possible to obtain high-resolution spectra (at frequencies of 360–500 MHz) of underivatised samples of glycopeptides or oligosaccharides. High-resolution ^{1}H-NMR, together with methylation analysis, has been used to determine the complete carbohydrate primary structures of many asparagine-linked oligosaccharides and several serine/threonine-linked oligosaccharides (Vliegenthart et al., 1980, 1981, 1983). The technique has the important advantage that results can be obtained in a short time compared with most other methods for the structural analysis of glycoproteins. However, instrumentation necessary for this approach to structure determination is available at only a limited number of centres. It is therefore necessary to arrange a collaborative study or to have samples run on a service or commercial basis.

High-resolution ^{1}H-NMR can give a structural 'identity card' characteristic for a particular compound which can be compared with spectra obtained from carbohydrate units of known structure. From the NMR spectrum it is possible to obtain qualitative and quantitative analysis of monosaccharide composition, the anomeric configuration of glycosidic linkages and information about the positions of glycosidic linkages. If the NMR spectra of the appropriate reference compounds are available, it may be possible to establish the primary structure of an oligosaccharide completely on the basis of its high-resolution ^{1}H-NMR spectrum. In other cases information from techniques such as methylation analysis and compositional analysis may be required to complement the NMR spectrum in order to obtain the structure. In contrast to many other techniques for structural analysis, which give misleading results when applied to samples in which the oligosaccharide units show structural heterogeneity, ^{1}H-NMR can be used to characterise samples which show limited heterogeneity (microheterogeneity).

Because of the low intensities of ^{1}H-NMR spectra the quantity of sample available can be a factor limiting the application of the technique. The minimum amount of glycopeptide which can be analysed by 360 MHz ^{1}H-NMR spectroscopy is of the order of 0.25 μmol (Dorland, 1979) but the development of 500-MHz machines and progress in data analysis have decreased the minimum sample size from which a spectrum can be obtained to about 25 nmoles (Vliegenthart et al., 1981). As the technique is non-destructive it is possible to recover the sample (from D_2O solution) and apply other analytical procedures.

Purified glycopeptides with only one or two amino acids are very suitable for structural analysis by ^{1}H-NMR. Oligosaccharide alditols (van Halbeck et al., 1980b) or oligosaccharides can also be analysed by this method. Some complication of the spectra of oligosaccharides with free reducing groups arises from the contributions of both α and β anomers of the reducing terminal group to the spectrum (Vliegenthart et al., 1981). Samples must be free of insoluble particles and impurities (e.g. paramagnetic ions) arising from buffers and chromatographic media.

The interpretation of ^{1}H-NMR spectra of asparagine-linked glycopeptides was developed initially by studies at 360 MHz (Dorland et al., 1977a,b; Fournet et al., 1978). Examination of the spectra in D_2O of model compounds and a series of oligosaccharides of known structure indicated that the resonances of certain protons were particularly useful for structural analysis. These protons have been called 'structural reporter groups' (Vliegenthart et al., 1980).

1. The resonances of protons attached to anomeric carbon atoms (H-1) are usually well resolved and occur at relatively high chemical shifts (δ) compared to other protons. These signals can be used to identify and quantitate the anomeric carbon atoms. The coupling constants ($J_{1,2}$) for the anomeric proton indicate whether the anomeric configuration is α or β (Dorland et al., 1977a,b).

2. The resonances of the protons of the methyl groups of the acetyl substituents of *N*-acetylglucosamine, *N*-acetylgalactosamine and *N*-acetylneuraminic acid are well separated from other protons in the spectrum at low δ (about 2 ppm) and can give information about the primary structures in which these residues occur.
3. The resonances of mannose H-2 protons lie apart from the other non-anomeric carbohydrate protons. These peaks can be used to count the number of mannose residues. Selective irradiation of the mannose H-2 resonances can be used to establish which H-1 resonance belongs to the same mannose residue.
4. Other resonances can give information about the positions of glycosidic linkages involving particular monosaccharide residues. For example the H-3 axial and H-3 equatorial protons of NeuAc are sensitive to the position to which sialic acid is linked. The chemical shifts of the H-1, H-5 and H-6 protons can give information about the linkage of fucose residues. The H-3 signal of mannose M-4 is senstitive to the substitution of this residue and consequently it is possible to detect the introduction of a new branch in complex-type structures by β1–4 substitution of mannose M-4 (Vliegenthart et al., 1981). In the carbohydrate units of *O*-linked glycoproteins the resonance of the H-3 of galactose residues can be shifted by substitution of the hydroxyl attached to C-3 (Van Halbeck et al., 1980a).

The 360 MHz ^{1}H-NMR spectrum of a biantennary glycopeptide is shown in Figure 6.6 and the chemical shift values assigned to the H-1, H-2 (mannose) and *N*-acetyl protons for this structure and some related compounds are given in Table 6.6. In the spectrum the HOD peak obscures the anomeric Man-3 resonance when the spectrum is scanned at 25°C but at 60°C the HOD peak is shifted (from 4.8 to 4.3 ppm), revealing the Man-3 peak.

The chemical shift values of anomeric and other protons can be influenced by the pattern of substitution of the oligosaccharide chain. Detailed consideration of these effects, and the structural information which can be deduced from them, is given in a series of publications on asparagine-linked (Dorland et al., 1977a,b; Fournet et al., 1978; Vliegenthart et al., 1980, 1983) and serine/threonine-

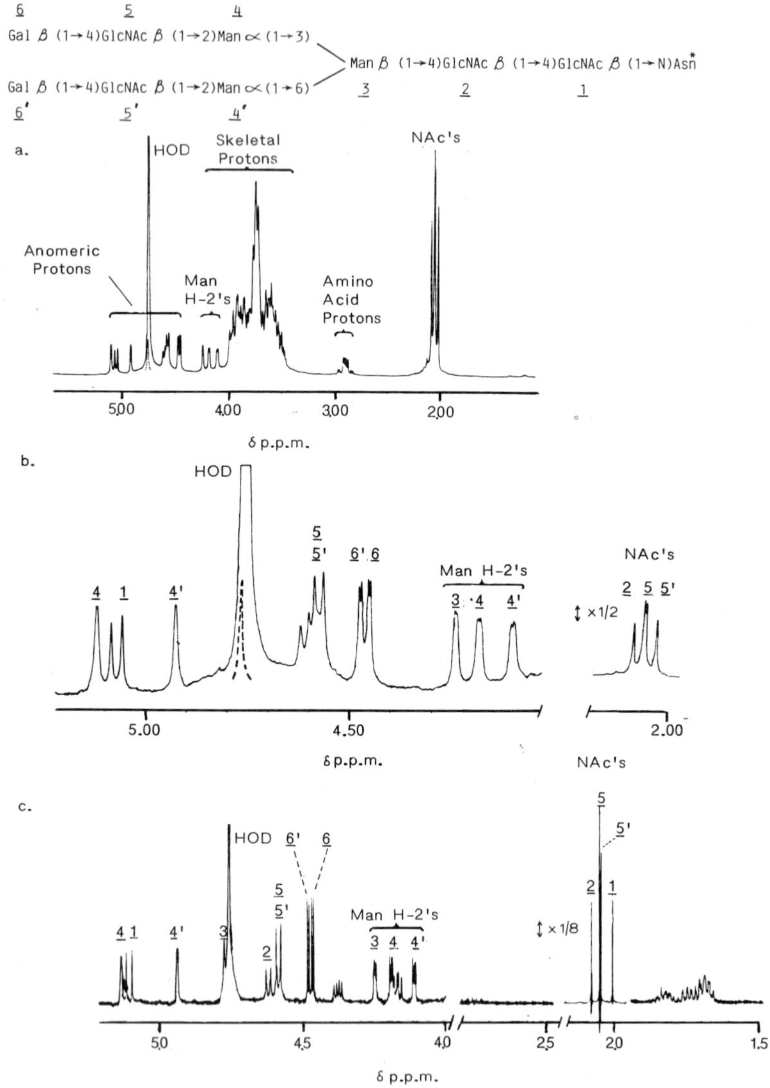

Fig. 6.6. ^{1}H-NMR spectra of a biantennary glycopeptide from transferrin. a and b show the 360 MHz spectrum of the desialylated glycopeptide whose structure is given at the top of the figure (Dorland et al., 1977). c is the 500 MHz spectrum of a similar glycopeptide (Vliegenthart et al., 1981) having an additional residue of lysine. The assignment of peaks to particular structural reporter groups is indicated. Note the greatly increased resolution obtained at 500 MHz.

TABLE 6.6

^{1}H-NMR data relating to anomeric protons, mannose -H-2 protons and *N*-acetyl methyl protons for glycopeptides isolated from human transferrin (see Fig. 6.6 for structure) and some related compounds (Dorland et al., 1977)

Compound	H-1 of residue									H-2 of residue			NAc of residue			
	1	2	3	4	4′	5	5′	6	6′	3	4	4′	1	2	5	5′
Asialo-glycan-Asn(-Lys)	5.07	4.61	4.78	5.12	4.93	4.58	4.58	4.47	4.47	4.24	4.18	4.11	2.01	2.08	2.05	2.05
	(9.6)	(7.6)	(0.9)	(1.3)	(1.7)	(8.0)	(8.0)	(8.0)	(8.0)							
Asialo-agalacto-glycan-Asn-Lys	5.07	4.63	4.77	5.12	4.92	4.59	4.59			4.24	4.19	4.10	2.01	2.08	2.05	2.05
	(9.2)	(7.6)	(<1)	(~1.5)	(~1.5)	(7.8)	(7.8)									
Man[4′]α(1–6)Man[3]β(1–4)GlcNAc[2]β(1–4)	5.07	4.69	4.77		4.92					4.08		3.96	2.02	2.09		
[Fucα(1–6)]GlcNAc[1]β1–Asn	(9.5)	(6.8)	(<1)		(1.2)											
GlcNAc[1]β1–Asn[4]	5.09												2.02			
	(9.8)															
Man[4]α(1–3)Man[3]β(1–4)GlcNAc[2]$_{\alpha}^{\beta}$		4.73	4.79	5.12						4.25	4.09			2.04		
		(7.8)	(<1)		(1.7)											
		5.21														
		(2.6)														

Chemical-shifts are given in ppm downfield from sodium 2,2-dimethyl-2-silapentane-5-sulphonate; the values in brackets are coupling constants $J_{1,2}$ in Hz.
The numbering of the monosaccharide units in the reference compounds corresponds to that in the asialo-glycan-Asn (Fig. 6.6).

linked carbohydrate units (van Halbeck et al., 1980a; Vliegenthart, 1981).

A considerable improvement in resolution and sensitivity can be achieved using a 500-MHz instrument (Vliegenthart et al., 1981, 1983). This is illustrated in Fig. 6.6c, which shows the high-resolution spectrum of the same biantennary structure illustrated in Fig. 6.6b. The increased resolution obtained at 500 MHz has been particularly valuable in the analysis of the high-mannose type of asparagine-linked carbohydrate chains. These compounds can now all be characterised in terms of the signals obtained from their structural reporter groups.

As well as the information obtained from chemical shifts, intensities and coupling constants, the line width of resonances can also give pertinent information on primary structure, the non-reducing terminal residues giving sharper peaks than more internal residues. This factor aids in distinguishing the resonances of CH_3 protons of *N*-acetylglucosamine residues in peripheral branches of complex oligosaccharide units from core *N*-acetylglucosamine residues (van Halbeck et al., 1980b).

The conformations of asparagine-linked oligosaccharides in solution have been examined by ^{1}H-NMR . The parameters analysed were chemical shifts and nuclear Overhauser enhancements (Brisson and Carver, 1983). Preferred conformations were evaluated by potential energy calculations.

Natural abundance ^{13}C-NMR spectroscopy has been applied to some asparagine-linked glycopeptides (Dijkstra et al., 1983). It was possible to assign all resonances of proton-bearing carbon atoms. Studies of ^{13}C-NMR are particularly suited to the examination of molecular dynamics by determination of spin-lattice (T_1) relaxation time, the spin-spin (T_2) relaxation time and the nuclear Overhauser enhancement (NOE) factor. ^{13}C-NMR requires substantially larger quantities of sample than ^{1}H-NMR.

Sample preparation and analysis Samples must be free of insoluble particulate material and of paramagnetic ions (which cause line-broadening). Other substances giving ^{1}H-NMR signals (such as acetate) must be removed.

Glycopeptides, oligosaccharides or oligosaccharide-alditols are dissolved in D_2O and adjusted to pH 7 if necessary. Deuterium exchange is achieved by lyophilising and redissolving in D_2O five times, finally using 99.96 atom% deuterated D_2O (Aldrich). Atmospheric moisture must be rigorously excluded. For analysis at 360 MHz 2–30 mM solutions of compounds in 0.5 ml D_2O were employed (Fournet et al., 1978). Analysis in a 500-MHz instrument has been performed on 0.1–3.0 mM solutions of compounds in 0.4 ml D_2O (Vliegenthart et al., 1981).

Spectra at 360 MHz were recorded in Fourier transform mode over a 30-min period. Probe temperatures of 25 and 60°C were selected (Fournet et al., 1978). Vliegenthart et al. (1981) have given details of analyses carried out with a 500-MHz instrument; this was operated in pulsed Fourier transform mode. Resolution enhancement was achieved by Lorenzian to Gaussian transformation. A few hundred acquisitions for each sample were accumulated. Probe temperature was 300°K.

Proteoglycans High-resolution NMR spectroscopy has also proved to be a valuable technique for studying the structures of glycosaminoglycans (Perlin, 1976; Gorin, 1981). Both ^{1}H and ^{13}C spectra have been used to characterise different glycosaminoglycan types (Perlin et al., 1970; Honda et al., 1974; Hamer and Perlin, 1976), to examine the nature of repeating units (Perlin et al., 1972) and structural heterogeneity (Hamer and Perlin, 1976). As for glycoproteins it is advantageous to employ an instrument capable of operating at high frequencies ($\geqslant$200 MHz for ^{1}H-NMR) to obtain adequate resolution of resonances for detailed structural interpretation. ^{13}C spectra have the advantage of a wider range of chemical shifts giving better separation of signals than those obtained with ^{1}H-NMR (Perlin, 1976) but the low natural abundance of ^{13}C necessitates the use of high concentrations of sample (or the accumulation of spectra over a prolonged period). The application of this methodology is, however, still limited to glycosaminoglycans which are available in milligram quantities.

The characteristics of the ^{1}H- and ^{13}C-NMR spectra of glycosaminoglycans have been discussed by Perlin (1976) and Gorin (1981) and will be treated only in outline here. Three main categories of 'structural reporter groups' are associated respectively with the anomeric centres, with *N*-substituted hexosamines and with uronic acids. Resonances arising from H-1 (at about 5 ppm) and C-1 (at about 100 ppm) are usually well separated from non-anomeric signals and are indicative of the monosaccharide residues present in the polymer. Integration of these signals can give the relative amounts of the constituents present. Whether the configuration of the anomeric linkage is α or β can be determined by the splitting pattern of the H-1 signal (Perlin, 1976). The methyl protons of the acetyl groups of *N*-acetylhexosamines are well separated from other signals (at about 2.2 ppm) and can be integrated to quantitate *N*-acetyl substituents relative to other signals in the spectrum. The H-2 resonances (3.2–3.5 ppm) and C-2 resonances (52–57 ppm) of hexosamine, whether acetylated or sulphated, are displaced upfield relative to the bulk of such signals from other nuclei. For uronic acid residues the H-5 signal is usually observed between 4 and 5 ppm. Assignment of this peak can be established by acidifying the solution, which leads to a downfield displacement of about 0.3 ppm. Glucuronic and iduronic acids can be distinguished by estimating their different splitting of the H-5 resonance.

NMR spectra can also give information about the presence of ester sulphate substituents in glycoconjugates. Substitution with *O*-sulphate results in a downfield shift of the resonance of the proton attached to the esterified carbon atom by about 0.3 ppm. Such resonances occur in the region below 4.0 ppm (Perlin, 1976). Deshielding of ^{13}C nuclei attached to ester sulphate by 5 ppm or more also occurs (Perlin, 1976). The ^{13}C-NMR spectrum of chondroitin 4-sulphate can readily be distinguished from that of chondroitin 6-sulphate (Honda et al., 1974).

The NMR spectra of glycosaminoglycans show the line-broadening characteristics of substances of high molecular weight. To obtain sharp signals it is necessary to obtain spectra at high temperature (e.g. 80°C). ^{13}C spectra are usually obtained using proton decoupling, which results in less complex spectra with sharper resonances.

The line broadening of glycosaminoglycans can itself be examined to investigate the restrictions to the mobility of carbohydrate chains. This type of study, involving measurements of relaxation times and NOE, has also been extended to proteoglycans and samples of connective tissue (Brewer and Kaiser, 1975; Torchia et al., 1977).

Samples can be prepared for ^{1}H or ^{13}C-NMR spectroscopy by repeatedly exchanging with D_2O as described for glycoproteins. D_2O is not essential for ^{13}C-NMR but is used for instruments operating with a deuterium lock system. Spectra are generally obtained in the Fourier transform mode. Particularly for ^{13}C-NMR, signals must be repeatedly acquired, accumulated and enhanced to give acceptable signal-to-noise ratios. The amounts of sample used to obtain NMR spectra have usually been very large (e.g. 200 mg for ^{13}C-NMR and 50 mg for ^{1}H-NMR; Perlin, 1977) but with modern instrumentation and data-acquisition techniques it should be possible to obtain spectra with substantially smaller samples. ^{1}H-NMR is the more sensitive technique and requires less instrument time.

6.9.4. *Methods for partial degradation*

6.9.4.1. Enzymic digestion The most important group of methods for investigating the oligosaccharide structure of glycoproteins and proteoglycans by partial degradation are those that involve digestion with specific enzymes. Because of the linear structures of glycosaminoglycan chains enzymes cleaving internal linkages are most valuable for these glycoconjugates. The short, branched structures of glycoproteins can be cleaved sequentially by the exoglycosidases considered in this section. Endoglycosidases can also be employed to release oligosaccharide units and to obtain structural information (Section 6.6).

Among the attractive features of enzymic digestion methods is their applicability to very small quantities of material and the information they reveal about the anomeric configuration of glycosidic linkages. Strategies for the structural analysis of carbohydrate moieties involving enzymic digestion have been outlined (Section 6.9) and tactical details relating to glycoproteins and proteoglycans are given below.

Glycoproteins — cleavage by exoglycosidases The specific release of monosaccharides from the non-reducing terminal ends of oligosaccharides or glycopeptides by digestion with exoglycosidases is of particular value in determination of the carbohydrate sequences of glycoproteins. As well as giving information about the order of monosaccharides the anomeric configuration of glycosidic linkages can be determined unequivocally from the known specificity of purified glycosidases. The complete structures of many glycoprotein carbohydrate units have been deduced by combining the information obtained by sequential digestion with exoglycosidases of different specificity with linkage analysis by methylation (Section 6.9.1) and periodate oxidation (Section 6.9.4). Enzymatic hydrolysis can be carried out on minute quantities of labelled sample allowing the structural analysis of only a few micrograms of oligosaccharide or glycopeptide.

Several exoglycosidases have been isolated which act on intact glycoproteins as well as on substrates of low molecular weight. These enzymes can be employed in the investigation of the functional role of the carbohydrate units of soluble or membrane-associated glycoproteins.

The specificities of exoglycosidases are directed primarily towards recognition of the terminal non-reducing monosaccharide residue and the configuration of the glycosidic linkage of this residue. However, in some instances, the position of linkage and the aglycone structure can be of great importance. There is absolute dependence on the anomeric configuration of the glycosidic bond attacked (Flowers and Sharon, 1979) so that β-galactopyranosides, for example, are distinguished from the α-anomers. In most cases exoglycosidases act only on a single type of terminal sugar so that α-fucosidase acts only on α-linked glycosides of fucose. However, several hexosaminidases are known which will cleave glycosides of either *N*-acetyl-D-glucosamine or *N*-acetyl-D-galactosamine. There is considerable variation in the extent to which positional isomerism affects the action of exoglycosidases. For example the α-fucosidases from *Torbo cornutus* and *Charonia lampas* cleave all fucosyl linkages so far reported in glycoproteins. However, the α-fucosidases from *Bacillus fulminans* and *Clostridium perfringens* cleave only the Fuc(α1–2)Gal linkage, while α-fucosidase 1 from almond emulsin hydrolyses Fuc(α1–3)GlcNAc and Fuc(α1–4)GlcNAc linkages but not Fuc(α1–2)Gal and Fuc(α1–6)GlcNAc (Ogata et al., 1977). These differences in specificity of exoglycosidases can be made use of to distinguish positional isomers. Table 6.7 lists some of the exoglycosidases which have been employed in structural studies of glycoproteins and gives an indication of their specificities for particular types of linkage. In addition to the position of linkage other structural features of the aglycone can have an important effect on the action of exoglycosidases. For example jack bean β-galactosidase cleaves Gal(β1–4)GlcNAc readily but Galβ1–4(Fucα1–3)GlcNAc is not hydrolysed, suggesting that the α-fucosyl residue sterically hinders approach of the enzyme to the oligosaccharide. More complex cases of steric hindrance, involving monosaccharides separated by several residues from the site of action of the exoglycosidase, have also been reported (Kobata, 1979). Hence the failure of a monosaccharide to be released by exoglycosidase treatment does not exclude the possibility that it is present in a non-reducing terminal position.

Exoglycosidase may hydrolyse glycosidic linkages in which the aglycone is not a carbohydrate. Cleavage of *O*-glycosidic glycopeptide bonds such as the *N*-acetylgalactosaminyl-serine, galactosyl-serine and xylosyl-serine linkages has been reported (Spiro, 1972).

A considerable variety of exoglycosidases have been characterised (Flowers and Sharon, 1979). Several of these enzymes are available commercially or they may be purified by the methods referred to in Table 6.7. The careful removal of contaminating glycosidase activities is essential.

In sequencing the oligosaccharide unit of a glycoprotein advantage

TABLE 6.7

Exoglycosidases employed in the structural analysis of the oligosaccharide units of glycoproteins

Enzyme	Source	Specificity	Refs.
α-L-Fucosidase	*Charonia lampas*	Very broad aglycone specificity	1, 2
	Turbo cornutus	Very broad aglycone specificity	3, 2
	Aspergillus niger	Only splits Fucα1–2Gal	4
	Bacillus fulminans	Only splits Fucα1–2Gal	5
	Clostridium perfringens	Only splits Fucα1–2Gal	6
	Almond emulsin	α-Fucosidase I splits Fucα1–4GlcNAc and Fucα1–3GlcNAc including Galβ1–4 (Fucα1–3) GlcNAcβ1–, but Fucα1–2Gal and Fucα1–6GlcNAc are not split	7
α-D-Galactosidase	*Aspergillus niger*		8
	Coffee bean		9
	Fig	Broad aglycone specificity	10
β-D-Galactosidase	Jack bean	Rate of hydrolysis Galβ1–6GlcNAc > Galβ1–4GlcNAc > Galβ1–3GlcNAc (can distinguish 1–3 linkage at low enzyme concentration)	11
	Aspergillus niger		8
	Clostridium perfringens		12
	Bovine testes	β1–3 and β1–4-linked Gal residues hydrolysed rapidly. Galβ1–6GlcNAc split slowly.	33
	Streptococcus (Diplococcus) pneumoniae	Splits Galβ1–4GlcNAc (but not when GlcNAc has Fuc at position 3). No cleavage of Galβ1–3GlcNAc or Galβ1–6GlcNAc	14
α-D-Mannosidase	Jack bean	Rate of hydrolysis of Manα1–2Man = Manα1–6Man > Manα1–3Man. Can be used to distinguish	

		linked 1–6 to core β-Man of 'complex type' unit is not split but a single Man linked α1–3 to the core β-Man is cleaved (16). Enzyme can distinguish substituents on α1–3 and α1–6 arms of complex units.	17
	Aspergillus niger	Splits Manα1–2Man but not 1–3, 1–4 or 1–6 linkage in di- and tri-saccharides. Not active on intact glycoproteins.	18
	Aspergillus niger	A second enzyme from this organism splits Manα1–4 and Manα1–6 links to Man or GlcNAc. Manα1–2 and α1–3 links cleaved slowly. Acts on intact glycoproteins.	19
	Aspergillus saoti	Splits Manα1–2Man but not Manα1–3Man or Manα1–6Man. Useful for identifying 'high-mannose type' oligosaccharides	20, 21
β-D-Mannosidase	Hen oviduct	Splits $ManGlcNAc_2Asn$	22
	Pineapple	Splits $ManGlcNAc_2Asn$	23
	Snail (*Achatina fulica*)	Splits $ManGlcNAc_2Asn$	24
	Turbo cornutus	Splits $ManGlcNAc_2Asn$	25
β-*N*-Acetylgalactosaminidase	*Aspergillus niger*	Acts on low and high MW substrates. Cleaves GalNAc-Ser link in glycoproteins when GalNAc is terminal	26
	Clostridium perfringens		12
β-*N*-Acetylhexosaminidase	Jack bean	Acts on β-linked GlcNAc and GalNAc in low MW substrates and glycoproteins	17
	Aspergillus niger		8
	Streptococcus (Diplococcus) pneumoniae	Splits GlcNAcβ1–2Man but not GlcNAcβ1–4Man or GlcNAcβ1–6Man. Steric effects can hinder hydrolysis, e.g. GlcNAcβ1–6(GlcNAcβ1–2)Man not split. In *O*-linked units GlcNAcβ1–3Gal and GlcNAcβ1–6Gal are split	13, 27

Neuraminidase	*Arthrobacter urifaciens* (enzymes I and II)	Broad specificity, splits NeuAcα2–3Gal, NeuAcα2–4Gal and NeuAcα2–6GlcNAc at similar rates.	33, 28, 29
	Clostridium perfringens	Cleaves NeuAcα2–3Gal, NeuAcα2–6Gal and NeuAcα2–6GlcNAc at similar rates.	30
	Vibrio cholerae (comma)	Rate of cleavage of NeuAcα2–3Gal > NeuAcα2–4Gal > NeuAcα2–6Gal. Requires Ca^{2+} for activity.	31, 29
	Influenza virus	Splits NeuAcα2–3Gal	32, 29

1. Iijima, Y. and Egami, F. (1971) J. Biochem. (Tokyo) *70*, 75; 2. Nishigaki, M., Muramatsu, T., Kobata, A. and Maeyama, K. (1974) J. Biochem. (Tokyo) *75*, 509–517; 3. Iijima, Y., Muramatsu, T., and Egami, F. (1971) Arch. Biochem. Biophys. *145*, 50–54; 4. Bahl, O.P. (1972) Methods Enzymol. *28*, 738–742; 5. Kochibe, N. (1973) J. Biochem. (Tokyo) *74*, 1141; 6. Aminoff, D. (1972) Methods Enzymol. *28*, 763–769; 7. Kobata, A. (1982) Methods Enzymol. *83*, 625–631; 8. Bahl, O.P. and Agrawal, K.M.L. (1972) Methods Enzymol. *28*, 728–734; 9. Courtois, J.E. and Pertek, F. (1966) Methods Enzymol. *8*, 565–571; 10. Li, Y. -T. and Li, S. -C. (1972) Methods Enzymol. *28*, 714–720; 11. Li, S. -C., Mazzotta, M.Y., Chien, S. -F. and Li, Y. -T. (1975) J. Biol. Chem. *250*, 6786–6791; 12. McGuire, E.J., Chipowski, S. and Roseman, S. (1972) Methods Enzymol. *28*, 755–763; 13. Glasgow L.R., Paulson, J.C. and Hill, R.L. (1977) J. Biol. Chem. *252*, 8615–8623; 14. Paulson, J.C., Prieels, J., Glasgow, L.R. and Hill, R.L. (1978) J. Biol. Chem. *253*, 5617–5624; 15. Tai, T., Yamashita, K., Ogata-Arakawa, M., Koide, N., Muramatsu, T., Iwashita, S., Inoue, Y. and Kobata, A. (1975) J. Biol. Chem. *250*, 8569–8575; 16. Yamashita, K., Tachibana, Y., Nakayama, T., Kitamura, M., Endo, Y. and Kobata, A. (1980b) J. Biol. Chem. *255*, 5635–5642; 17. Li, Y. -T. and Li, S. -C. (1972) Methods Enzymol. *28*, 702–713; 18. Swaminathan, N., Malta, K.L. and Bahl, O.P. (1972) J. Biol. Chem. *247*, 1775–1779; 19. Matta, K.L. and Bahl, O.P. (1972) J. Biol. Chem. *247*, 1780–1787; 20. Yamashita, K., Ichishima, E., Arai, M. and Kobata, A. (1980) Biochem. Biophys. Res. Commun. *96*, 1335–1342; 21. Ichishima, E., Arai, M., Shigematsu, Y., Kumagai, H. and Sumida-Tanaka, R., (1981) Biochim. Biophys. Acta *658*, 45–53; 22. Sukeno, T., Tarentino, A.L., Plummer, T.H. and Maley, F. (1972) Methods Enzymol. *28*, 777–782; 23. Li, Y. -T. and Lee, Y. -C. (1972) J. Biol. Chem. *247*, 3677–3683; 24. Sugahara, K., Okumura, T. and Yamashita, I. (1972) Biochim. Biophys. Acta *268*, 488–496; 25. Muramatsu, T. and Egami, F. (1967) J. Biochem. (Tokyo) *62*, 700; 26. McDonald, M.J. and Bahl, O.P. (1972) Methods Enzymol. *28*, 734–738; 27. Yamashita, K., Ohkura, T., Yoshima, H. and Kobata, A. (1981) Biochem. Biophys. Res. Commun. *100*, 226–232; 28. Uchida, Y., Tsukuda, Y. and Sugimori, T. (1974) J. Biochem. (Tokyo) *82*, 1425–1433; 29. Kobata, A. (1979) Anal. Biochem. *100*, 1–14; 30. Cassidy, J.T., Jourdian, G.W. and Roseman, S. (1965) J. Biol. Chem. *240*, 3501–3506; 31. Ada, G.L., French, E.L. and Lind, P.E. (1961) J. Gen. Microbiol. *24*, 409–421; 32. Drzeniek, R., Seto, J.T. and Rott, R. (1966) Biochim. Biophys. Acta *128*, 547–558; 33. Uchida, Y., Tsukada, Y. and Sugimori,

is taken of the fact that only non-reducing terminal residues are removed. For example, if the carbohydrate chain terminates with the sequence NeuAc(α2–6)Gal(β1–4)GlcNAc(β1–2)Man(α1–) treatment with neuraminidase will release sialic acid but other exoglycosidases will be without effect. Subsequent sequential digestion with β-galactosidase, β-*N*-acetylhexosaminidase and α-mannosidase can establish the sequence as NeuAcα-Galβ-GlcNAcβ-Man.

The effects of exoglycosidase digestion on an oligosaccharide substrate can be followed by assay of the monosaccharide which is released using gas-liquid chromatography, paper chromatography or enzymatic assays (Sections 5.5–5.10). An alternative, and very much more sensitive, procedure is to radioactively label the oligosaccharide (by reduction with NaB^3H_4) and to analyse its behaviour by paper chromatography or gel permeation chromatography (on Bio-Gel P-4) before and after enzymatic digestion. Changes in chromatographic behaviour indicate whether cleavage has occurred and from gel permeation data the number of residues removed can be deduced. Because the label is incorporated at the 'reducing' end of the oligosaccharide the radioactive oligosaccharide can be repeatedly reisolated and used for subsequent rounds of treatment with exoglycosidases. An alternative approach for following digestion with exoglycosidases has been suggested by Wenn (1975). In this method a glycopeptide containing a single amino acid is converted to an ultraviolet fluorescent dansyl derivative. The change in molecular weight of the glycopeptide following glycosidase digestion is estimated from measurement of mobility on paper electrophoresis. This method is simple and fairly sensitive but has not been widely applied.

The practical importance in structural analysis of employing glycosidases which are free of contaminating glycosidase activities cannot be overemphasised. Often prolonged digestion with relatively high concentrations of enzyme are necessary to bring about the complete release of a terminal monosaccharide. Under these conditions even levels of contaminating activities which may appear low when assayed for a few minutes with synthetic substrates can have a major effect on oligosaccharide substrates.

Procedures for exoglycosidase digestion

Labelled oligosaccharide method (Kobata, 1979) This extremely sensitive method can be applied to oligosaccharides released from glycoproteins by hydrazinolysis (*N*-linked units), endoglycosidase digestion, or alkaline borohydride treatment (*O*-linked units). The oligosaccharides are labelled by reduction of the reducing terminal monosaccharide to an alditol with tritiated sodium borohydride. Oligosaccharide alditols obtained by hydrazinolysis should be re-*N*-acetylated (Section 6.5.2). Alternatively, glycopeptides labelled by *N*-acetylation can be employed.

TABLE 6.8
Experimental conditions for the digestion of glycoprotein oligosaccharide units with exoglycosidases (Matsumoto et al., 1983)

Enzyme	Amount	pH	Buffer
1. *Charonia lampas* α-fucosidase	0.2 U	4.0	0.1 M acetate-0.1 M NaCl
2. Almond emulsion α-fucosidase I	40 U	5.0	0.1 M citrate-phosphate
3. *Bacillus fulminans* α-fucosidase	17.5 μg	6.6	0.05 M phosphate
4. Jack bean β-galactosidase	0.7 U	3.5	0.1 M citrate-phosphate
5. *Diplococcus pneumoniae* β-galactosidase	0.02 U	6.0	0.1 M citrate-phosphate
6. Jack been β-*N*-acetylhexosaminidase	0.7 U	5.0	0.1 M citrate-phosphate
7. *Diplococcus pneumoniae* β-*N*-acetylhexosaminidase	4 mU	6.0	0.1 M citrate-phosphate
8. Jack bean α-mannosidase	0.5 U	4.5	0.1 M acetate
9. *Aspergillus saoti* α-mannosidase	0.15 μg	5.0	0.1 M acetate
10. Snail β-mannosidase	0.02 U	4.5	0.5 M citrate

Radioactive oligosaccharide (1-2 nmole, $1–2 \times 10^4$ cpm) is incubated with one of the above reaction mixtures in a total volume of 50 μl (or 30 μl for 3, 9 and 10). Reaction products are examined by gel filtration on Bio-Gel P4.

Labelled oligosaccharides (2×10^4 cpm) are incubated with glycosidases at 37°C for 18 h in a buffer at the pH optimum of the enzyme in a total volume of 50 μl. Suitable enzyme amounts (Matsumoto et al., 1983) and buffers are indicated in Table 6.8. Toluene (50 μl) is included to inhibit bacterial growth. The reaction is terminated by heating in a boiling water bath for 2 min. Reaction products are analysed by passing them through a column of Bio-Gel P-4 (1 m × 2 cm i.d.) calibrated with a standard mixture of glucose oligomers (Yamashita et al., 1982). An increase in the elution volume of the oligosaccharide following enzymic digestion indicates that one or more sugar units has been released by the exoglycosidase. The actual number of sugar units released can be determined by reference to the column calibration with glucose oligomers. For each radioactive oligosaccharide the elution volume is interpolated on the calibration curve of elution volume versus number of glucose units to obtain a number of glucose units equivalent to the oligosaccharide. The change in equivalent number of glucose units resulting from exoglycosidase digestion can then be determined. The numbers of glucose units equivalent to many oligosaccharides of known structure have been determined (Yamashita et al., 1982). From these empirical observations it has been shown that removal of one α-linked fucose usually results in a change in elution equivalent to about 0.5 glucose units, β-galatose is equivalent to about 1 glucose unit and a residue of β-*N*-acetylhexosamine is equivalent to 2 glucose units. These figures are usually sufficiently precise for the number of terminal monosaccharide units released by the exoglycosidase to be determined. Whenever possible it is desirable to compare the elution of a radioactive oligosaccharide with an authentic standard of the same structure.

Mizuochi et al. (1979) applied this procedure to determine the carbohydrate sequence of oligosaccharides released by hydrazinolysis of bovine prothrombin. After reduction the oligosaccharides were fractionated by paper electrophoresis, desialated by digestion with neuraminidase and fractionated by paper chromatography. The neutral oligosaccharides were then digested sequentially with exoglycosidases and chromatographed on Bio-Gel P4 (Fig. 6.7). When incubated with jack bean β-galactosidase, two galactose residues were

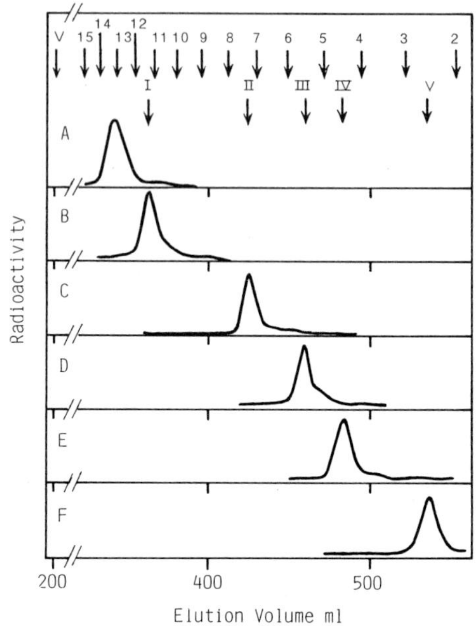

Treatment	Monosaccharides released
A. Untreated oligosaccaharide	–
B. Oligosaccharide treated with β-galactosidase	2 Gal
C. Peak from B treated with β-N-acetylhexosaminidase	2 GlcNAc
D. Peak from C treated with α-mannosidase	2 Man
E. Peak from D treated with β-mannosidase	1 Man
F. Peak from E treated with β-N-acetylhexosaminidase	1 GlcNAc

Fig. 6.7. Sequential exoglycosidase digestion. An oligosaccharide (N1) released by hydrazinolysis (of prothrombin) was reduced with NaB^3H_4 and was chromatographed on a column of Bio-Gel P4 before and after sequential treatment with exoglycosidases. The column was calibrated with glucose oligomers (2–15 glucose units, upper arrows) and with oligosaccharide of known structure (lower row of arrows) I, $GlcNAc_2 \cdot Man_3 \cdot GlcNAc \cdot GlcNAcol$; II, $Man_3 \cdot GlcNAc \cdot GlcNAcol$; III, Man · GlcNAc · GlcNAcol; IV, GlcNAc · GlcNAcol; V, GlcNAcol (*N*-acetylglucosaminitol). The number of monosaccharides released was deduced from the change in elution volume (Mizuochi et al., 1979).

removed from the oligosaccharide. The original oligosaccharide was resistant to β-*N*-acetylhexosaminidase treatment but the β-galactosidase digest liberated two *N*-acetylglucosamine residues when incubated with β-*N*-acetylhexosaminidase. The major component at this stage showed the same mobility as authentic Man_3GlcNAcGlcNAcol. Two residues of mannose were released from the radioactive peak (Fig. 6.7 C) by α-mannosidase. Subsequent treatment with β-mannosidase released another mannose residue. After β-*N*-acetylhexosaminidase digestion radioactive *N*-acetylglucosaminitol was obtained. These results indicated that the structure of the neutral glycopeptides could be written as (Galβ1–GlcNAcβ1–Manα1–$)_2$Manβ1–GlcNAcβ1–GlcNAcol. Two isomeric oligosaccharides, N1 and N3, both appeared to have the same sugar sequence if the initial digestion with jack bean β-galactosidase was carried out at high enzyme concentration. At lower enzyme concentration, however, the galactose of N3 was liberated but only slow hydrolysis of N1 was observed. Under these conditions jack bean β-galactosidase cleaves Gal(β1–4)GlcNAc much more rapidly than Gal(β1–3)GlcNAc linkages. Treatment of N1 and N3 with diplococcal β-galactosidase produced cleavage of N3 but not N1, indicating that the Gal(β1–4) linkage was present in the former oligosaccharide. The presence of Gal(β1–3) linkages in N3 was also demonstrated by other techniques (Mizouchi et al., 1979).

Glycosaminoglycans — enzymic degradation Different types of glycosaminoglycans can be distinguished by their susceptibility to fragmentation by enzymes specific for the cleavage of particular oligosaccharide sequences. The products of digestion can be analysed to determine the nature of the repeating units present in the glycosaminoglycan chain. Structural modifications occurring within glycosaminoglycan chains may also be detected.

Enzymes acting on glycosaminoglycans which are valuable in structural analysis are listed in Table 6.9. *Streptomyces* hyaluronidase is a specific enzyme useful in the characterisation of hyaluronate. Chondroitinase ABC and AC can be used to distinguish the chondroitin sulphates from dermatan suphate, which is digested by the former

TABLE 6.9

Enzymes used in the structural analysis of glycosaminoglycans (* other endo-β-galactosidases acting on keratan sulphate are included in Table 6.4)

Enzyme	Source	EC number	Description of activity	Substrates	Products	Reference
Testicular hyaluronidase	Bovine testes	EC 3.2.1.35	Endo-*N*-acetylhexosaminidase, random hydrolysis of β1–4 link to GlcA. Also catalyses transglycosidation.	Hyaluronate, chondroitin 4 and 6 sulphates, dermatan sulphate (in part)	Mainly tetrasaccharides $(GlcAGlcNAc)_2$	1, 2
Leech hyaluronidase	*Hirondo medicinalis*	EC 3.2.1.36	Endo-β-glucuronidase. No transglycosidation has been detected.	Hyaluronate and oligosaccharides (hexasaccharide and larger) derived from it.	Mainly tetrasaccharides $(GlcNAcGlcA)_2$	1
Streptomyces hyaluronidase	*Streptomyces hyaluroniticus*	EC 4.2.2.1	Lyase	Hyaluronate	Tetra and hexasaccharides with Δ4,5 unsaturated non-reducing terminal	3, 1
Chondroitin ABC lyase (chondroitinase ABC)	*Proteus vulgaris*	EC 4.2.2.4	Splits *N*-acetylhexosamine β1–4 uronic acid linkage with formation of Δ4,5 unsaturated product	Chondroitin 4 and 6 sulphates, dermatan sulphate and hyaluronate (slowly)	Disaccharides with Δ4,5 unsaturated non-reducing terminal uronic acid residue	4
Chondroitin AC lyase (chondroitinase AC)	*Flavobacterium heparinum*	EC 4.2.2.5	Splits *N*-acetylhexosamine β1–4 glucuronic acid linkage with formation of Δ4,5 unsaturated product	Chondroitin 4 and 6 sulphates and hyaluronate (slowly)	Disaccharides with Δ4,5 unsaturated non-reducing terminal uronic acid residue	4
Chondroitin AC_{II} lyase (chondroiti-	*Arthrobacter aurescens*	EC 4.2.2.5	Splits *N*-acetylhexosamine β1–4 glucuronic acid linkage with for-	Chondroitin 4 and 6 sulphates and hyaluronate (slowly)	Disaccharides with Δ4,5 unsaturated non-reducing termi-	5

			acid linkage with formation of Δ4,5 unsaturated product		ed non-reducing terminal uronic acid	
Keratan sulphate* endo-β-galactosidase	*Pseudomonas*		Hydrolyses β-linkage between non-sulphated galactose and GlcNAc	Keratan sulphate (and some glycoproteins)	Oligosaccharides of varying sizes	7
	Escherichia freundii			Keratan sulphate (and some glycoproteins)	Oligosaccharides of varying sizes	8
	Flavobacterium keratolyticus			Keratan sulphate (and some glycoproteins)	Oligosaccharides of varying sizes	9
Chondro-4-sulphatase	*Proteus vulgaris*	EC 3.1.6.9	Cleaves sulphate at C-4 of disaccharides	Unsaturated disaccharide Δ4,5-GlcAβ1–3GalNAc (4-sulphate) or saturated derivative	Free sulphate and disaccharide	10
Chondro-6-sulphatase	*Proteus vulgaris*	EC 3.1.6.10	Cleaves sulphate at C-6 of disaccharides	Unsaturated disaccharide Δ4,5-GlcAβ1–3GalNAc (6-sulphate) or saturated derivative	Free sulphate and disaccharide	10
Arylsulphatase B			Cleaves sulphate	4-*O*-sulphate groups in chondroitin 4-sulphate and dermatan sulphate	Free sulphate	11

1. Meyer, K. (1971) in The Enzymes (Boyer P.D., ed.), 3rd Edn., Vol. 5, pp. 307-320, Academic Press, New York and London; 2. Borders, C.L. and Raftery, M.A. (1968) J. Biol. Chem. *243*, 3756–3762; 3. Ohya, T. and Kaneko, Y. (1970) Biochim. Biophys. Acta *198*, 607–609; 4. Suzuki, S. (1972) Methods Enzymol. *28*, 911–917; 5. Hiyama, K. and Okada, S. (1975) J. Biol. Chem. *250*, 1824–1828; 6. Hovingh, P. and Linker, A. (1970) J. Biol. Chem. *245*, 6170–6175; 7. Li, Y.-T., Nakagawa, H., Kitamikado, M. and Li, S-C. (1982) Methods Enzymol. *83*, 610–619; 8. Kitamikado, M., Ito, M. and Li, Y.-T. (1982) Methods Enzymol. *83*, 619–625; 9. Nakazawa, K. and Suzuki, S. (1975) J. Biol. Chem. *250*, 912–917; 10. Suzuki, S. (1972) Methods Enzymol. *28*, 917–921; 11. Flaharty, A.L. and Edmond, J. (1978) Methods Enzymol. *50*, 537–547.

but not the latter enzyme. Endo-β-galactosidase hydrolyses keratan sulphate but not other glycosaminoglycans.

The cleavage of glycosaminoglycan chains to small fragments can be followed by observing the loss of high molecular weight material on gel filtration (Hart, 1976) or by the decrease in material precipitated with cetylpyridinium chloride (Hart, 1976; Hunter et al., 1983). Low molecular weight products can be characterised by paper chromatography or electrophoresis (Murata, 1980) and the position of sulphate substituents can easily be determined. Several of the enzymes acting on glycosaminoglycan substrates are lyases rather than hydrolases and cleavage of a glycosidic linkage results in the formation of an unsaturated oligosaccharide.

Two examples of procedures for the characterisation of glycosaminoglycans by the use of specific enzymic cleavage are given below. The general strategies involved in the use of these methods have been summarised in Section 6.9.

Procedure for estimating the proportions of glycosaminoglycans in a mixture (Hart, 1976, 1978) This method was developed to measure the relative amounts of biosynthetically labelled glycosaminoglycans synthesised by cultured embryonic corneas. It can be applied to small amounts of sample (up to a maximum of 2 mg each GAG) and differentiates between heparan sulphate (plus heparin), hyaluronate, chondroitin sulphates (plus dermatan sulphate) and keratan sulphate.

After labelling with [^{3}H]glucosamine and [^{35}S]sulphate (see Chapter 8) tissue (4.9–9.5 mg dry wt.) and culture medium are digested with Pronase to release glycosaminoglycans (Section 6.4.1). After precipitation with 5% trichloroacetic acid and centrifuging, the supernatant is extracted with chloroform:methanol (2:1 vol./vol.) and the aqueous phase is lyophilised. The sample dissolved in chromatography solvent (0.1 M sodium acetate containing 20% v/v ethanol), is applied to a column (1 × 200 cm) of Sephadex G-50 (fine) equilibrated with the same solvent. The glycosaminoglycan fraction eluted in the void volume of the column is counted, pooled and lyophilised.

Heparan sulphate is degraded selectively by treating the sample with nitrous acid as described in Section 6.9.4.5. The neutralised reaction mixture is lyophilised, redissolved in chromatography solvent and chromatographed on Sephadex G-50 and the undegraded glycosaminoglycans in the void volume are counted, pooled and lyophilised. The amount of heparan sulphate plus heparin in the original sample can be calculated from the loss of counts in the glycosaminoglycan peak.

Hyaluronate is degraded by dissolving the glycosaminoglycan fraction in 3 ml 0.02 M acetate buffer, pH 5.0, and digesting with *Streptomyces* hyaluronidase (3.3 turbidity reducing units/ml; Calbiochem-Behring) for 24 h at 37°C with addition of fresh enzyme after 12 h. Undegraded GAGs are again isolated by Sephadex G-50 chromatography.

Chondroitin sulphates (plus dermatan sulphate) are degraded by digesting the glycosaminoglycan fraction, dissolved in 3 ml 0.01 M Tris-HCl, pH 8.0, with chondroitinase ABC (0.133 units/ml; Miles) for 24 h at 25°C. The isolation of undegraded glycosaminoglycans is repeated.

Keratan sulphates are digested with keratan sulphate-endo-β-galactosidase (2.25 units/ml) in 0.05 M Tris-HCl, pH 7.2, for 48 h, followed by gel filtration of the product. The enzyme can be isolated from *Pseudomonas* sp IFO-13309 (Nakazawa and Suzuki, 1975) or alternatively the endogalactosidases from *E. freundii* or *F. keratolyticus* (Table 6.9) could be used with suitable modification of the digestion conditions.

The decrease in labelled GAG after each degradative step is used to determine the proportion of each type of labelled glycosaminoglycan species present. Sequential treatment of the sample minimises the amount of material required. Each of the steps should be checked to ensure that complete degradation of standard glycosaminoglycans occurs. This can be done by subjecting the standards (2 mg each) to the conditions of the digestion and adding cetylpyridinium chloride 1% w/v. The absence of turbidity following digestion indicates effective degradation has occurred. Precipitation with cetylpyridinium chloride in the presence of carrier glycosaminoglycan can be employed to follow the degradation of labelled glycosaminoglycans (see Hunter et al., 1983, for example).

This type of procedure can be modified to estimate other glycosaminoglycans of interest in a particular investigation. For example, if analysis of dermatan sulphate was required digestion with chondroitinase AC could be included prior to digestion with chondroitinase ABC. In addition the digestion products can be further examined.

Analysis of unsaturated disaccharides from chondroitin 4-sulphate, chondroitin 6-sulphate, dermatan sulphate and hyaluronate Separation and quantitation of the unsaturated disaccharides resulting from the digestion of glycosaminoglycans with chondroitinase ABC and chondroitinase AC provides a sensitive and specific method for characterisation and measurement of several of these carbohydrate polymers. Saito et al. (1968) developed procedures for estimating chondroitin 4-sulphate. Hyaluronate can also be measured by a recent modification of the method (Mason et al. 1982). It is possible to measure the chondroitin sulphates, dermatan sulphate and hyaluronate in the presence of other glycosaminoglycans. Examinations of the disaccharides resulting from chondroitinase digestion of sulphated glycosaminoglycans are also important in establishing the extent of sulphation and the positions of substitution with ester sulphate groups.

The reactions catalysed by chondroitinase ABC, chondroitinase AC, chondro-4-sulphatase and chondro-6-sulphatase are indicated in Table 6.9 and can be summarised as follows:

$$\text{Chondroitin 6-sulphate} \xrightarrow{\text{chondroitinase ABC}} n\Delta\text{Di-6S}$$

$$\text{Chondroitin 4-sulphate} \xrightarrow{\text{chondroitinase ABC}} n\Delta\text{Di-4S}$$

$$\text{Dermatan sulphate} \xrightarrow{\text{chondroitinase ABC}} n\Delta\text{Di-4S}$$

Hyaluronate $\xrightarrow{\text{chondroitinase ABC}}$ $n\Delta$Di-OS(H)

Chondroitin 6-sulphate $\xrightarrow{\text{chondroitinase AC}}$ $n\Delta$Di-6S

Chondroitin 4-sulphate $\xrightarrow{\text{chondroitinase AC}}$ $n\Delta$Di-4S

ΔDi-4-S $\xrightarrow{\text{chondro-4-sulphatase}}$ ΔDi-OS + SO_4^{2-}

ΔDi-6-S $\xrightarrow{\text{chondro-6-sulphatase}}$ ΔDi-OS + SO_4^{2-}

where *n* indicates the number of repeating units. Hyaluronate is cleaved only slowly by chondroitinase ABC compared to chondroitin and dermatan sulphates and special conditions are required to drive this reaction to completion.

The basis of the technique is to digest the glycosaminoglycan mixture exhaustively, separately with chondroitinase ABC and chondroitinase AC. The unsaturated disaccharides are then separated by paper or thin-layer chromatography and quantitated. From the known specificities of the enzymes it is possible to deduce the repeating structures present in the glycosaminoglycan mixture. The sulphated disaccharides can be further characterised by digestion with chondrosulphatases and by paper electrophoresis (Murata, 1980).

The procedure described below, due to Murata (1980), is a modification of the method of Saito et al. (1968). It is, however, still well worth reading the original paper by Saito et al. (1968), which discusses several enzymic assays of glycosaminoglycans, one of which can be used to determine only 3 μg of chondroitin or dermatan sulphate in a mixture. The smallest amount of glycosaminoglycan which can be analysed by characterisation of the unsaturated disaccharides released by chondroitinase digestion as described below is about 30 μg. The same methodology can be applied to radioactively labelled samples with the addition of unlabelled 'carriers'.

Procedure for analysis of chondroitin 4-sulphate, chondroitin 6-sulphate and dermatan sulphate This method requires chondroitinase ABC from *Proteus vulgaris*, chondroitinase AC from *Flavobac-*

terium heparinium, chondro-4-sulphatase and chondro-6-sulphatase from *Proteus vulgaris* (Suzuki, 1972a,b; Murata et al., 1968) which are commercially available. The specificities of the enzymes should be checked with authentic samples of chondroitin and dermatan sulphates or the sulphated unsaturated disaccharides derived from them.

Glycosaminoglycans are isolated by exhaustive proteolysis of proteoglycan (10 mg), followed by inactivation of the protease and purification of glycosaminoglycan chains (Section 6.4.1).

Three samples are prepared, one containing chondroitinase ABC, one chondroitinase AC and a control containing no enzyme. Each sample should contain approximately 500 μg glycosaminoglycan (sodium salt) in 50 μl water. Chondroitinase ABC, 1 unit dissolved in 20 μl of Tris-HCl buffer, pH 8.0, (prepared by dissolving 3 g Tris, 2.4 g sodium acetate, 1.46 g NaCl and 50 mg crystallised bovine serum albumin in 100 ml of 0.13 M HCl) is added to one tube and chondroitinase AC (1 unit) dissolved in 20 μl of Tris-HCl buffer, pH 7.0, (otherwise prepared as for pH 8.0 buffer) is added to another tube. Digestion is carried out at 37°C for 2 h and the enzymes are inactivated by heating in a boiling water bath for 10 min.

Samples of the digests, containing unsaturated disaccharides, are applied as bands (2.5 cm for a 500 μg sample) to Whatman No. 1 filter paper sheets 55 cm long. Standards of digested chondroitin 4-sulphate and chondroitin 6-sulphate are also applied. Desalting of the samples is carried out by descending chromatography in 1-butanol-ethanol-water 13:8:4 (v/v) for 24 h at room temperature. The paper is allowed to dry and then the disaccharides are separated by chromatography in 1-butanol-acetic acid-1 M ammonium hydroxide 2:3:1 (v/v) for 24 h or in *n*-butyric acid-0.5 M ammonium hydroxide 5:3 (v/v) for 72 h.

The separated unsaturated disaccharides are located by means of an ultraviolet lamp and are identified by comparison of their mobilities with standards. The distance from the origin for the disaccharides decreases in the order: unsaturated non-sulphated, 4-sulphated, 6-sulphated and disulphated disaccharides. Disaccharides are then eluted from the paper and quantitated. If the original glycosaminoglycan preparation was highly purified this estimation can be done by measuring ultraviolet absorbance due to the double bonds in the disaccharides. The regions of paper containing disaccharides and equivalent areas of paper to act as controls are excised, cut into small pieces and extracted in 2.0 ml of 0.01 M HCl in a sealed centrifuge tube at 50°C for 10 min. After centrifugation the absorbance of the supernatant solution is measured at 232 nm and the absorbance from the paper blank subtracted. Saito et al. (1968) reported the millimolar absorption coefficients of ΔDi-4S and ΔDi-6S to be 5.1 and 5.5 respectively when measured in 0.01 M HCl. It is thereforepossible to calculate the amounts of glycosaminoglycans present in the original sample in the following way:

$E_3 \times \frac{2}{5.1}$ = amount of chondroitin 4-sulphate as micromoles of repeating unit in the assay mixture.

$(E_1 - E_3) \times \frac{2}{5.1}$ = amount of dermatan sulphate as micromoles of repeating unit in the assay mixture.

$E_2 \times \frac{2}{5.5}$ = amount of chondroitin 6-sulphate as micromoles of repeating unit in the assay mixture.

when E_1, E_2 and E_3 are the absorbances for ΔDi-4S in the chondroitinase ABC digest, for ΔDi-6S in the chondroitinase ABC digest and for ΔDi-4S in the chondroitinase AC digest respectively. This assay is suitable for quantities of disaccharide in the range 0.05–0.2 μmol.

An alternative procedure which should be used if there is any evidence for the presence of UV-absorbing contaminants is to elute the disaccharides with water (1.0–1.5 ml) and assay them by the carbazole reaction (Section 5.6).

The identity of the ΔDi-4S and ΔDi-6S disaccharides can be confirmed by incubating them with chondro-4-sulphatase and chondro-6-sulphatase. Desulphation produces ΔDi-OS, which can be characterised by paper chromatography.

In glycosaminoglycan digests the presence of disulphated disaccharide is indicative of the occurrence of 'oversulphated' residues in the polymer. A spot corresponding to the unsulphated unsaturated disaccharide ΔDi-OS results from the digestion of chondroitin. However, in the paper chromatography system described above the disaccharide produced by the action of chondroitinases on hyaluronate (ΔDi-OSH) is not separated from ΔDi-OS. It is possible, however, to separate ΔDi-OSH from ΔDi-OS by cellulose thin-layer chromatography (Mason et al., 1982). Conditions for the complete digestion of hyaluronate by chondroitinase ABC with subsequent assay of the released ^{35}S-labelled disaccharide have been described (Mason et al., 1982). This methodology is, however, only suitable for the assay of labelled glycoconjugate mixtures in the presence of unlabelled carrier.

The assay of glycosaminoglycans by measuring disaccharides released by chondroitinase digestion is valuable because it gives a specific identification of the repeating units present. This method does not, however, take account of the heterogeneity of the carbohydrate chains of glycosaminoglycans. Thus it cannot be assumed that all disaccharide units of a particular type arise from homogeneous chains with that repeating unit throughout. However, if the proteoglycan sample is highly purified before enzymic digestion the disaccharides released can give information about the extent of structural heterogeneity of the polymer.

6.9.4.2. Partial acid hydrolysis Examination of the products of limited cleavage of glycosidic bonds by acid hydrolysis is an important general technique for analysing the structures of polysaccharides (Lindberg et al., 1975; McNeil et al., 1982). Information about the sequence of sugars in the carbohydrate units of glycoproteins (Marshall and Neuberger, 1972) and proteoglycans can be obtained by this method.

Limited acid hydrolysis of glycoproteins or glycopeptides has frequently been employed for the selective removal of non-reducing terminal residues of sialic acid or fucose. More vigorous hydrolysis produces a mixture of oligosaccharides arising from the cleavage of glycosidic bonds in different locations within the carbohydrate moiety of the glycopeptide. Separation and structural analysis of these oligosaccharides permit the identification of structural units present in the original polymer. When the carbohydrate polymer contains a linear repeating unit, as in the poly(glycosyl) glycopeptides of erythrocyte membranes or the carbohydrate units of proteoglycans, partial acid hydrolysis can be employed to identify the repeating structural unit in the carbohydrate chain. It is often possible to obtain further structural information by changing the major sites of hydrolytic cleavage of carbohydrate polymers by chemical modifications such as de-*N*-acetylation of amino sugars or carboxyl reduction of uronic acids.

The rates of hydrolysis of glycosidic linkages found in heterosaccharides depend on several variables, including the configurations and substituent groups of the monosaccharides involved and the positions and anomeric configurations of their linkages (BeMiller, 1967; Marshall and Neuberger, 1972). Rates of hydrolysis of particular glycosidic linkages in carbohydrate polymers cannot be precisely predicted from the behaviour of simple monosaccharide derivatives. In general, however, it has been found that glycosidic linkages involving sugars in the furanoside form and 2-deoxy sugars are rapidly hydrolysed while, in contrast, linkages involving hexuronic acids and 2-deoxy-2-aminohexoses (but not their *N*-acetylated derivatives) are particularly resistant to cleavage by acid (Section 5.3).

The conditions required for the selective release of sialic acid and fucose residues from glycoproteins or glycopeptides have been considered in Section 5.3 and 5.7. These residues can often be removed essentially completely without significant cleavage of other glycosidic linkages. This is a consequence of the lability of their glycosidic linkages and because they occur in non-reducing terminal positions. The positions of linkages between terminal residues of fucose or sialic acid and the penultimate sugar in a glycopeptide can be determined by methylation analysis before and after treatment with acid (Krusius and Finne, 1978).

Measurement of the rates of release of monosaccharides on hydrolysis of a glycoprotein has occasionally been employed to obtain preliminary evidence about the carbohydrate sequence (Spiro, 1966b). A more generally useful procedure is to isolate and characterise oligosaccharides released from glycopeptides by treatment with acid. For example, the presence of the disaccharides Gal(β1–4)GlcNAc and Gal(β1–3)GlcNAc was demonstrated in a hydrolysate of rat brain glycopeptides after hydrolysis in 0.1 M HCl at 100°C for 75 min (Krusius and Finne, 1978). *N*-Acetyllactosamine had previously been isolated, in a yield of 10% (based on galactose content), after hydrolysis of the glycoprotein fetuin in 0.025 M H_2SO_4 at 120°C for 2 h in a sealed tube (Spiro, 1962) and this disaccharide sequence is widespread in glycoproteins.

The uronic acid residues of proteoglycans slow the hydrolysis of glycosidic bonds involving these residues. Prior reduction of the carboxyl group of the uronic acid to an alcohol by treatment with $LiAlH_4$ or with $NaBH_4$ (or $NaBD_4$) in the presence of water-soluble carbodiimide (Section 5.6) can be employed to increase the susceptibility of glycosidic bonds involving the modified residue to acid hydrolysis. Glycosides of 2-deoxy-2-acetamido-hexoses are generally fairly readily cleaved by acid hydrolysis but the 2-deoxy-2-aminohexosidic linkage is resistant to hydrolysis. Deacetylation of *N*-acetylhexosamines in glycosaminoglycans, by treatment with hydrazine, can be employed to redirect hydrolysis away from the hexosaminidic linkage. Loss of the *N*-sulphate group in heparin similarly leads to resistance of the glycosidic linkages involving hexosamine to acid hydrolysis.

Reaction conditions for acid hydrolysis should be chosen to maximise the yield of the desired product and minimise side-reactions. Because of the reactions which can occur between amino acids and reducing sugars (Section 5.4) it is preferable that the amount of peptide material present should be as small as possible (i.e. glycopeptides are preferable to intact glycoproteins) and that the concentration of the sample should be low (1–2 mg/ml). A low sample concentration will also minimise the possibility of transglycosidation reactions.

Following partial acid hydrolysis the reaction mixture is neutralised, desalted and the resulting oligosaccharides are separated. Earlier methodology employed cation-exchangers for the selective absorption of products containing 2-amino-2-deoxyhexose residues and paper chromatography or adsorption chromatography on carbon columns to fractionate oligosaccharides. In more recent studies of the oligosaccharides produced by hydrolysis of glycoproteins the reduced, permethylated oligosaccharide derivatives have been fractionated by GLC. Selective detection of particular products (e.g. those containing *N*-acetylhexosaminitol) can be achieved by mass fragmentography at appropriate *m/e* values.

Powerful methodology for the structural analysis of polysaccharides by partial acid hydrolysis coupled with separation of oligosaccharides (as reduced permethylated derivatives) by HPLC has been described (McNeil et al., 1982). This type of approach could also find application in glycoprotein and proteoglycan carbohydrate units.

Procedure for partial acid hydrolysis. The selective release of sialic acid and fucose residues from glycopeptides is discussed in Sections 5.3 and 5.7 and the isolation of the GlcNAc-Asn linkage released by partial acid hydrolysis is described in Section 6.8.1. Earlier methodology for the graded acid hydrolysis of glycoproteins has been reported by Spiro (1962, 1966b). The following procedure was applied by Krusius and Finne (1978) to determine the mode of linkage between galactose and *N*-acetylglucosamine in rat brain glycopeptides.

Glycopeptides are hydrolysed in 2 ml of 0.1 M HCl for 75 min, neutralised with 1.0 M NaOH and reduced with 500 μl of 1 M $NaBH_4$ at room temperature for 3 h. After destruction of $NaBH_4$ by addition of glacial acetic acid the reduced oligosaccharides are desalted by passing through a column of Dowex-50 (0.8 cm × 4 cm) and dried on a rotary evaporator. Borate is removed by repeated evaporation with methanol/acetic acid (1000/1, v/v). Reduced oligosaccharides are methylated (Section 6.9.1) and separated by GLC on OV-225 at 245°C and on QF-1 at 225°C. If GLC-MS facilities are available differently linked hexosyl-hexosaminitols can be detected at *m/e* values of 218, 219, 276 and 378 (Mononen et al., 1977). The reduced disaccharides Gal(β1–4)GlcNAcol, Gal(β1–3)GlcNAcol and Gal(β1–6)GlcNAcol can be distinguished by this technique.

Further examples of the application of partial acid hydrolysis to the structural analysis of glycopeptides and glycosaminoglycans are given below.

Partial hydrolysis of the poly(glycosyl) chains of erythrocyte glycopeptides. The structure of the repeating unit of the poly(glycosyl) glycopeptides obtained by protease digestion of delipidated human erythrocyte ghosts was determined using acid hydrolysis of both the native and *N*-deacetylated glycopeptides (Krusius et al., 1978). Partial acid hydrolysis of the glycopeptides and analysis of the products was performed by the method described above. The sole hexosyl-hexosamine present in the hydrolysate was identified as Gal(β1–4)GlcNAc, which occurred in a yield corresponding to 11% of the galactose present in the poly(glycosyl) chains. To examine the linkage between hexosamine and other sugars the poly(glycosyl) glycopeptide was de-*N*-acetylated (by hydrazinolysis) and then subjected to acid hydrolysis in 0.5 M HCl at 100°C for 16 h followed by re-*N*-acetylation.

After reduction and permethylation gas-liquid chromatography (with total ion detection) revealed one major peak in the disaccharide region. The mass-spectrum of this peak was consistent with an *N*-acetylhexosaminyl (1–3 or 4) hexitol. From methylation data obtained from the original disaccharide which indicated that the only 3- or 4-substituted hexose present was 3-*O*-substituted galactose the disaccharide was identified as GlcNAc(1–3)Gal. Taken together with other evidence a repeating unit of –3Gal(β1–4)GlcNAc (β1–) with branch points at the C-6 of galactose residues was proposed (Krusius et al., 1978).

Partial hydrolysis of chondroitin sulphate. The structural analysis of disaccharides obtained by partial acid hydrolysis of glycosaminoglycans has provided information about the repeating units of these molecules. Davidson and Meyer (1954) hydrolysed the barium salt of chondroitin sulphate in 0.5 M H_2SO_4 for 4 h at 100°C. Following neutralisation with barium hydroxide and removal of barium sulphate the hydrolysate was applied to a column of Dowex-50 (H^+).

The ninhydrin-positive disaccharide 'chondrosamine' was eluted in 5 mM acetic acid and was found to be homogeneous by paper chromatography. Structural analysis indicated that the disaccharide, which was obtained in 67% yield, was GlcA(β1–3)GalN.

Hydrolysis of carboxyl-reduced and desulphated chondroitin for 2.25 h at 100°C in 0.5 M H_2SO_4 followed by re-*N*-acetylation led to the isolation of the disaccharide Glc(β1–3)GalNAc (Wolfrom and Juliano, 1960). If the desulphated carboxyl-reduced derivative of chondroitin sulphate is de-*N*-acetylated (with hydrazine) the hexosaminic linkage is resistant to acid hydrolysis and cleavage occurs preferentially at the reduced uronic acid linkage. This was demonstrated by Onodera et al. (1964), who isolated the disaccharide GalN(β1–4)Glc from a hydrolysate of desulphated, carboxyl reduced, deacetylated chondroitin sulphate. Together these studies establish the repeating structural unit of (desulphated) chondroitin as –4)GlcA(β1–3)GalNAc(β1.

6.9.4.3. Acetolysis Like partial acid hydrolysis, acetolysis provides a means of cleaving glycosidic linkages of glycoproteins or proteoglycans to produce oligosaccharides. Structural analysis of these oligosaccharides gives information about the sequence of the original polymer. As the types of glycosidic linkage cleaved preferentially by acetolysis are often different from those split most rapidly by acid hydrolysis the two methods are complementary. For example, 1–6 linkages between hexopyranosides are relatively stable to acid hydrolysis but they are preferentially cleaved by acetolysis. The reagent usually employed for acetolysis is an acetic anhydride-acetic acid-H_2SO_4 mixture. In acetolysis the attacking species is $[CH_3CO]^+$ (Guthrie and McCarthy, 1967) and the reaction yields peracetylated carbohydrate derivatives.

Glycosidic linkages involving sialic acid are much more stable to acetolysis than to acid hydrolysis. Oligosaccharides containing sialic acid have been obtained by acetolysis of oligosaccharides and from glycopeptides (Bayard and Montreuil, 1972). Cleavage of the carbohydrate units of glycopeptides isolated from the egg white glycoprotein, ovomucoid, by acetolysis yielded thirteen neutral oligosaccharides, twelve of which had mannose as the non-reducing terminal residue (Bayard et al., 1972). The *N*-glycosylamine linkage between *N*-acetylglucosamine and asparagine was found to be relatively stable to acetolysis (Bayard and Montreuil, 1972). No transglycosylation was detected on acetolysis of glycopeptides (Bayard and Montreuil, 1972). However, anomerisation may occur during acetolysis (Lindberg et al., 1975).

Separation of the products of acetolysis, after de-*O*-acetylation, can be carried out by chromatography on carbon-Celite and by paper chromatography (Bayard and Montreuil, 1972). Much smaller quantities of sample can be analysed by GLC-MS of the permethylated derivatives of the de-*O*-acetylated acetolysis products (Fournet et al., 1980). Procedures which have recently been developed for the separation of acetylated oligosaccharides by reversed-phase HPLC (Wells et al., 1982) could also be applied to the separation of acetolysis products.

Procedures for the acetolysis of glycopeptides or oligosaccharides. Detailed methodology for the acetolysis of glycopeptides and the fractionation of the oligosaccharides produced have been described by Bayard and Montreuil (1972). The procedure given here is based on work from the same group in which acetolysis products of glycopeptides were separated and characterised by GLC and mass spectrometry (Fournet et al., 1980). Dried glycopeptide (1 mg/ml) is cleaved by acetolysis for 3 days at 20°C in a mixture of acetic acid, acetic anhydride and concentrated sulphuric acid (10/10/1, v/v). The reaction mixture is poured into five volumes of ice water and sodium bicarbonate powder is added to bring the pH to 4.5. Peracetylated oligosaccharides are extracted into chloroform and the solvent is removed by rotary evaporation. The *O*-acetyl groups are removed by treatment at 4°C for 30 min with a mixture of 0.2 M NaOH and acetone (1/1, v/v). The oligosaccharides are desalted by passing them through a column of Dowex 50 (H^+). If the oligosaccharides are to be analysed by GLC-MS they are then reduced (with $NaBD_4$) and permethylated as described in Section 6.9.1. Suitable conditions for the GLC of acetolysis products are described by Chambers et al. (1973) and by Fournet et al. (1980). Interpretation of the electron impact and chemical ionisation mass spectra is discussed in detail by Fournet et al. (1980).

Applications of acetolysis to structure determination in glycoproteins. Bayard and Montreuil (1972) applied partial acetolysis to a mixture of glycopeptides obtained by proteolytic digestion of ovomucoid. The acetolysate after de-*O*-acetylation was fractionated by passing it through Dowex 50- × 2 (H^+) and Dowex 1- × 4 (formate). Neutral oligosaccharides were not absorbed. A 'basic fraction' of glycopeptide material bound to Dowex 50 and could be eluted with 5% aq. ammonia. The 'basic fraction' contained oligosaccharides which remained attached to amino acids and this fraction therefore contains the 'inner core' of the carbohydrate. Thus the neutral oligosaccharides obtained mainly arise from the exterior portion of carbohydrate units. Structural analysis of thirteen different neutral oligosaccharides indicated the variety of linkages present in carbohydrate units of ovomucoid. Similar information can be obtained more rapidly and with much smaller amounts of sample by GLC-MS (Fournet et al., 1980). With this technique the monosaccharide sequences in the oligosaccharide chains can be obtained unambiguously but the determination of glycosidic linkages becomes more difficult as the number of residues in the oligosaccharides increases.

The application of partial acetolysis to investigate the structure of the carbohydrate units of soybean agglutinin has been described by Lis and Sharon (1978). Acetolysis of a purified (but microheterogeneous) glycopeptide of composition $Man_9GlcNAc_2Asn$ yielded approximately equimolar amounts of Man(α1–2)Man, Man(α1–2)Man-(α1–2)Man and a glycopeptide approximating in composition to $Man_5GlcNAc_2Asn$. These findings taken together with methylation and other results were interpreted as indicating the preferential cleavage of a Man(α1–6)Man linkage releasing a manobiosyl or manotriosyl unit.

6.9.4.4. Periodate oxidation and Smith degradation The modification of vicinal diols by oxidation with periodate has long been established as a technique for the structural analysis of carbohydrates (Dyer, 1956). Periodate oxidation of glycopeptides or oligosaccharides released from glycoproteins or proteoglycans can yield information about the positions of glycosidic linkages between sugars, the location of substituents such as ester sulphate and the sequence of monosaccharide residues. Serial periodate oxidation, reduction and mild acid hydrolysis (Smith degradation) can be valuable in sequencing the branched oligosaccharide chains of glycoproteins. Periodate oxidation techniques are quite simple to apply but care in the interpretation of the results obtained is necessary because of the occurrence of side reactions.

Periodic acid reacts with polyols according to the general equation:

$$\begin{array}{l} CH_2OH \\ | \\ (CHOH)n + (n+1)HIO_4 \rightarrow 2HCHO + nHCOOH \\ | \\ CH_2OH \qquad\qquad\qquad\qquad\qquad\qquad + H_2O + (n+1)HIO_3 \end{array}$$

Secondary hydroxyl groups give rise to formic acid, while primary alcohols give formaldehyde. On treatment with periodic acid the anomeric carbon atom of glucose is oxidised to formic acid. When a single free hydroxyl group is adjacent to a glycosidic linkage, an *O*-acetyl, *O*-sulphate or an *N*-acetylamido group it is resistant to periodate. In contrast, the bond between a carbon atom substituted with a hydroxyl group and an adjacent carbon bearing a free amino group is cleaved by periodate. The reaction between periodate and a carbohydrate derivative can be monitored by quantitation of the periodate consumed and the formaldehyde and formic acid produced and inferences can be drawn about the number of primary and secondary hydroxyl groups in the molecule which was oxidised. This approach has been successfully applied to structural analysis of monosaccharide derivatives and simple oligosaccharides. When more complex oligosaccharides or glycopeptides are oxidised with periodate, difficulties in the interpretation of periodate consumption can arise. For these types of molecule a clearer picture can often be obtained by quantitation of the destruction of sugars by periodate or by use of the Smith degradation technique.

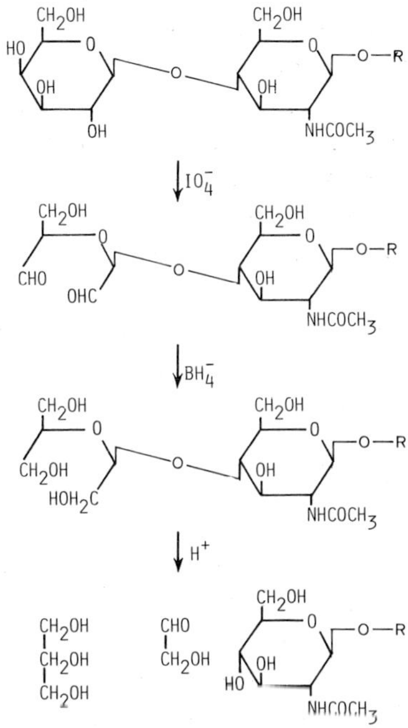

Fig. 6.8. Smith degradation of an oligosaccharide. Only the first cycle of the degradation of an oligosaccharide with non-reducing terminal Gal(β1–4)GlcNAc is shown.

The steps involved in the Smith degradation procedure are illustrated in Fig. 6.8. In this example periodate oxidation of a glycoside containing *N*-acetyllactosamine cleaves the ring of the terminal galactose residue with release of formic acid. The aldehyde groups of the product are then reduced to alcohols with sodium borohydride. Selective hydrolysis of the acetal resulting from ring cleavage can be performed under mildly acidic conditions without splitting glycosidic linkages (Goldstein et al., 1965). This sequence of reactions results in the destruction of galactose and the replacement of the glycosidic linkage at position 4 of *N*-acetylglucosamine with a hydroxyl group. Thus the position of linkage between galactose and *N*-acetylglucosamine could be established by methylation analysis before and after Smith degradation. As the *N*-acetylglucosamine residue of the oligosaccharide or glycopeptide produced by Smith degradation has a vicinal diol the periodate oxidation, reduction and hydrolysis procedure can be repeated. If the carbohydrate structure is suitable (i.e. only sugars in the non-reducing terminal positions are destroyed by periodate) this process of sequential removal of sugars may be continued for several cycles. It must be borne in mind, however, that this procedure relies on the specificity of the periodate oxidation and the selectivity of the hydrolysis step. Structural information can also be obtained by examination of the low molecular weight borohydride-reduced fragments which are released by hydrolysis during each round of the degradation.

The rates of reaction with periodate of different diol groups in glycoconjugates vary considerably. Steric factors are clearly of importance, *cis*-diols generally being cleaved considerably more rapidly than *trans*-diols. Oxidation of acyclic glycols is very fast compared to that of reducing sugars. Charge effects and hydrogen bonding of uronic acid residues influence the rates of oxidation of glycosaminoglycans. In addition to the reaction of periodate with vicinal diols discussed above other reactions termed 'overoxidation' or 'non-Malapradian' oxidation occur. It is important that these side-reactions should be minimised by an appropriate choice of reaction conditions. These reactions cannot, however, be entirely eliminated and they may become significant when the rate of diol oxidation is low.

Periodate can react with certain free amino acids and with polypeptides (see Marshall and Neuberger, 1972b, for review). For some purposes periodate oxidation can usefully be applied to intact glycoproteins as in the periodic acid-Schiff staining procedure (Section 4) and the use of Smith degradation to prepare glycoprotein derivatives to act as acceptors in glycosyltransferase assays. However, for structural analysis it is preferable that the amount of peptide present in the sample should be low. Either glycopeptides or oligosaccharides released by β-elimination, hydrazinolysis, endoglycosidases or glycosylamino-hydrolase are suitable for periodate oxidation. Oligosaccharides are usually reduced with borohydride before treatment with periodate. The polyol produced from the reducing terminal monosaccharide residue is likely to have free vicinal diol groups and will, in that case, be cleaved by periodate. When this terminal residue is an amino sugar identification of the amino sugar alcohol derivative resulting after periodate oxidation, borohydride reduction and acid hydrolysis can give information defining the position of the linkage between this residue and the penultimate residue from the reducing end (Spiro, 1972).

The conditions employed for the periodate oxidation of glycopeptides and oligosaccharides isolated from glycoproteins have been reviewed in detail by Marshall and Neuberger (1972b). To minimise side-reactions concentrations of periodate below 0.1 M should be employed. An excess of periodate is essential, the amount of periodate added being at least 2.5–5 times the quantity consumed in the reaction. For the complete oxidation of low concentrations of oligosaccharides or glycopeptides it may be advantageous to employ a considerably greater excess of periodate in the range 10–100 moles of periodate per mole of monosaccharide residue (Yoshima et al., 1981; Krusius and Finne, 1981). Specificity of oxidation is enhanced at low temperature (0–4°C) and by the use of a slightly acid pH (usually pH 4.5). Metaperiodate decomposes giving rise to ozone in the presence of light but solutions of the reagent are stable in the dark.

The reaction of glycosaminoglycans with periodate can be influenced by the presence of uronic acid residues and sulphate groups. Overoxidation associated with the carboxyl group of hexuronic acids and some loss of sulphate can occur as side-reactions (Kennedy, 1979). The rates of reaction of glycosaminoglycans, including hyaluronate, chondroitin sulphate and heparin, with periodate are low when compared with an uncharged polysaccharide such as dextran. This low rate of reaction is attributable, in part, to exclusion of negatively charged periodate ions from the polysaccharide domain by electrostatic repulsion as the reaction rate can be accelerated significantly by increasing the ionic strength (e.g. by addition of 0.2 M sodium perchlorate) or by lowering the pH (Scott and Harbinson, 1968). Even in the presence of perchlorate ions, however, the oxidation of chondroitin sulphate, hyaluronate and heparin requires more vigorous conditions (e.g. temperatures of 20–37°C) than does the oxidation of dermatan sulphate (Scott, 1968). Reaction conditions can be chosen so that the iduronic acid residues of dermatan sulphate are oxidised selectively, leaving the glucuronic acid residues which are also present in this polymer largely unmodified (Fransson, 1974). The lower reactivity of glucuronic acid residues than iduronic acid may result from the formation of hydrogen bonds between the carboxyl group of glucuronic acid and the hydroxyl group at position three in this residue (Scott, 1968).

Periodate oxidation procedures. A typical procedure for following the periodate oxidation of a glycopeptide or oligosaccharide derived from a glycoprotein, modified from Spiro (1966b), is given here. Examples of methodology suitable for small quantities of radioactively labelled oligosaccharides or to particular glycosaminoglycans are outlined later in this section.

The sample to be oxidised (2–5 mg/ml) in 0.05 M sodium acetate buffer, pH 4.5, containing 0.05 M sodium metaperiodate is incubated at 4°C in the dark. Aliquots of the reaction mixture are removed at intervals up to 72 h and are analysed to determine the sugars remaining unaffected by periodate. The time course of the reaction can be followed by measuring periodate consumption. If an appropriate excess of periodate is not present at the end of the reaction conditions must be modified. A rapid and sensitive colourimetric method for determination of periodate has been described by Avigard (1969). Alternatively, periodate consumption can be monitored by measuring the decrease in absorbance at 222.5 nm (Aspinall and Ferrier, 1957). The effects of light on periodate and the possibility of side-reactions, particularly of glycopeptides, giving rise to UV-absorbing products, give cause for reservations about the latter procedure. Formaldehyde can be determined by the chromotropic acid reagent after reduction of periodate with arsenite (Dyer, 1956).

To measure the destruction of sugars in samples taken from the reaction mixture periodate is first destroyed by reaction with an excess of ethylene glycol, the sample is desalted by gel filtration, hydrolysed or cleaved by methanolysis, and the constituent sugars are analysed by one of the methods described in Chapter 5. The destruction of sugars by periodate oxidation of a glycopeptide can be interpreted on the basis of the following considerations (Spiro, 1966). Any non-reducing terminal monosaccharide will be destroyed. Glycosidically linked hexoses in the pyranoside form which are monosubstituted will be destroyed unless the substituent is at position 3. Di-substituted hexoses will be stable if substituted at positions 2 and 4 or position 3 and any other. Glycosides of monosubstituted *N*-acetylamino sugars will be stable unless they are substituted at position six. The possible presence of other substituents which will block oxidation such as *O*-sulphate should be borne in mind. Substitution of sialic acid residues with an *O*-acetyl group at position 8 blocks oxidation with periodate. Formaldehyde is released when sialic acid which is not blocked in position 8 or 9 is oxidised.

Smith degradation procedure. The sample is oxidised with periodate as described previously to produce complete destruction of susceptible sugars. Excess periodate is destroyed by addition of ethylene glycol and allowing the reaction to proceed for 1 h at room temperature in the dark. A large excess of sodium borohydride, freshly dissolved in 0.1 M sodium borate buffer, pH 8.0, is added and left to react for 5 h. Excess borohydride is destroyed by acidifying the solution to pH 5 with acetic acid and the reduced glycopeptide is isolated by gel filtration on Sephadex G-25 in 0.05 M pyridine acetate buffer, pH 5.0. The glycopeptide is concentrated and the buffer removed by lyophilising twice. Mild acid hydrolysis can be carried out in 0.025 M H_2SO_4 at 80°C for 1 h (Spiro, 1966) and the reaction is stopped by neutralising with 1 M NaOH. More complete cleavage of internal mannose residues in certain glycopeptides can be achieved with 0.1 M HCl at 80°C for 1 h (Krusius and Finne, 1981). Further rounds of Smith degradation can be carried out on the products.

The sugar composition of the desalted borohydride-treated sample can be examined after hydrolysis or methanolysis. Polyols produced can be separated, identified and quantitated by GLC (Clamp, 1974).

Examples of the application of periodate oxidation techniques

Carbohydrate structures of Sendai virus glycoproteins. Yoshima et al. (1981) have applied Smith degradation to very small quantities of labelled oligosaccharides. These carbohydrate units were released from *N*-glycosidic linkage by hydrazinolysis and were subsequently re-*N*-acetylated and labelled by reduction with NaB^3H_4. Oligosaccharides (5×10^4cpm, 3 nmol) dissolved in 100 μl of 0.04 M acetate, pH 4.6, containing 0.6 μmol $NaIO_4$ were incubated at 4°C for 40 h in the dark. Ethylene glycol, 10 μl, was added and allowed to react 1 h at room temperature. Aldehyde residues were reduced with $NaBH_4$ (5 mg) in 0.1 M sodium borate (pH 8.0) for 5 h. The solution was acidified with acetic acid and passed through a column (3 × 0.5 cm) of Bio-Rad AG-50 (H^+ form). The eluate was evaporated to dryness and the residue freed from borate by repeated evaporation with methanol. Hydrolysis was carried out in 0.3 ml of 0.025 M H_2SO_4 at 80°C for 1 h. Oxidised oligosaccharides were examined by descending paper chromatography in ethyl acetate-pyridine-acetic acid-water (5:5:1:3) and were characterised by methylation analysis. A second cycle of periodate oxidation, reduction and hydrolysis was also carried out. Oligosaccharide HN-1 had been shown by exoglycosidase digestion and methylation analysis to have one or other of the following structures:

```
Manα1–6Manα1–6Manβ1–4GlcNAcβ1–4GlcNAcol      (1)
            3       3
            |       |
        Manα1   Manα1
```

or

```
Manα1–6Manα1–3Manβ1–4GlcNAcβ1–4GlcNAcol      (2)
            3       6
            |       |
        Manα1   Manα1
```

Smith degradation resulted in:

$$\text{Man}\alpha 1\text{–}6\text{Man}\beta 1\text{–}4\text{GlcNAc}\beta 1\text{–}4\text{XylNAcol} \qquad (3)$$

A second Smith degradation gave:

$$\text{GlcNAc}\beta 1\text{–}4\text{XylNAcol} \qquad (4)$$

These findings indicated that the correct structure was (1). The formation of *N*-acetylxylosaminitol in structures (3) and (4) indicates that the original *N*-terminal residue of the glycopeptide, reduced to *N*-acetylglucosaminitol in (1), was in 1–4 linkage with the penultimate residue. Identification of the various aminoalditols which could be produced from a reducing terminal *N*-acetylhexosamine involved in different types of linkages has been discussed in detail by Spiro (1972).

Application of Smith degradation coupled with methylation analysis to glycopeptides in a study of the branching pattern of the complex-type carbohydrate units of glycoproteins has been described by Krusius and Finne (1981).

Periodate oxidation of heparin. Heparin contains both glucuronic acid and iduronic acid residues. The relative stabilities of these residues towards periodate oxidation was found by Axelsson and Lindahl (1971) to be influenced by their substitution with *O*-sulphate groups.

Purified heparin (200 mg) in water (25 ml) was mixed with an equal volume of 0.1 M sodium metaperiodate and left in the dark for 96 h at room temp. Ethylene glycol (0.25 ml) was added and allowed to react for 1 h. Aldehydes were reduced with KBH_4 (400 mg) for 14 h, the solution was acidified, concentrated, and desalted on Sephadex G-25.

It was found that periodate destroyed about 30% of the total uronic acid of heparin and almost all of the glucuronic acid. The results indicated that the glucuronic acid was essentially non-sulphated, unlike the iduronic acid, which was found to be the major sulphated uronic acid of heparin and consequently did not have an available vic-diol group. In these experiments oxidation was carried out at room temperature; the glucuronic acid of glycosaminoglycans is cleaved only very slowly at 4°C (Fransson, 1974).

Selective cleavage of iduronic acid residues in dermatan sulphate by Smith degradation. Iduronic acid residues in glycosaminoglycans can be oxidised by periodate in conditions under which reaction of glucuronic acid residues is negligible (Fransson, 1974). Dermatan sulphate can be cleaved selectively at its iduronic acid residues by Smith degradation. Glucuronic acid residues remain intact (Fransson and Carlstedt, 1974).

The dermatan sulphate (2 mg/ml) was oxidised in 50 mM sodium periodate in 50 mM sodium citrate (pH 3.0) at 4°C for 24 h in the dark. After destruction of excess periodate by reaction with mannitol the sample was dialysed. The oxidised polysaccharide (100 mg) dissolved in water (10 ml) was reduced with 200 mg KBH_4 for 3 h at room temperature. After dialysis and lyophilisation the polysaccharide was dissolved in 25 ml of 5 mM H_2SO_4 and heated at 60°C for 3 h. The reaction was terminated by neutralising with NaOH and the products were fractionated on Sephadex G-50 in 0.2 M pyridine acetate buffer, pH 5.0.

The oligosaccharides produced by this procedure were consistent with dermatan sulphate being a copolymer containing sequences of D-GlcA-GalNAc-4-OSO_3^-. The polyol, L-threonic acid, resulting from cleavage of iduronic acid was identified by GLC. Periodate oxidation at low temperature and relatively low pH enhances the oxidation of iduronic acid residues relative to glucuronic acid.

6.9.4.5. Deamination with nitrous acid The carbohydrate units of proteoglycans and glycoproteins can be cleaved selectively by nitrous acid at glycosidic linkages involving hexosamine residues in which the amino groups are either *N*-sulphated or unsubstituted. During the reaction residues of D-glucosamine are converted to 2,5-anhydro-D-mannose (Fig. 6.9). This type of cleavage is of great importance in the characterisation of heparin and heparan sulphates, both of which contain *N*-sulphate residues. Nitrous acid has also been used to degrade the oligosaccharide chains of asparagine-linked carbohydrate units of glycoproteins after removal of *N*-acetyl substituents from glucosamine by hydrazinolysis.

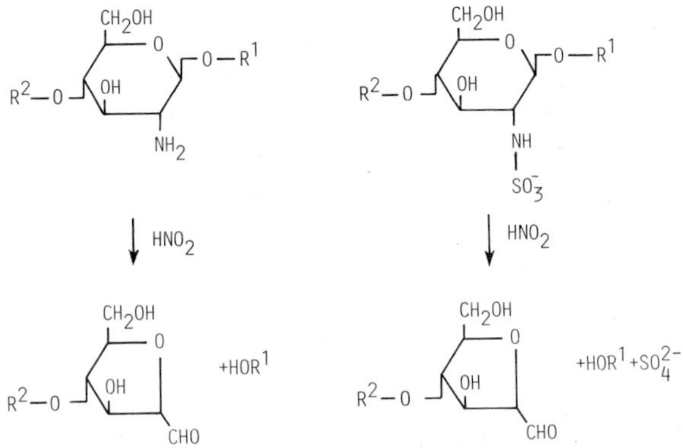

Fig. 6.9. Deamination of glucosamine and glucosamine *N*-sulphate residues with nitrous acid.

Glycosaminoglycans. The depolymerisation of *N*-sulphated glycosaminoglycans with nitrous acid has been extensively investigated (Cifonelli, 1968; Cifonelli and King, 1972; Shively and Conrad, 1976). The reaction is pH-dependent; only low pH (<2.5) conditions favour the direct deamination of *N*-sulphated glucosamine residues without prior desulphation. In contrast, *N*-unsubstituted amino sugars are deaminated more rapidly at pH 4 than at pH 2 or pH 6. Solutions of nitrous acid are unstable at room temperature, losing about 1/4 to 1/3 of their capacity to deaminate amino sugars in 1 h. The reagent must therefore be prepared immediately prior to use. Two alternative low-pH procedures for cleavage of heparin and heparan sulphate which use different means of generating nitrous acid are described below.

In the procedure of Shively and Conrad (1976), *N*-sulphated glycosaminoglycans (or their carboxyl-reduced derivatives) dissolved in water (50 μl) are mixed with 200 μl of a nitrous acid solution prepared as follows: 1 ml of 0.5 M H_2SO_4 at $-5°C$ is thoroughly mixed with a solution containing 114 mg of barium nitrite (Fluka AG) in 1 ml water at $-5°C$. Precipitated $BaSO_4$ is removed by centrifugation for 2 min at 1000 rpm and the supernatant solution of nitrous acid is drawn off with a Pasteur pipette. The reagent has a pH of 1.5 and is stable for several hours when stored at $-5°C$. The nitrous acid and sulphated glycosaminoglycan solutions are mixed and allowed to warm to room temperature. Cleavage of carboxyl-reduced heparin under these conditions is complete in 10 min. The reaction mixture is neutralised and the products can be examined by gel filtration or paper chromatography.

An alternative method of forming nitrous acid is to employ butyl nitrite (Cifonelli and King, 1972; Hart, 1976). In this procedure the sample of heparin or heparan sulphate in 2 ml water is mixed with 1 ml 1 M HCl and 1 ml freshly prepared 10% *n*-butyl nitrite (v/v) (Eastman-Kodak) in absolute ethanol and the reaction is allowed to proceed for 2 h at room temperature. Selective and complete degradation of *N*-sulphated glycosaminoglycans has been reported. The reaction is stopped by neutralisation and the solution lyophilised. This cleavage procedure has been used as part of a general method for the quantitative estimation of different classes of proteoglycans (Hart, 1976, 1978; Section 6.9.4.1).

Another nitrous acid procedure which has been used widely is that of Cifonelli (1968).

The reaction products can be examined by gel filtration to determine their size. Hart (1976, 1978) measured the amount of labelled heparin plus heparan sulphate present in a mixture of labelled glycosaminoglycans by the decrease in radioactivity excluded from Sephadex G-50 after nitrous acid treatment. Examining the size of the oligosaccharides released by gel filtration on Bio-Gel P-10 can give information about the frequency of *N*-acetyl and *N*-sulphate groups in the glycosaminoglycan chain (Yanagishita and Hascall, 1983). Colourimetric analysis of dehydromannose can be done by the method of Dische and Borenfreund (1950); this procedure is described in Section 6.10.

Glycoproteins. Specific cleavage at glucosamine residues by treatment with nitrous acid after de-*N*-acetylation has found a place in the structural analysis of asparagine-linked oligosaccharide units of glycoproteins since its introduction by Isemura and Schmid (1971). Oligosaccharide units free of *N*-acetyl groups can be obtained by hydrazinolysis of glycoproteins (Section 6.5.2). This procedure cannot be applied to *O*-linked carbohydrate units which are often degraded during hydrazine treatment. The fragments obtained by deamination of de-*N*-acetylated oligosaccharides can be analysed by paper chromatography (Bayard and Fournet, 1976) or by GLC-MS (Krusius and Finne, 1978; Strecker et al., 1981). Deamination has been of particular value in studying the linkages of monosaccharides or oligosaccharides attached to glucosamine residues. For example, the mannose residues in the 'core' of *N*-linked oligosaccharides (Bayard and Fournet, 1976) and the attachment of fucose residues to glucosamine either in the core chitobiose unit or fucose attached to glucosamines in antennary positions (Strecker et al., 1981) have been studied.

The following procedure can be used to characterise oligosaccharides resulting from hydrazinolysis-nitrous acid deamination of *N*-linked glycopeptides (Strecker et al., 1981). Glycopeptides (0.2–1 mg) are dissolved in anhydrous hydrazine (Pierce Chemical Co.) and heated at 105°C for 24 h and then dried under nitrogen. Traces of hydrazine are removed in a vacuum desiccator containing H_2SO_4. The dried residue is dissolved in 1 ml of 0.5 M acetic acid and 5 mg of $NaNO_2$ is added. After standing at room temperature excess HNO_2 is destroyed by addition of ethylamine, the solutions are degassed under nitrogen and then adjusted to pH 9 with 1 M NaOH. The oligosaccharides containing 2,5-anhydromannose are reduced for 2 h with 5 mg of sodium borodeuteride or sodium borohydride and then neutralised with Dowex 50 × 8 (25–50 mesh; H^+ form) and boric acid is removed by repeated evaporation with methanol. The reaction mixture is lyophilised, dried over P_2O_5 and methylated (Section 6.9.1).

Reduced oligosaccharides are separated by gas chromatography using packed columns of 3% SE-30 on Chromosorb W-HMDS (3 m × 0.2 cm i.d.; temperature-programmed from 140 to 310°C at 4°/min) and capillary columns of OV-101 (60 m × 0.4 mm; 140 to 320°C, 5°/min) and are detected by total ionisation current and by mass fragmentography. Mass spectra of oligosaccharides were obtained (Strecker et al., 1981) and could be interpreted to identify the positions of linkage of sugar residues to the reduced derivative of 2,5-anhydromannose.

6.9.4.6. Oxidation with CrO_3 The anomeric configuration of glycosidic linkages in oligosaccharides or glycopeptides can be determined by NMR spectroscopy (Section 6.9.3), by digestion with glycosidases or by oxidation with chromium trioxide. This latter method depends on the difference in stability towards oxidation by CrO_3 in acetic acid of fully *O*-acetylated derivatives of α- and β-linked glycosides. It is a straightforward procedure which has been applied to several glycoproteins (Krusius and Finne, 1978).

The method is based on the finding that the pyranoside form of pentose and hexose residues in which the aglycone occupies an equatorial position in the most stable chair form (i.e. most β-linkages) are

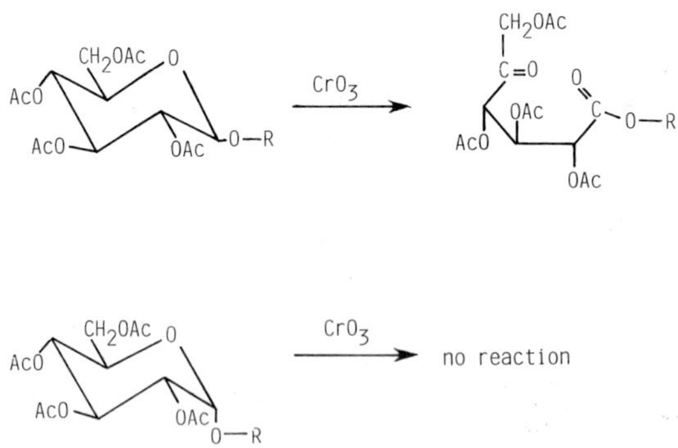

Fig. 6.10. Use of CrO_3 oxidation in acetic acid to distinguish between α and β glycosides.

readily oxidised by CrO_3 in acetic acid. In contrast, corresponding residues in which the aglycone occupies an axial position (most α-linkages) are more resistant to oxidation. The product arising from oxidation of a β-linked hexopyranoside is the corresponding 5-hexulosonic acid (Fig. 6.10). Thus the reaction is accompanied by destruction of the β-linked sugar (Hoffman et al., 1972).

Before carrying out the reaction oligosaccharides with free reducing groups should be reduced to the corresponding alditol. The oligosaccharide or glycopeptide is then fully peracetylated.

Procedure (Krusius and Finne, 1978). Glycopeptides were mixed with a known, approximately equimolar, amount of inositol as internal standard and taken to dryness. Samples were acetylated with 0.2 ml of pyridine:acetic anhydride (1:1, v/v) at 100°C for 1 h. Toluene (1 ml) was added and the mixture taken to dryness under reduced pressure. Acetylated glycopeptide, in 0.1 ml glacial acetic acid, and powdered CrO_3 (10 mg) were incubated at 50°C in an ultrasonic bath for 2 h. The mixture was diluted with water (2 ml) and extracted three times with 2-ml portions of chloroform. The combined chloroform layers were evaporated to dryness and analysed for sugars after hydrolysis or methanolysis (Chapter 5). Control samples were treated similarly without addition of CrO_3. Recoveries of individual sugars were calculated relative to inositol which, when fully acetylated, is resistant to oxidation by CrO_3.

6.10. *Location of substituents*

The monosaccharide residues of glycoproteins may be substituted with a variety of groups which include acetyl, glycolyl, sulphate and phosphate. Proteoglycans are extensively substituted with ester sulphate and both heparin and heparan sulphate are *N*-sulphated. In this section methods for the determination of the amounts of *O*- and *N*-linked sulphate and the location of these substituents will be considered. Determination of the acetyl content of glycoproteins (Section 5.11) and the characterisation of the various acetyl and glycolyl derivatives of sialic acid (Section 5.7) are described elsewhere.

Total sulphate determination The sample of glycosaminoglycan or glycoprotein is hydrolysed with acid and released sulphate is determined turbidometrically as barium sulphate. Samples (1–3 mg) of sulphated proteoglycan or glycoprotein are hydrolysed in 1 ml of 1 M HCl at 100°C for 6 h. After cooling the hydrolysates are taken to dryness on a rotary evaporator and the residue is redissolved in 1 ml of water. Aliquots are analysed for sulphate as described in Section 5.11.

Hydrolysis under these conditions releases both acid-labile *N*-sulphate and ester sulphate constituents.

O-Sulphate Several different procedures are available for determination of the site of attachment of *O*-sulphate groups (Rodén et al., 1972).

Infra-red spectra of sulphated polysaccharides have a strong band at 1240 cm^{-1} (S–O bond stretching), which is characteristic for all sulphate esters, and also show absorbance in the 800–850 cm^{-1} region (C–O–S vibration) which can be diagnostic for the site of attachment of sulphate. A peak at 810–820 cm^{-1} is characteristic of a 6-sulphate (with the hexose in the $^4C_1$ conformation), a peak at about 830 cm^{-1} can be attributed to a sulphated group occupying a secondary equatorial position (e.g. glucose 4-sulphate) and at 850–860 cm^{-1} to a sulphate in an axial, secondary position (e.g. galactose 4-sulphate; Turvey et al., 1965). Spectra can be obtained by preparing samples (e.g. 1–2 mg of sulphated glycosaminoglycan) as a KBr disc (Wood, 1957).

Another spectroscopic approach which defines the position of sulphation precisely is NMR. This approach is described in Section 6.9.3.

For the location of ester sulphate substituents in the chondroitin sulphates and dermatan sulphate a simple and sensitive method is to examine the sulphated unsaturated disaccharides resulting from digestion with chondroitinase ABC. The procedure is described in Section 6.9.4.1. The disaccharides can be further characterised by digestion with chondro-4-sulphatase or chondro-6-sulphatase.

Another method, which has occasionally been used to investigate the location of ester sulphate groups in sulphated polysaccharides, is based on the relative rates of hydrolysis of ester sulphate groups linked to equatorial and axial secondary hydroxyl groups and primary hydroxyl groups (Rees, 1963). Because of difficulties in calculating half-lives of sulphate esters when several sulphate groups are present in the sample this approach has been largely superseded by other methods.

N-Sulphate The presence of *N*-sulphate in glycosaminoglycans is indicated by the release of sulphate by mild acid hydrolysis or by solvolysis with aqueous dimethyl sulphoxide and by the formation of anhydromannose on treatment of the glycosaminoglycan with nitrous acid.

Hydrolysis of heparin in 0.04 M HCl at 100°C for 90 minutes produces complete *N*-desulphation (Durant et al., 1962). The release of inorganic sulphate from a carbohydrate polymer under these conditions suggests the presence of *N*-sulphate. However, this is not definitive evidence as some release of ester sulphate also occurs under these conditions (Nagasawa and Inoue, 1974). Furthermore the presence of partially hydrolysed glycosaminoglycans can interfere with the assay for inorganic sulphate. When sulphate estimations are carried out by the barium sulphate turbidometric procedure prior precipitation of glycosaminoglycans with cetylpyridinium chloride is advisable (Saito et al., 1968).

An alternative procedure for the selective release of sulphate from hexosamine *N*-sulphate has been proposed by Nagasawa and Inoue (1974). The pyridinium salt of the sulphated polymer is treated with 95% dimethyl sulphoxide/5% water (v/v) at 50°C for 2 h and free sulphate is measured. Little release of *O*-sulphate occurs under these conditions. The pyridinium salts of glycosaminoglycans can be obtained by gel filtration on Sephadex G-50 (8 ml bed volume for a

0.5-ml sample) equilibrated and eluted with 1 M pyridine acetate, pH 5.0, followed by lyophilising the excluded peak (Yanagishita and Hascall, 1983b).

Treatment of polysaccharides containing *N*-sulphated glucosamine with nitrous acid results in the formation of 2,5-anhydro-D-mannose residues with concurrent cleavage of adjacent glycosidic linkages (Section 6.9.4.5). The oligosaccharides containing reducing terminal 2,5-anhydromannose can be estimated colourimetrically (Dische and Borenfreund, 1950) or by chromatography after reduction with NaB^3H_4 (Shively and Conrad, 1976).

Several procedures have been employed for nitrous acid deamination of *N*-sulphated glycosaminoglycans. The method of Lagunoff and Warren (1962) involves treatment of the glycosaminoglycans with sodium nitrite in acetic acid at room temperature for 90 min. Under these conditions anhydromannose is formed from hexosamines with a free amino group (including D-glucosamine) as well as from *N*-sulphated hexosamine. In addition the conversion of *N*-sulphated hexosamine in heparin to anhydromannose is far from complete (Shively and Conrad, 1976). The procedure of Cifonelli (1968) is specific for *N*-sulphated hexosamine but requires substantial quantities of sample and is time-consuming. The method described below, due to Cifonelli and King (1972), is much simpler to carry out.

The solution of glycosaminoglycan (2–10 μg hexosamine) in 0.2 ml is mixed with freshly prepared 5% butyl nitrite in ethanol (0.2 ml) and 2 M HCl (0.2 ml). (Butyl nitrite (Distillation Products) is stored at −20°C). After incubation of the reaction mixture at room temperature for 1 h, ammonium sulphamate (12.5%, 0.2 ml) is added and the mixture shaken occasionally for 1 h. Following the addition of 1 ml 5% v/v HCl and 0.2 ml of 0.5% w/v indole in ethanol the tubes are covered and heated for 5 min in a boiling water bath. The tubes are cooled and ethanol (1.2-ml) is added and the colour measured at 492 nm against the appropriate blank. Standard curves should be prepared with fully *N*-sulphated heparin.

6.11. Deglycosylation of glycoproteins and proteoglycans

The removal of all or part of the carbohydrate portion of glycoproteins or proteoglycans can be advantageous for several purposes, including protein sequence determination, studying the function of carbohydrate units or preparing substrates for glycosyltransferases. Several methods used for deglycosylation are listed in Table 6.10.

TABLE 6.10
Methods for producing deglycosylated peptide chains

		Section
Anhydrous HF		
0°C, 1 h	Partial removal of Ser/Thr- and Asn-linked units and cleavage of Gal-Hpr link	6.11
23°C, 3 h	Removal of Ser/Thr-linked units and all but linkage GlcNAc of Asn-linked units	6.11
Trifluoromethane sulphonic acid	Removal of most Ser/Thr-linked units and all but linkage GlcNAc of Asn-linked units, Xyl-Ser linkage cleaved	6.11
HF pyridine	Xyl-Ser linkage cleaved	6.11
Smith degradation	Removal of peripheral residues (occasionally complete removal possible for *O*-linked units), peptide-chain damage likely	6.9.4.4
Exoglycosidases	Removal of peripheral residues	6.4
Endoglycosidases	Removal of carbohydrate units, all but linkage Asn of GlcNAc-Asn-linked units	6.6
Tunicamycin	Prevents formation of Glc-Asn-linked units	8.2.8

Amino acid sequence studies have been carried out on several glycoproteins which have been treated with anhydrous HF to remove most of their carbohydrate moieties (Mort and Lamport, 1977). This procedure requires special apparatus to deal with the highly toxic and corrosive HF. Suitable Kel-F coated vacuum equipment (Peninsular Laboratories Inc., Santa Carlos, California, U.S.A.) may be available in laboratories carrying out chemical synthesis of peptides. The sample (up to 100 mg) is thoroughly dried and then treated with 1 ml of anisole. HF (10 ml) is added by distillation and the reaction mixture is incubated for 3 h at 25°C or 1 h at 0°C. Anisole is added as a scavenger to remove fluorinated compounds such as glucosylfluoride which are generated in the reaction and which readily alkylate aromatic compounds. The reaction is terminated by removal of HF in vacuo and the deglycosylated glycoprotein is purified by gel filtration. Under the more severe conditions (25°C) cleavage of *O*-glycosidic linkages approaches completion but peptide bonds and the *N*-glycosylamine linkage between Asn and GlcNAc remain intact. Treatment at 0°C for 1 h results in removal of most of the hexose and peripheral sugars from *N*-linked units but much of the 'core' *N*-acetylglucosamine is retained. The protein-linked galactosamine of *O*-linked carbohydrate units is largely resistant to HF under these conditions but most sugars are cleaved from plant glycoproteins with the galactose-hydroxyproline linkage (Mort and Lamport, 1977).

An alternative procedure for deglycosylation using trifluoromethane sulphonic acid which does not require the specialised apparatus needed for safe handling of HF has been described by Edge et al. (1981). Anisole (1 ml) and 2 ml trifluoromethane sulphonic acid (Aldrich) are mixed in a glass tube with a Teflon-lined screw cap and cooled to 0°C. (Trifluoromethane sulphonic acid should be stored desiccated, under N_2 below 0°C; it is moisture-sensitive and hygroscopic. This reagent causes burns and contact with the skin or inhalation of vapour must be avoided.) The sample (5–10 mg) of thoroughly dried glycoprotein or proteoglycan is dissolved in 1 ml of the anisole-trifluoromethane sulphonic acid mixture in a 2 ml Reactivial (Pierce Chemical Co.) and nitrogen is bubbled for 30 s followed by magnetic stirring at 0°C in an ice water bath for a period of 2–3 h (the optimal time varies for different glycoproteins). The deglycosylated sample is recovered by the addition of two volumes of diethyl ether, precooled to −40°C, followed by 3 ml of ice-cold 50% vol./vol. aqueous pyridine. The precipitate of pyridinium salt is redissolved by vortexing and the ether phase discarded. After repeated ether extraction the aqueous phase is dialysed against 2 mM pyridine acetate pH 5.5 (4 l) and lyophilised. Further purification by gel filtration or affinity chromatography can be carried out. Cleavage under these conditions gives products with an intact peptide chain, there is rapid cleavage of peripheral sugars from glycoproteins, slow loss of serine/threonine-linked GalNAc and retention of *N*-glycosidically linked GlcNAc (Edge et al., 1981). Cleavage of the *O*-glycosidic protein-carbohydrate linkages in bovine nasal cartilage proteoglycan also occurs. Harrison (1983) has described a simple alternative method for neutralisation of trifluoromethane sulphonic acid after solvolysis by addition of anhydrous Na_2CO_3. The reaction mixture is then diluted with water, dialysed and lyophilised. Molecular weight analysis by SDS-gel electrophoresis of human Cl^- inhibitor deglycosylated in this way has been reported (Harrison, 1983).

Another procedure for deglycosylation of proteoglycan has been described by Coudron et al. (1980). Chondroitin sulphate proteoglycan (75–100 mg) was dried thoroughly, placed in a polyethylene vial, and 1 ml of anisole scavenger and 10 ml of 70% polyhydrogen fluoride in pyridine (PRC Inc., Gainsuille, FL) was added. After stirring 8 h at room temperature the reaction mixture was diluted into 100 ml water and dialysed, concentrated and chromatographed on Sephadex G-200. The product had a high xylosyl-transferase acceptor activity, indicating that cleavage of protein-carbohydrate linkages had occurred.

Despite the apparent severity of the conditions required for chemical deglycosylation, it has been found that immunological and other biological activities depending on protein conformation can survive treatment such as incubation with HF at 0°C for 1 h. The partially deglycosylated products have been employed in studies of the functional role of carbohydrate in glycoproteins. For example, it has been shown that human chorionic gonadotrophin deglycosylated with HF (the scavenger anisole was omitted in these experiments) binds to cellular receptors with equal or higher affinity than native hCG but that the deglycosylated hormone did not show other activities of the native hormone such as the stimulation of cAMP formation (Manjurath and Sairam, 1982; Rebois and Fishman, 1983). The effect of chemical deglycosylation (with trifluoromethane sulphonic acid) on the reassembly of subunits of hCG has been examined by Kalyan and Bahl (1983).

Partially deglycosylated glycoproteins suitable for studies of the functional consequences of loss of carbohydrate can also be obtained by digestion with endoglycosidases (Section 6.6) or with combinations of exoglycosidases (Section 6.4), as discussed elsewhere in this book. Enzymic methods should, in principle, be less likely than chemical methods to produce denaturation of glycoproteins. However, in some cases quite prolonged incubations with glycosidases are required and trace impurities with proteases can pose difficulties. The use of the antibiotic tunicamycin, which inhibits the addition of Asn-linked carbohydrate units to protein, has been very valuable in producing, in biosynthetic systems, small amounts of unglycosylated peptide chains.

In preparing substrates for glycosyl transferases by partially, or completely, deglycosylating glycoproteins or proteoglycans, Smith degradation (Section 6.9.4.4) or exoglycosidases have been widely used. Chemical deglycosylation of proteoglycans to produce xylosyltransferase acceptor 'core protein' by the HF-pyridine (Coudron et al., 1980) or the trifluoromethane sulphonic acid methods may also be valuable.

CHAPTER 7

Lectin techniques

7.1. Introduction

A lectin is a sugar-binding protein of non-immune origin which agglutinates cells or precipitates glycoconjugates (Goldstein et al., 1980; Dixon, 1981). Agglutination of cells results from reversible non-covalent interactions between lectins, which contain at least two carbohydrate-binding sites per molecule, and the carbohydrate chains of cell surface glycoproteins or glycolipids. The ability of lectins to bind specifically and reversibly to carbohydrates has been exploited in a wide variety of techniques used to study soluble glycoproteins, glycopeptides and oligosaccharides as well as the glycoconjugates of intact cellular membranes (Table 7.1). The ready availability of several highly purified lectins of well-defined specificity has led to the widespread adoption of lectin techniques.

Because lectins do not usually enter cells they can be used as probes to give information about the location, abundance and function of glycoconjugates at cell surfaces. Methods for measuring the binding of lectins to cells, or other membrane bounded systems, are discussed in Section 7.2. The agglutination of cells by lectins (Section 7.3) can be used to obtain information about the carbohydrate determinants present at the cell surface or as an assay for lectin activity. Inhibition of lectin-induced agglutination by glycoproteins, glycopeptides or oligosaccharides can be used as a means of investigating either lectin specificity or the structure of carbohydrate units.

The specific, reversible interactions between lectins and oligosaccharides form the basis for important affinity chromatography me-

TABLE 7.1

Lectin techniques used to study the structure, cellular location and functions of glycoconjugates

	Reference
Detection and quantitation of cell surface carbohydrate	
Agglutination of cells	7.3
Lectin binding assays	7.2
Localisation of cell surface carbohydrates	
Fluorescently labelled lectin binding	Bittiger and Schebli (1976)
Binding of conjugated lectins which have electron-dense markers for localisation by electron microscopy	Bayer et al. (1982)
Cell fractionation	
Lectin affinity chromatography	Reisner and Sharon (1980)
Selection of cells with altered surface carbohydrate	
Cells selected by growth in presence of toxic lectins	Stanley (1980, 1983), Baker et al. (1982)
Isolation, fractionation and structural analysis	
Lectin affinity chromatography for:	
1. Separation of glycosylated from non-glycosylated proteins	7.4.2
2. Fractionation of glycoproteins with microheterogeneous carbohydrate units	7.4.3
3. Membrane glycoprotein ('lectin receptor') isolation	7.4.4
4. Isolation and structural characterisation of glycopeptides	7.4.5
Detection of glycoconjugates	
Staining gels with labelled lectins	7.4
Detection of glycoproteins by precipitation methods	7.6 and 7.7
Studies of the functional role of glycoconjugates	
Examination of effect of lectins on cellular processes (e.g. mitogenic action of lectins on lymphocytes)	

thods (Section 7.4) in which insolubilised lectins are used to isolate glycoproteins, investigate glycoprotein heterogeneity and fractionate glycopeptides. Understanding of the structural requirements for binding of oligosaccharides to insolubilised lectins has now developed to a stage at which considerable structural information can be deduced from their chromatographic behaviour in this type of system.

Labelled lectins can be employed to detect glycoproteins following separation by gel electrophoresis (Section 7.5). An alternative approach to the detection of glycoproteins on an analytical scale is to precipitate (labelled) glycoprotein specifically by the addition of lectin and anti-lectin antibody (Section 7.6). Lectin-glycoprotein interactions can also be detected by precipitation reactions (7.7) including precipitation in gels ('lectin immunodiffusion', Section 7.7.3) or by electrophoresis through gels containing an insolubilised lectin (7.7.4).

Because lectins bind reversibly to glycoconjugates it is not always possible to establish directly which cell surface component is acting as the 'receptor' for the lectin. One approach to identification of lectin receptors by covalent cross-linking is discussed in Section 7.8.

There is not space in this chapter to discuss all techniques employing lectins and consequently some methods for studying the interactions of lectins with cells have been omitted. These include studies of the mitogenic action of lectins, techniques for assessing the localisation and distribution of lectin receptors by fluorescence and electron microscopy, the fractionation of cell populations with lectins and the selection of lectin-resistant cell lines. References to these techniques are given in Table 7.1.

Lectins can bind to specific carbohydrate determinants in proteoglycans as well as in other glycoproteins (Toda et al., 1981). However, this type of interaction has not been investigated extensively to date and the emphasis in this chapter will therefore be placed on glycoproteins other than proteoglycans. The interactions of lectins with glycolipids and proteoglycans must, however, be kept in mind when considering the interaction of lectins with intact cells.

The interactions between glycoconjugates and lectins have a great deal in common with the interactions of carbohydrate-specific antibodies with glycoconjugates. There are especially close parallels between lectins and monoclonal antibodies which, like the lectins, are composed of a homogeneous molecular population. Many of the techniques discussed in this chapter could also be applied to carbohydrate-specific antibodies. The production of carbohydrate-specific antibodies has been discussed in Methods Enzymol. *50* (1978) 155–174, and the production of monoclonal antibodies has already been described in a book in this series (Campbell, 1984).

7.1.1. Isolation of lectins

Lectins have been isolated from microorganisms (Eshdat and Sharon, 1982), invertebrates, vertebrates (Simpson et al., 1978; Barondes, 1981; Ashwell and Harford, 1982) and from a variety of plant tissues (Lis and Sharon, 1980). The seeds of many plants contain substantial quantities of lectins and most lectins used as tools for studying glycoproteins are prepared from this source. Lectins are usually extracted from powdered seeds after preliminary defatting with an organic solvent. Modern isolation procedures generally employ affinity chromatography on an insoluble carbohydrate derivative (Goldstein and Hayes, 1978; Brown and Hunt, 1978; Lis and Sharon, 1980; also see Methods in Enzymology, Volumes 24, 28, 30 and 83). It should be borne in mind that other proteins which bind to carbohydrate, such as glycosyl transferases and glycosidases, may co-purify with lectins. A number of lectins occur in more than one molecular form. This may arise from proteolytic nicking of the peptide chain or from the existence of 'isolectins' which are closely related peptide chains differing in primary structure, or from multimeric lectins containing different combinations of subunits. Where different molecular forms have identical binding characteristics it may be acceptable to use a mixture. Iso-lectins can be separated by conventional protein fractionation techniques such as ion-exchange chromatography (Allen et al., 1973). Many lectins are now available commercially in a pure state as judged by SDS-polyacrylamide gel electrophoresis.

Suppliers offering a range of lectins include Boehringer, Calbiochem-Behring, Miles, P-L Biochemicals, Pharmacia, Pierce, Polysciences, Sigma and Vector (British Drug Houses in the U.K.).

7.1.2. Molecular properties

The lectins are a diverse group of proteins varying widely in molecular weight, number of peptide chains, metal ion requirements and the presence or absence of disulphide bonds. Many, but not all, are glycoproteins. Some properties of widely used lectins are given later in this section.

Agglutination. The activity of lectins was first recognised by their ability to agglutinate red blood cells. Agglutination occurs as a consequence of multivalent lectin forming crosslinks between glycoprotein (or glycolipid) determinants on the erythrocyte membrane in a process analogous to the agglutination produced by antibodies. Agglutination requires cells to be brought together overcoming mutual repulsion due to charged groups at the cell surface. Thus binding of lectins to cells may occur without agglutination when the crosslinkages formed between cells are not sufficient to overcome this repulsion. The susceptibility of cells to agglutination by lectins can often be drastically modified by treating the cells with proteases or glycosidases. Agglutination is a complex process influenced by the nature of the lectin, the number and structure of surface receptors and cell membrane factors (Nicolson, 1974, 1976b; Section 7.3). The reversal or inhibition of agglutination by simple sugars or glycosides is indicative of a specific interaction between lectin and cell surface glycoconjugates.

Specificity. The specificities of lectins were originally defined in terms of the monosaccharides or simple glycosides which were capable of preventing the agglutination of cells or inhibiting the precipitation of glycoconjugates by the lectin. These studies showed that the specificity of many lectins was directed primarily at particular monosaccharide residues (Table 7.2) but that some cross-reactions occurred between sugars of similar structure. Thus *Ricinus communis* agglutinin interacts both with glycosides of galactose and *N*-acetylgalactosamine, and concanavalin A can be inhibited by glycosides of mannose, glucose or *N*-acetylglucosamine. It was found that the anomeric configuration of the sugar glycosides strongly influenced the binding

TABLE 7.2

Carbohydrate and blood group specificities of some lectins used to study glycoconjugates

Further information on specificity is given in the text and in Table 7.3

	Ref.	Sugar inhibitors	Human blood group specificity
D-Mannose (D-Glucose)			
1. Concanavalin A, *Canavalia ensiformis* (jack bean)	1	α-Man > α-Glc (see Table 7.3)	non-specific
2. *Lens culinaris* (lentil)	1	α-Man > α-Glc (see Table 7.3)	non-specific
3. *Pisum sativum* (pea)	1	α-Man > α-Glc (see Table 7.3)	non-specific
4. *Vicia faba* (fava bean)	1	α-Man > α-Glc	non-specific
N-Acetyl-D-glucosamine			
1. *Triticum vulgare* (wheat germ)	1	$[GlcNAc\beta1\text{–}4]_3 > [GlcNAc\beta1\text{–}4]_2$, NeuAc	non-specific
2. *Solanum tuberosum* (potato)	1	$[GlcNAc\beta1\text{–}4]_4 > [GlcNAc\beta1\text{–}4]_3$	–
3. *Ulex europeus* II (gorse)	1	$[GlcNAc\beta1\text{–}4]_2$	O (A_2 B)
4. *Bandeiraea simplicifolia* II	1	β-GlcNAc,α-GlcNAc	not A, B, O. T-activated cells agglutinated
5. *Cystisus sessilofolius*	1	GlcNAc(β1–4)GlcNAc	O, A_2
N-Acetyl-D-galactosamine			
1. *Dolichos biflorus* (horse gram)	1	α-GalNAc	$A_1 \gg A_2$
2. *Glycine max* (soybean)	1	α-GalNAc,β-GalNAc	A > O > B
3. *Helix pomatia* (snail)	1	α-GalNAc > α-GlcNAc	A
4. *Phaseolus lunatus syn limensis* I and II (lima bean)	1	α-GalNAc > α-Gal	A
5. *Sophora japonica* (japanese pagoda tree)	1	β-GalNAc > β-Gal	B, A, O, I antigen

1. *Ricinus communis* I (castor bean)	1	β-Gal	non-specific
2. *Bandeiraea simplicifolia* I	1	α-Gal > α-GalNAc	$B \gg A_1$
3. *Arachis hypogea* (peanut)	1	Gal(β1–3)GalNAc	no agglutination of ABO unless neuraminidase-treated
4. *Abrus precatorius* (jequirity bean)	1	β-Gal > α-Gal	B, O > A
L-Fucose			
1. *Lotus tetragonolobus* I, II and III (asparagus or winged pea)	1	α-L-Fuc	$O \gg A_2$
2. *Ulex europaeus* I (gorse)	1	α-L-Fuc	O A_2
3. *Anguilla anguilla* (eel serum)	1	α-L-Fuc	O A_2
4. *Griffonia simplicifolia* IV	4		Le
N-Acetylneuaraminic acid			
1. *Limax flavus* (slug)	5	NeuAc	
2. *Limulus polyphemus* (limulin, horseshoe crab haemolymph)	1	NeuAc, GlcA	
Other			
1. *Phaseolus vulgaris* erythroagglutinin (red kidney bean)	6	See Table 7.3	non-specific
2. *Phaseolus vulgaris* leucoagglutinin (red kidney bean)	1	See Table 7.3	non-specific
3. *Vicia graminea*	1		N

1. Goldstein, I.J. and Hayes, C.E. (1978) Adv. Carbohydr. Chem. Biochem. *35*, 127–340; 2. Lis, H. and Sharon, N. (1980) in The Biochemistry of Plants: A Comprehensive Treatise, Vol. 6 (Marcus, A., ed.), pp. 371–447; 3. Tollefsen, S.E. and Kornfeld, R. (1983) J. Biol. Chem. *258*, 5172–5176; 4. Shibata, S., Goldstein, I.J. and Baker, D.A. (1982) J. Biol. Chem. *257*, 9324–9329; 5. Miller, R.L., Collawn, J.F. and Fish, W.W. (1982) J. Biol. Chem. *257*, 7574–7580; 6. Cummings, R.D. and Kornfeld, S. (1982) J. Biol. Chem. *257*, 11230–11234.

of some lectins but not others. For example, concanavalin A binds to derivatives of mannose and glucose where the anomeric linkage is α rather than β. It is the preference of ConA for α-linked sugars which excludes its interaction with β-*N*-acetylglucosamine in mammalian glycoproteins. However, some lectins such as soybean agglutinin are almost completely devoid of anomeric specificity.

Many lectins interact with the non-reducing terminal sugar residues of oligosaccharides or glycoproteins. However, some lectins can tolerate substitution at certain positions of the sugar residue primarily determining their specificity. Concanavalin A, for example, can bind to oligosaccharides with internal α-linked mannose residues which are substituted at position 2.

However, the binding sites of many lectins interact with more than a single monosaccharide unit. For example, the agglutination of red blood cells by wheat germ agglutinin is inhibited by *N*-acetylchitotriose at one thousand-fold lower concentration than is required for inhibition by the monosaccharide *N*-acetyl-D-glucosamine (Allen et al., 1973). Some lectins are not inhibited by monosaccharides, whereas oligosaccharides or glycopeptides act as potent inhibitors (Allen and Neuberger, 1973). Inferences about the size and shape of the saccharide binding site of lectins have been drawn from specificity studies (Kabat, 1978).

The interactions between panels of oligosaccharides or glycopeptides of known structure and lectins have been compared by measurements of binding constants or assessed by affinity chromatography using insolubilised lectins (Ogata et al., 1975; Narasimhan et al., 1979; Baenziger and Fiete, 1979a,b; Kornfeld et al., 1981; Cummings and Kornfeld, 1982a; Debray et al., 1983). These studies have shown that subtle changes in oligosaccharide structure can have considerable influence on binding affinity and form the basis for the use of lectins in the fractionation of glycopeptides by affinity chromatography.

The strength of interaction between a lectin and a glycoprotein can be influenced by steric factors arising from the structure of the carbohydrate receptors within a carbohydrate unit or which are located at different sites on the peptide chain (Beeley et al., 1983).

Non-specific binding. As well as interacting specifically with carbohydrate groups of glycoproteins, lectins may also undergo non-specific binding to other proteins, cells or glass and plastic surfaces. Concanavalin A has a well-defined site which binds small hydrophobic molecules (Edelman and Wang, 1978) and the lectin from lima beans (*Phaseolus lunatus*) also has a hydrophobic site (Roberts and Goldstein, 1982). In carrying out experiments with lectins it is essential to use conditions which will minimise non-specific binding and to carry out adequate controls to show that any effects produced are specifically reversed by monosaccharides or glycosides known to bind to the lectin.

Membrane 'receptors' for lectins. The term 'lectin receptor' is frequently applied to the complex heterosaccharides of glycoproteins which interact with lectins. These structures are not necessarily in any physiological sense 'receptors' for lectins, but they happen to have carbohydrate sequences with which the lectin can interact. In some cases the structures of the cell surface glycoconjugates to which lectins bind have been determined (Thomas and Winzler, 1969; Kornfeld and Kornfeld, 1970; Cummings and Kornfeld, 1982a).

The binding of lectins to membrane glycoproteins or glycolipids is dependent both on the carbohydrate groupings present and their environments. Unmasking of cryptic carbohydrate determinants may occur following modification of the cell membrane by proteolysis, treatment with glycosidases, exposure to hypotonic conditions or shear stress (Greig and Brooks, 1979). Binding of lectin to membrane receptors may also be influenced by the mobility of receptors in the membrane which in turn reflects the viscosity of the membrane lipid or anchorage of membrane glycoproteins to cystoskeletal proteins.

When a heterogeneous population of glycoconjugates such as that occurring in the plasma membrane is examined it is not surprising that many different receptors for a single lectin may exist. In studies of the binding of lectins to cells it has been observed that there are classes of receptors with very different affinities for lectin (Feller et al., 1979; Cuatrecasas, 1973). Often only a small fraction of the total lectin binding sites may be associated with a particular functional effect of lectin binding (Jacobs and Cuatrecasas, 1976; Reisner et al., 1976).

Properties of individual lectins

Concanavalin A (ConA). This lectin is available commercially or can be isolated from jack bean (*Canavalia ensiformis*) using affinity chromatography on Sephadex (Liener, 1976).

At neutral pH ConA consists of a tetramer (M_r 102 000) of four identical subunits (M_r 25 000) each containing one binding site (Becker et al., 1975). Between pH 2 and pH 5.5 ConA exists as a dimer and above pH 9 aggregation occurs. Stabilisation of the tetrameric form is favoured by high ionic strength ($I = 1.0$). At an ionic strength of 0.3 and pH 7 comparable proportions of the dimeric and tetrameric forms co-exist (McKenzie et al., 1972). Low temperatures promote dissociation to the dimer (Huet et al., 1974). The position of the dimer-tetramer equilibrium can have a pronounced effect on the biological properties of ConA. Divalent ConA can be produced by chemical modification of the lectin with succinic anhydride (Gunther et al., 1973).

Each subunit of ConA contains a binding site for Mn^{+} and for Ca^{++} (Becker et al., 1975). Removal of metal ions (e.g. by dialysis at low pH) inactivates the lectin.

ConA binds α-D-mannopyranoside and its derivatives. Any modification or substitution in positions C-3, C-4 or C-6 results in drastically decreased binding. The α-pyranosyl forms of glucose and *N*-acetylglucosamine also bind to ConA. Certain sugars containing the five-membered furanoside ring also bind to the lectin. These include α- and β-D-arabinofuranoside and α- and β-D-fructofuranoside (So and Goldstein, 1969).

ConA forms precipitates with several branched-chain polysaccharides, including α-glucans, α-mannans and β-fructans, with certain glycoproteins and with a glycopeptide from ovalbumin (Brewer, 1979). Binding can be to non-reducing terminal monosaccharides but also to internal 2-*0*-substituted α-D-mannopyranosyl units. The 'core' structures of asparagine-linked glycopeptides containing two α-linked mannose residues are bound strongly to ConA, and binding may be influenced by substituents remote from the α-linked mannose residues (Baenziger and Fiete, 1979a). The widespread occurrence in glycoproteins of carbohydrate groups which bind concanavalin A has contributed to the extensive use made of this lectin in studies of glycoproteins.

The interaction between ConA and sugars results in conformational change (Hardman and Ainsworth, 1976). As well as interacting with carbohydrates ConA binds small hydrophobic molecules (Edelman and Wang, 1978) but at a location different from the carbohydrate binding site.

Wheat germ agglutinin (WGA). This lectin can be obtained commercially or isolated from defatted wheat germ (*Triticum vulgare*) by conventional methods (Nagata et al., 1974) or by affinity chromatography (Block and Burger, 1974; Bouchard et al., 1976). The pure protein has a molecular weight of 36 000 and contains two similar peptide chains each having two binding sites for sugars (Goldstein and Hayes, 1978). The high stability of the protein has been attributed to the presence of a large number of intrachain disulphide bridges. Wheat germ agglutinin is a highly basic protein.

The lectin binds specifically to *N*-acetyl-D-glucosamine and its β1–4-linked oligomers (Allen et al., 1973; Privat et al., 1974; Goldstein et al., 1975). Sialic acid (Greenaway and LeVine, 1973) and glycoproteins containing *N*-acetylneuraminic acid (Bhavanandan and Katlic, 1979) also bind to the lectin. There was at one time uncertainty as to whether this binding was specific or resulted from electrostatic interactions. The finding that *N*-acetylneuraminic acid but not *N*-glycolyneuraminic acid inhibits agglutination of red blood cells supports the proposal that interaction is specific (Bhavanandan and Katlic, 1979). Binding sites for sialic acid and *N*-acetylglucosamine have been identified in the three-dimensional structure of the lectin by X-ray crystallography (Wright, 1980a,b). There now seems little doubt that much of the binding of wheat germ agglutinin to cell surfaces is a result of its interaction with sialic acid (Cuatrecasas, 1973; Baxter, 1974).

Ricinus communis agglutinin and toxin (RCA_I and RCA_{II}). Two carbohydrate-binding proteins can be prepared from castor beans (*Ricinus communis*) by affinity chromatography and gel filtration

(Nicolson et al., 1974; Olsnes et al., 1974) although adequate safety precautions must be observed because of the high toxicity of the material. *Ricinus communis* agglutinin can be obtained commercially. The agglutinin RCA_I (or RCA_{120}) has a molecular weight of about 120 000 and is composed of 2 A chains (each of M_r 29 500) and 2 B chains (M_r about 37 000) which are associated by non-covalent interactions. The toxin, RCA_{II} (or RCA_{60}) has very little haemagglutinating activity, has a molecular weight of a little over 60 000 and is made up of one A chain (M_r 29 500) and one B chain (M_r about 34 000) which are covalently joined by a single disulphide band. The toxicity of RCA_{II} results from enzymatic inactivation of the 60 S subunit of eucaryotic ribosomes by the B chain. Both RCA_I and RCA_{II} are glycoproteins. They differ in their carbohydrate specificities.

RCA_I has affinity for β-linked galactose residues. Binding studies with labelled glycopeptides indicate that RCA_I binds to glycopeptides of the complex type which contain β-galactose and to glycopeptides containing Galβ1–3GalNAc. The presence of sialic acid substituents on galactose reduces or abolishes binding (Baenziger and Fiete, 1979b).

RCA_{II} binds to β-galactose residues in glycopeptides. Binding occurs with equal or lower affinity than RCA_I to complex-type glycopeptides but Galβ1–3GalNAc-containing glycopeptides are bound more strongly by RCA_{II}. Unlike RCA_I, RCA_{II} can bind to glycopeptides containing GalNAc but no galactose (Nicolson et al., 1974; Baenziger and Fiete, 1979b).

Some controversy exists over whether the toxin from *Ricinus communis* (ricin) should be described as a lectin. Until recently it seemed that ricin had only one carbohydrate-binding site and so lacked one of the properties required of a lectin. However, evidence is now available suggesting that there are in fact two galactose-binding sites per B chain of ricin (Houston and Dooley, 1982) and the toxin may show a low level of agglutinating activity.

Lentil lectin (LCA). The *Lens culinaris* agglutinin can be purified easily by affinity chromatography on Sephadex (Howard and Sage, 1969; Sage and Green, 1973) and is commercially available. Molecular weights in the range 42–63 000 have been reported and the molecule is made up of two α subunits (M_r 6 000) and two M_r 18 000 β subunits (Foriers et al., 1978) with two binding sites per mole. Mn^{++} is required for activity.

Monosaccharide specificity resembles concanavalin A in that mannose and glucose derivatives are bound, but equilibrium dialysis studies indicated the affinity to be considerably lower ($K_a = 100\ M^{-1}$ for methyl α-D-glucopyranoside: Stein et al., 1971).

While there is some overlap between the glycopeptides bound to lentil lectin and concanavalin A there are also interesting differences in specificity. For high-affinity binding of Asn-linked glycopeptides to lentil lectin-Sepharose the presence of fucose attached to the Asn-linked GlcNAc is essential (Kornfeld et al., 1981). An additional requirement for binding is the presence of two α-linked mannose residues. Substitution of the mannose residues at C-2 does not prevent binding (c.f. ConA). Substitution of one Man at both C-2 and C-4 does prevent binding but substitution of one Man at C-2 and C-6 does not. Binding of glycopeptides to lentil lectin is enhanced by exposure of terminal GlcNAc residues.

Differences have also been observed in the glycoproteins bound by ConA and LCA. Findlay (1974) showed that the erythrocyte glycoprotein glycophorin binds to lentil lectin but not to ConA.

Soybean agglutinin (SBA). This lectin can be isolated from extracts of soybean (*Glycine max*) meal by affinity chromatography on *N*-acetylgalactosamine linked covalently to CH-Sepharose (Allen and Neuberger, 1975). Four 30 000 molecular weight subunits are associated to give a molecule of 120 000 which has two binding sites for sugars (Lotan et al., 1974). Soybean agglutinin is a glycoprotein (Lis and Sharon, 1980) and Mn^{++} is required for activity (Jaffe et al., 1977). The lectin has a monosaccharide specificity for *N*-acetylgalactosamine and galactose.

Soybean agglutinin has a tendency to aggregate on storage, or aggregates can be formed by cross-linking the lectin with glutaraldehyde (Lotan et al., 1973) and the aggregated lectin shows increased agglutination of certain cells because of its increased size and valency.

7.2. Lectin binding

7.2.1. Introduction

The binding of lectins to cells, subcellular components or isolated membranes has been investigated to obtain information about the number and affinity of available receptor sites (for reviews see Brown and Hunt, 1978; Sharon and Lis, 1975; Nicolson, 1976a,b). Changes in membrane glycoconjugates accompanying growth, transformation, differentiation, induced by the selection of lectin-resistant cell clones, or induced by perturbation with proteases or glycosidases, can be examined.

The binding of lectins to receptors either in solution or in membrane-bound form has many similarities to the binding of antibody to antigen. Unlike antibodies (other than monoclonals), however, lectins are usually homogeneous, which greatly simplifies the theoretical treatment of binding.

The reversible bimolecular reaction between a lectin (L) and receptor (R) to form a lectin-receptor complex (LR) can be described by the equation

$$\mathrm{L} + \mathrm{R} \underset{k_\mathrm{d}}{\overset{k_\mathrm{a}}{\rightleftharpoons}} \mathrm{LR}$$

where k_a and k_d are respectively the association and dissociation rate constants. The equilibrium (association constant), K, for the reaction can be written as

$$K = \frac{k_\mathrm{a}}{k_\mathrm{d}} = \frac{[\mathrm{LR}]}{[\mathrm{L}][\mathrm{R}]}$$

Where [LR] is the molar concentration of the complex, [L] the concentration of the free lectin and [R] that of free receptor.

Cells, or subcellular organelles, carry multiple glycoprotein (and glycolipid) receptors for lectins. To simplify the analysis of binding the assumption is made that although lectins are multivalent they *behave* as though they are monovalent when binding to cellular receptors. This assumption, originally proposed in analysis of binding of antibodies to red blood cells (Wurmser and Fillitti-Wurmser, 1957), follows from the geometrical consideration that, for a multivalent binding protein, the number of molecules bound to two cells will be small compared with the total number bound by a single valency to a single cell. (Although useful, this assumption may not always be valid, because a multivalent lectin may be able to form bridges between different carbohydrate chains on the surface of the same cell.)

When increasing quantities of lectin are added to a constant amount of cells it is usually found that the amount of lectin specifically bound reaches a saturating level. The number of molecules of lectin required to saturate the receptors can be equated directly with the number of cellular binding sites provided that the lectin behaves as though it were monovalent.

Measurement of the affinity of the lectin (i.e. the apparent intrinsic association constant K) and the number of lectin receptor sites per cell, n, can be made by determination of free and bound lectin over a range of lectin concentrations and analysis of the results using one of the equations given below. These relationships can be derived from application of the Law of Mass Action to the multiple equilibria involved in binding an effectively monovalent lectin to independent and identical binding sites.

$$r = \frac{n[L]}{1 + K[L]} \tag{1}$$

Where $r = \frac{[LR]}{[C]}$; [LR] is the bound lectin concentration (in moles per litre); [C] is the number of cells per litre $\div 6.03 \times 10^{23}$. For soluble glycoproteins or glycopeptides [C] is the molar concentration of the carbohydrate determinants which bind to lectin.

This expression can also be written in reciprocal form (Steck and Wallach, 1965)

$$\frac{1}{r}=\frac{1}{n}+\frac{1}{nK[\mathrm{L}]} \tag{2}$$

By plotting $\frac{1}{r}$ against $\frac{1}{[\mathrm{L}]}$ a straight line is obtained if sites are independent and identical.

Extrapolation to $\frac{1}{r}=0$ yields the $\frac{1}{[\mathrm{L}]}$ intercept equal to $-K$. The third and most widely applied method for determination of K and n makes use of the expression derived by Scatchard (1949).

$$\frac{r}{[\mathrm{L}]}=Kn-Kr \tag{3}$$

A graph of $\frac{r}{[\mathrm{L}]}$ against r gives a straight line with intercept n and slope $-K$. These types of graph are shown in Fig. 7.1.

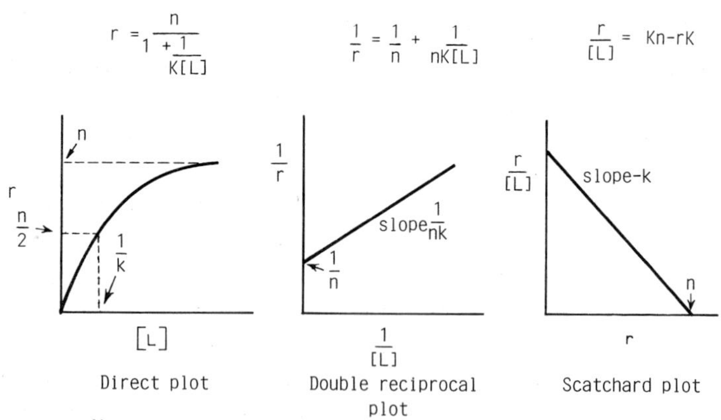

Fig. 7.1. Graphical methods for analysing lectin-binding measurements.

The reciprocal and Scatchard plots will give straight lines for non-interacting binding sites which all have the same affinity for lectin. The Scatchard plot is more sensitive than the reciprocal plot to deviations from linearity resulting from heterogeneity of binding constants in the population of receptors. The existence of classes of binding sites differing in affinity for lectins has been reported for several cell types and is perhaps not surprising because each lectin may bind to a number of different glycoconjugates each of which is likely to display microheterogeneity. Deviation of reciprocal or Scatchard plots from linearity may also arise from the existence of sites interacting with positive or negative co-operativity.

However, interpretations of non-linear binding curves must be treated with caution. The Scatchard plot in particular emphasises data obtained at the extremities of the binding curve. Minor non-specific interactions can distort the curve and give the impression of a second class of receptors of low affinity and high capacity. Further, when applied to high-affinity systems studied under conditions where only a small fraction of the total sites are occupied, binding of lectin may be a linear function of lectin concentration. Scatchard analysis in this range of lectin concentrations would give the misleading impression of an infinite number of binding sites (Chang and Cuatrecasas, 1976).

The results of binding experiments can be expressed as the number of lectin binding sites per molecule of glycopeptide or glycoprotein or per cell. For membrane receptors it is useful to indicate the number of receptor sites per unit area of membrane. The surface area of cells can be estimated from measurements of cell size or volume if it can realistically be assumed that they are spherical in shape. However, many cells are not spherical and/or have convoluted membranes. To avoid the problem of measuring surface area the number of binding sites is sometimes related to cell protein content, which is presumed to be dependent on cell size. In practice, the quantitative determination of the number of receptors for a lectin or the affinity of lectin for receptors both require measurement of the specifically bound (or free) lectin concentrations. Thus labelled lectins are usually required together with methods for separating free and bound lectin. Careful choice of reaction conditions such as temperature and time of incubation is also necessary.

7.2.2. Radioactive labelling of lectins

Binding studies require lectins labelled to high specific activity with the least possible change in properties of the lectin. The isotope most commonly used has been ^{125}I, which emits γ-rays and has a half-life of 60 days. Labelling may be carried out using chloramine T (Hunter and Greenwood, 1962; Greenwood et al., 1965) to oxidise I^- to I_2. Damage to the protein can be minimised by the use of a short reaction time and a low ratio of oxidant to protein. Enzymatic iodination using lactoperoxidase and hydrogen peroxide (Marchalonis, 1969) has been applied to lectins (Gurd, 1977), or hydrogen peroxide can be generated in situ using glucose oxidase and glucose (Hubbard and Cohn, 1972; Meager et al., 1976). Preparations of lactoperoxidase and glucose oxidase coupled to an insoluble support which facilitate removal of the enzymes from the reaction mixture are available (Bio-Rad). A chemical method of labelling using iodine chloride which avoids oxidising conditions but requires prior oxidation of any SH group (McFarlane, 1958) has been applied to lectins (Nicolson et al., 1975).

Concanavalin A labelled by acylation with [^{3}H]acetic anhydride or [^{14}C]succinic anhydride has been used in binding studies (Noonan and Burger, 1973; Gunther et al., 1973) and these derivatives have the advantage that the isotopes are long-lived. Acetylated and succinylated ConAs exist in the dimeric form at pH 7.4, whereas the native lectin is tetrameric (Gunther et al., 1973). Extensive acylation can, however, alter the affinity without change of specificity (Chang and Cuatrecasas, 1976).

Concanavalin A has been radioactively labelled by substituting ^{63}Ni for the coordinately bound Mn (Inbar and Sachs, 1969). However, the molecular weight and binding properties of this derivative differ from those of the native lectin and use of this labelling technique is inadvisable (Cline and Livingstone, 1971; Ozanne and Sambrook, 1971).

During labelling an inhibitory sugar may be included in the reaction mixture to protect the combining site. The sugar must be removed prior to purification of the lectin by affinity chromatography.

After labelling excess reagents can be removed by dialysis, gel filtration, precipitation of the lectin or by binding it to an affinity column. It is highly desirable to re-purify labelled lectin by affinity chromatography to remove any inactivated material.

Labelled lectin should be examined to determine whether (1) the haemagglutination (or precipitating) activity is unchanged, (2) the label is covalently bound to protein (i.e. can be precipitated by TCA), (3) the label co-migrates with protein on SDS-gel electrophoresis, (4) binding of labelled lectin is inhibited by an excess of unlabelled lectin, (5) dilution of labelled lectin with unlabelled lectin produces a proportionate decrease in bound radioactivity. If these criteria are satisfied it can be assumed that the number of counts bound accurately reflects the amount of lectin bound. A number of discrepancies found in binding data reported in the literature may have arisen in part through the use of lectins whose affinities have been altered during labelling (Sandvig et al., 1976; Nicolson et al., 1975).

^{125}I-labelling procedures

(a) Chloramine T. Lectin (1–2 mg) in 200 μl 0.1 M sodium phosphate buffer is added to 100 μl of 0.25 M sodium phosphate buffer, pH 7.5, containing an inhibitory sugar or glycoside (0.1 M) and 0.2–2 mCi of carrier-free $Na^{125}I$. Chloramine T (1 mg) dissolved in 100 μl of water is added, mixed, and after 30 s at room temperature 100 μl (1 mg) of sodium metabisulphite is added. The reaction mixture is cooled on ice and unreacted iodine and inhibitory sugar are removed by gel filtration or dialysis.

The labelled lectin is then isolated by affinity chromatography. Concanavalin A can be absorbed onto a 5-ml column of Sephadex G-75 equilibrated with 50 mM Tris, pH 7.5, and containing 1 mM $CaCl_2$ and 1 mM $MnCl_2$. Labelled lectin can be selectively eluted with 0.3 M methyl α-D-mannoside in 0.1 M Tris-HCl, pH 8.5. Bound methyl α-mannoside is removed by dialysis.

The specific activities of labelled lectin obtained lie in the range 0.1–1 mCi/mg. Increased specific activity can be obtained using the higher amounts of radioactive iodide or decreased amounts of protein. When smaller amounts of protein are used losses due to non-specific adsorption can be reduced by adding albumin (0.1% w/v) to the lectin after termination of iodination with metabisulphate (Cuatrecasas, 1973). This labelling technique gives satisfactory recoveries of active lectin with ConA and WGA. High specific activities can be obtained. Appreciable inactivation of some lectins (e.g. *Ricinus communis* agglutinin) occurs.

(b) Lactoperoxidase and H_2O_2 (Marchalonis, 1969; Sandvig et al., 1976). The reaction mixture contains lectin, such as RCA I or II (1 mg), $Na^{125}I$ (500 μCi) diluted with NaI to give about 1 atom of iodine per protein molecule and lactoperoxidase (5 μg) in a total volume of 200 μl buffered with 0.05 M sodium phosphate, pH 7.3, containing 0.15 M NaCl. Hydrogen peroxide (5 μl), 4.9 mM, is added and the mixture incubated at 20°C for 45 minutes. Free iodide is removed by passing the solution through a Sephadex G-25 column and the iodinated lectin is isolated by affinity chromatography.

(c) Lactoperoxidase and glucose oxidase (Hubbard and Cohn, 1972; Meager et al., 1976). Ricin (1 mg) to be labelled is dissolved in phosphate-buffered saline (2 ml) containing 2.5 mM glucose, 110 mM galactose, 20 μg of lactoperoxidase/ml, 0.2 units of glucose oxidase/ml and 500–1000 μCi of carrier-free $Na^{125}I$. After incubation at room temperature for 15 min the reaction mixture is diluted with ice-cold water and dialysed extensively against phosphate-buffered saline at 4°C for 2–3 days. The dialysed mixture (5 ml) is applied to a column (1 × 10 cm) of Sepharose 6B equilibrated with phosphate-buffered saline. After washing with buffer to remove enzyme, reagents and free iodide, the lectin, which binds specifically to agarose, is eluted with 0.1 M galactose. Fractions containing labelled lectin are dialysed to remove galactose. Albumin, 50 μg/ml, is added and the lectin stored frozen until used for binding assays.

Lectins can also be iodinated with Enzymobeads (Bio-Rad), which contain lactoperoxidase and glucose oxidase bound covalently to polymer beads. The beads are removed from the reaction mixture by centrifugation.

7.2.3. *Measurement of lectin binding*

Quantitation of binding requires measurement of the amount of lectin bound specifically to carbohydrate receptors of glycopeptides, glycoproteins or membranes. Non-specific binding of lectins to cells, glass and plastic surfaces also occurs. Specific binding is distinguished by its reversibility by simple monosaccharide or glycoside inhibitors. Such inhibitors added before or after lectin should produce complete reversal of binding at concentrations of the order of 100 mM. Binding not reversed under these conditions is assumed to be non-specific or due to uptake of lectins into cells. Specific binding of labelled lectin should also be inhibitable by dilution with unlabelled lectin. Binding of lectin to receptors is a temperature-dependent process.

Both the rates of association and dissociation generally increase with temperature. Even at 4°C equilibrium is usually attained between lectin and cell-bound receptors within minutes although the time required is a complex function of the concentration of both lectin and receptor sites. Equilibrium in the binding of glycopeptides to lectins in solution can occur within a few seconds (Baenziger and Fiete, 1979a).

Cells with endocytotic or pinocytotic activity are liable to internalise lectin (Philips et al., 1974; Nicolson et al., 1975). Internalisation of lectin by cells incubated at 25°C has been detected by failure of sugar inhibitors added *after* lectin to fully reverse binding despite preventing binding when added prior to lectin (Philips et al., 1974). The uptake of lectins into cells has been confirmed by electron microscopy (Nicolson et al., 1975). For this reason it is advantageous to carry out binding experiments at 4°C using the shortest incubation time necessary for equilibrium, or steady state, conditions to be obtained. At low temperatures membrane fluidity is decreased. This may reduce the mobility of membrane receptors and decreases the probability of cross-linking receptors within the membranes. In the case of concanavalin A decreased temperature promotes dissociation of the lectin to its dimeric form (Huet et al., 1974).

The concentration of lectin also affects the association-dissociation equilibrium of ConA. Cuatrecasas and Hollenberg (1976) emphasise the importance of using lectin of high specific activity to allow the detection of high-affinity binding sites present in low concentrations but which may be of biological significance.

Divalent cations required for the stability of several lectins should be added to reaction mixtures in which lectin binding is examined. An exception has to be made when divalent cations would damage or perturb cells. In this situation it is important to check that the lectin is not inactivated under the conditions used for measuring binding.

The methods used for separation of free and bound lectin and the interpretation of results depend on whether glycopeptides, glycoproteins or membrane-bound carbohydrates are being investigated.

Membranes and cells. Binding studies on membrane-bound glycoconjugates involve separation of bound lectin by washing, centrifugation through a cushion or filtration.

The simplest method is to remove unbound lectin by washing at 0°C. This procedure may be carried out on layers of cells attached to the surface of a Petri dish (Feller et al., 1979) or on detached cells, subcellular particles or membranes by washing in a centrifuge. Valid results can be obtained if the dissociation of lectin is slow enough to be negligible during the time taken for washing. This must be established in preliminary experiments. In some cases appreciable loss of lectin may occur during washing (Philips et al., 1974). Another potential problem is the non-specific binding of lectin to plastic and glassware. Siliconisation of surfaces (where adhesion of cells is not required), the inclusion of albumin in the reaction mixture or prior saturation of plastic surface with albumin can reduce adsorption of labelled lectin.

Many of the difficulties associated with washing procedures can be obviated by centrifugation of cells through a layer of immiscible oil or albumin in a microfuge (Philips et al., 1975; Philips and Furmanski, 1976).

Filtration methods allow the separation of free from bound lectin in a short time with little opportunity for dissociation of bound lectin. Careful choice of filters is necessary to minimise binding of lectin. Cuatrecasas (1973) found that nylon filters gave low background absorption of ^{125}I-labelled wheat germ agglutinin, while Teflon filters were suitable for concanavalin A. Adsorption of radioactivity to these filters was unaffected by *N*-acetylglycosamine or methyl α-mannoside. However, cellulose or cellulose ester filters bound large amounts of lectin and binding was reversed by monosaccharide inhibitors.

The following procedures illustrate different approaches to measurement of the binding of lectin to cells.

Lectin binding to cultured fibroblasts (Feller et al., 1979). Lectins (concanavalin A, pea and lentil lectins) were ^{125}I-labelled with lactoperoxidase and contained $2–3 \times 10^4$ cpm/mg protein. Fibroblasts were cultured on 60-mm diameter plastic Petri dishes. The cell monolayers were cooled to 4°C and washed with cold phosphate-buffered saline (PBS), pH 7.4, (to remove serum glycoproteins). Cells were incubated for 1 h at 4°C in 2 ml buffer (PBS) containing labelled lectin (concanavalin A) in the presence or absence of 0.01 M methyl α-glucoside or 2 mg/ml unlabelled concanavalin A. Following the period of incubation cells were washed five times with cold PBS to remove unbound lectin. Cells were solubilised with 0.1 M NaOH and samples were counted to determine the amount of lectin bound to the cells and an aliquot was removed for protein determination. Specific binding was distinguished from non-specific binding (to the cells and plastic dishes) by its prevention by dilution with a large excess of cold lectin and by the addition of methyl α-glucoside.

The results obtainied for binding of ConA are shown in Fig. 7.2. The Scatchard plot of these results indicates two populations of receptors, 3.1×10^6 high-affinity sites with association constants of 2.2×10^9 M^{-1} and about 4.8×10^8 sites with low affinity with apparent intrinsic association constants of 2.4×10^6 M^{-1}.

It is also possible to examine the competition between different lectins for cellular receptors using this type of binding assay. Feller et al. (1979) showed that concanavalin A binding completely inhibit-

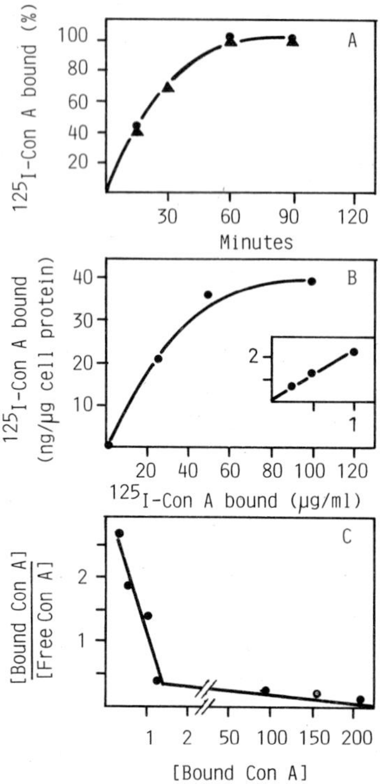

Fig. 7.2. Binding of ^{125}I-labelled concanavalin A to cultured fibroblasts (Feller et al., 1979). A. Rate of binding of ^{125}I-labelled ConA 0.1 μg/ml (●) or 50 μg/ml (▲) to human fibroblasts at 4°C. Unbound lectin was removed by washing the cell monolayer and the results are expressed as a percentage of the maximum specific binding obtained. B. The saturation curve for specific binding of ConA to fibroblasts is shown with data obtained at low ConA concentrations inset. Non-specific binding (not reversed by the addition of methyl α-D-mannoside) has been subtracted. C. Scatchard plot of the data indicating two classes of binding sites with high (2.2×10^9 M^{-1}) and low (2.4×10^6 M^{-1}) affinity.

ed subsequent binding of pea and lentil lectins (which have similar monosaccharide specificity) but prior binding of pea or lentil lectin inhibited concanavalin A binding by only 25%. Thus all pea and lentil binding sites are also concanavalin A sites but only one quarter of all ConA binding sites are also pea and lentil lectin sites. Binding of wheat germ agglutinin also produced limited inhibition of ConA binding, indicating that monosaccharide specificity alone is not sufficient to determine the sites to which a lectin binds at a cell surface.

Gurd (1977) has examined the binding of pairs of lectins to rat synaptic plasma membranes and to the glycoproteins isolated from the same membranes by SDS-gel electrophoresis. While competition could be demonstrated between receptors in the synaptic membrane no competition was observed in binding to the isolated glycoproteins. This suggests that topographical arrangement of receptors in the membrane may lead to competition for lectin binding.

Separation of cell-bound lectin using a microfuge. The procedure developed by Philips and Furmanski (1976) is to add cells ($1–10 \times 10^6$) in phosphate-buffered saline (0.5 ml) to labelled lectin in glass test tubes. After mixing and allowing the reaction to take place for 10 min at 4°C the cells are pelleted by centrifugation, the supernatant is removed, and the cells taken up in 50 μl of phosphate-buffered saline and layered on top of an albumin cushion. The cushion consists of 300 μl of a 5% w/v bovine serum albumin solution in phosphate-buffered saline and should have been given a preliminary 30-second spin to eliminate air locks. After layering the cells on, the microfuge (Beckman) is spun for 1 min at top speed. The tubes are rapidly frozen and the bottoms are cut off with a razor blade and counted.

Binding to simple saccharides, glycopeptides and glycoproteins. The affinities of monosaccharides and oligosaccharides for lectins have been measured using equilibrium dialysis (Olsnes et al., 1974; Greenaway and LeVine, 1973) or relative affinities have been estimated in a semi-quantitative way by inhibition of agglutination (Section 7.3).

Glycopeptide binding to lectins has been quantitated by direct Scatchard analysis of the binding curves obtained with labelled glycopeptides (Baenziger and Fiete, 1979a,b). The glycopeptides were labelled by the procedure of Bolton and Hunter (1973) and bound glycopeptides were separated by ammonium sulphate precipitation.

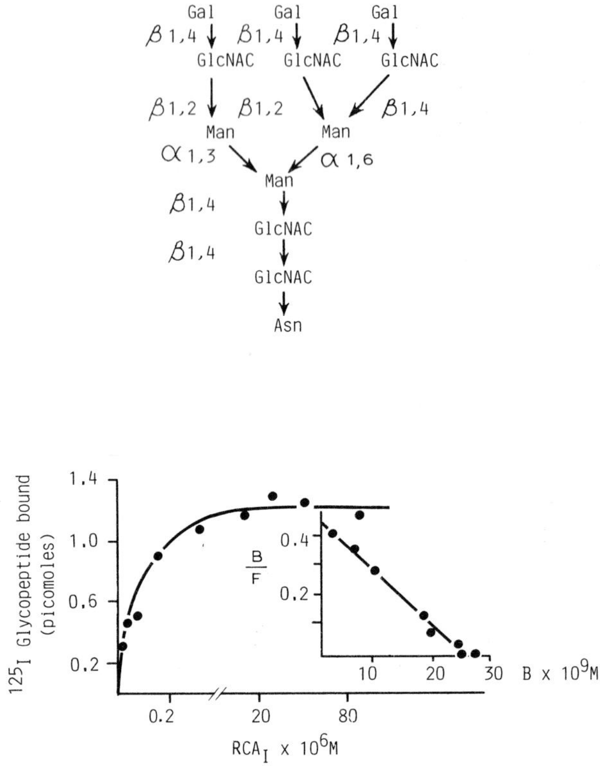

Fig. 7.3. Binding of a glycopeptide to *Ricinus communis* agglutinin (RCA_I). The saturation curve and Scatchard plot for the binding of the ^{125}I-labelled glycopeptide to RCA_I are illustrated (Baenziger and Fiete, 1979b). The structure of the desialylated glycopeptide isolated from fetuin is shown.

Results obtained by this approach are illustrated in Fig. 7.3. Measurement of equilibrium association constants for glycopeptides of known structure allows precise definition of the specificity of lectins. The binding constants obtained (Baenziger and Fiete, 1979b) for glycopeptides binding to *Ricinus communis* agglutinin I and II (about $1–15 \times 10^6\ M^{-1}$) fall within the range of values ($8 \times 10^6\ M^{-1}$ to $4.2 \times 10^8\ M^{-1}$) reported by Sandvig et al. (1976) for the iodinated lectins binding to erythrocytes and HeLa cells. The higher-affinity binding of some cellular receptors may arise from the presence of a different type of carbohydrate structure or from multivalent binding. Semi-quantitative binding measurements can also be performed using columns of insolubilised lectin (Section 7.4).

7.3. *Agglutination methods*

7.3.1. *Introduction*

Binding of lectins to the glycoproteins and glycolipids on cell surfaces may lead to agglutination. Agglutination methods have been used to detect carbohydrate determinants on the surface of cells, organelles and vesicles. Changes in agglutination of cells have been observed during differentiation and following transformation (for reviews see Brown and Hunt, 1978; Nicolson, 1974). Agglutination inhibition techniques can be employed to study glycoproteins, glycopeptides or other saccharides.

The process of agglutination requires binding of lectin to cellular receptors followed by formation of cross-linkages between cells. In addition to having the necessary specificity and affinity for binding, the valency and size of a lectin may determine its ability to form bridges between cells (Lotan et al., 1973) and whether it can interact with glycolipids as well as glycoproteins. Other factors which may influence agglutination include the nature, number, distribution, exposure and mobility of receptors and the deformability, fluidity and surface charge of the membrane (Nicolson, 1974).

Thus although agglutination measurements provide a simple method for the detection of changed cell surface properties the complexity of the factors affecting agglutination make it difficult to determine the nature of the change which has occurred.

Experimental factors affecting agglutination. Because of the complexity of the process of agglutination and the diversity of techniques used for its quantitation, variable and occasionally contradictory results have been reported. Careful attention to possible sources of experimental variation is therefore required (Schnebli, 1976).

Suspensions of blood or lymphoid cells can be readily prepared for agglutination assays in buffered isotonic saline. The properties of these cells remain stable for a few hours at 4°C and red cells can be used for certain purposes for several days. For agglutination studies it is preferable to grow tissue culture cells in suspension culture or release them from monolayers with EDTA-saline. Release of cells with proteases leads to altered agglutinability. Nucleated cells have a tendency to aggregate when stored in serum-free medium. Serum cannot be used in agglutination studies because of the glycoproteins which it contains. Clumping of cells due to leakage of DNA may be reduced by the addition of protease-free DNAase to the cell suspension.

Measurements of agglutination are often carried out under conditions where true equilibrium is not established. Thus the end point of the 'titration' of cells with serial dilutions of lectin changes with time.

The rate of agglutination is dependent on lectin concentration and cell density. In practice a cell density (10^5–10^8 cells/ml) is chosen that is convenient for measurement of agglutination.

Agglutination rate is also affected by temperature (Schnebli, 1976). In addition to possible effects on the mobility of membrane receptors the temperature determines the extent to which cells can internalise lectin (Philips et al., 1974; Nicolson et al., 1975). Possible effects of temperature on the association-dissociation equilibria of lectins should also be considered (Huet et al., 1974).

Probably the most widely used agglutination method, equivalent to passive haemagglutination, involves initial mixing of cells followed by a period in which the cells are allowed to settle under gravity. However, a variety of other arrangements involving different degrees and kinds of mixing have been used and this may have a profound influence on agglutination. Vigorous shaking can result in exposure of extra determinants (Greig and Brooks, 1979) so it is essential that the mixing procedure should be carefully standardised.

For agglutination to occur cells must approach each other sufficiently closely for lectin bridges to be formed. Occasionally cells have been centrifuged in the presence of lectin in order to force them together (Steck and Wallach, 1965). Decreasing the repulsive force due to surface charge may also promote agglutination. For example the negative charge of human erythrocytes is largely produced by sialic acid residues of glycoproteins (Eylar et al., 1962); removal of sialic acid with neuraminidase increases agglutinability by many lectins as does the removal of sialoglycopeptides by trypsinisation.

A buffered salts solution in which the cells are stable should be chosen for agglutination assays. Most lectins will work effectively in a medium tolerated by cells. The range of conditions in which agglutination assays are performed can be extended by fixing the cells with formaldehyde or glutaraldehyde (Section 7.3.3). Agglutination of fixed cells can be examined in the presence of detergent and over a much wider range of pH and ionic strength than untreated cells.

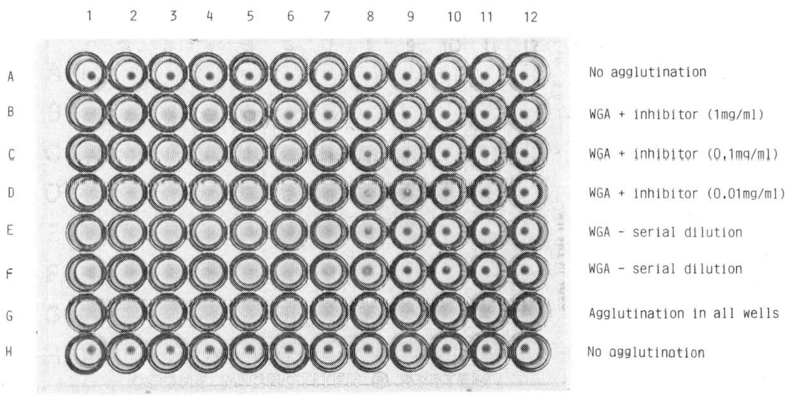

Fig. 7.4. Agglutination of erythrocytes by a lectin and the inhibition of agglutination. Haemagglutination was carried out with trypsinised human erythrocytes (0.5% by volume) in a total volume of 80 μl in the wells of aMicrotitre tray. Unagglutinated cells are present in rows A and H and fully agglutinated cells are present in all wells of row G. Rows E and F show the end point (well 8) when serial dilutions of wheat germ agglutinin (50 μg/ml in well 1) were added. Rows B, C and D show inhibition of agglutination by ovomucoid.

7.3.2. Quantitation of agglutination

Agglutination may be measured by titration assays in which the concentration of lectin is varied to obtain a certain proportion (usually 50%) of agglutinated cells. Alternatively the degree of agglutination under a given set of conditions may be assessed.

Titration methods usually employ two-fold dilutions of lectin and the concentration of lectin agglutinating approximately 50% of the cells present in a given time is measured. The length of the experiment is determined by the time taken for free cells to settle to the bottom of the well of an agglutination tray (Fig. 7.4) and become unavailable for reaction. Alternatively the degree of agglutination can be measured by counting free and agglutinated cells either with a microscope or using a particle counter (François-Gerard et al., 1979).

Agglutination of erythrocytes. Red blood cells have been widely used in agglutination experiments because they are readily obtained, easily visualised and their surfaces carry a range of different types of carbohydrate group. The variety of glycoprotein and glycolipid determinants can be extended by treatment of cells with proteases or glycosidases or by using erythrocytes from different species.

Progressive changes occur in stored erythrocytes as a consequence of metabolic depletion which results in loss of membrane, changes in cell shape, and altered agglutination behaviour. For this reason there may be some day-to-day quantitative variation in haemagglutination titres obtained with stored blood cells.

The conditions available for agglutination tests with fresh cells are restricted to approximately pH 5–8.5 and concentrations of salts close to isotonic and temperatures between 0 and 40°C. Outside these limits haemolysis may be extensive after relatively short incubations. Certain ions (e.g. Ca^{++}) are deleterious to the stability of cells. To extend the range of conditions in which agglutination tests can be performed cells 'fixed' with formaldehyde or glutaraldehyde may be employed. The use of fixed cells is valuable when studies are carried out on glycoproteins (or membrane-bound lectins) which have been solubilised in detergents or when it is necessary to use the same cells over an extended period of time. Fixed cells do, however, suffer the disadvantage of being less easily dispersed than fresh cells.

The sensitivity of agglutination assays can often be greatly enhanced by using neuraminidase- or trypsin-treated red blood cells. Cells can be fixed after enzyme treatment.

Preparation of erythrocytes. Fresh human (or rabbit) blood is collected into one half of its volume of Alsever's solution. This is prepared by dissolving glucose 2.05 g, anhydrous sodium citrate 0.80 g and NaCl 0.42 g, in 100 ml water and adjusting to pH 6.1 with 10% w/v citric acid. Other anticoagulants such as citrate-phosphate-dextrose or EDTA are also satisfactory. Cells should be used within 2–3 days.

The erythrocytes in Alsever's solution are centrifuged at 800 *g* for 10 min and the supernatant and 'buffy coat' of white cells which form a layer on top of the packed erythrocytes are removed using a Pasteur pipette connected by tubing to a vacuum aspirator. The cells are resuspended in at least 2 volumes of phosphate-buffered saline (PBS). This buffer contains 150 mM NaCl and 5 mM NaH_2PO_4 adjusted to pH 7.4 with 1 M NaOH.

The washing procedure is repeated five times. For haemagglutination assays the cells are diluted to 2% v/v in PBS.

Preparation of trypsinised erythrocytes. Crystalline bovine trypsin (10 mg/ml) is dissolved in 1 mM HCl. Washed erythrocytes are suspended in PBS at a concentration of 4% v/v. One volume of trypsin solution is added to 100 volumes of the diluted cell suspension and incubated at 37°C for 1 h. After centrifugation the trypsinised cells are washed three further times in PBS. The washed cells are resuspended at a concentration of 2% v/v in PBS for the agglutination assays.

Preparation of formalinised erythrocytes (Butler, 1963). Washed erythrocytes are diluted to 8% v/v in PBS, pH 7.4. One volume of cells is added to an equal volume of 3% v/v aqueous formaldehyde in a conical flask which is then agitated gently at 37°C for 18 h. The formalinised cells are spun down and washed 5 times in phosphate-buffered saline. Treated erythrocytes adhere together strongly and gentle mixing with a glass rod may be required to ensure adequate washing. Finally the cells may be stored as a 10% v/v suspension at 4°C in pH 7.4 buffer containing a trace of preservative (e.g. 1:10 000 merthiolate). Fixation of erythrocytes can be carried out by the same procedure using gluteraldehyde at a concentration of 1.5% v/v.

Haemagglutination assays. Agglutination of erythrocytes by lectins or antibodies is generally indicated by a complete carpet of cells covering the bottom of the well in an agglutination tray, while non-agglutinated cells slide down to form a compact button or ring at the centre of the curved (or conical) well (Herbert, 1978). This is usually graded as + + + (an even carpet of cells) + +, +, ± and − (a negative button). The end-point (+) is taken as an even carpet of cells with a ring at the edge (Fig. 7.4). The titre may be recorded as the dilution of the lectin at the end-point or as the reciprocal of this dilution, which is known as the agglutination index.

The assay can be carried out in the wells of a Perspex agglutination plate by adding 0.1 ml of a 2% suspension of cells to doubling dilutions of the lectin in 0.1-ml volumes of buffered saline. The agglutination trays should be covered to prevent evaporation and are usually allowed to develop at room temperature for a period between 0.5 and 24 h. Controls containing erythrocytes diluted in saline should also be included.

If haemagglutination assays are shaken the cells will become dispersed except at high lectin titres. It is, however, possible to carry out agglutination experiments in containers which are shaken continuously or periodically (Schnebli, 1976). In this case the agglutinated cells appear as a button in the centre of the well but non-agglutinated cells remain dispersed. The end-points obtained using shaken cells are likely to differ considerably from those found by allowing cells to settle. It is also possible that shearing forces may expose additional receptors. Variation in the technique used for agglutination assays may account for some of the variations in reported titres.

Agglutination assays may readily be carried out with small samples of lectin using microtitre plates (Fig. 7.4). The wells require only 20 μl of lectin and 20 μl of cells. Results of such assays should be regarded as semi-quantitative with an error of ±1 well. Where more precise quantitation is required it may be desirable to use a method for the estimation of agglutination with a particle counter (François-Gerard et al., 1979) or a fragiligraph (Marikovsky et al., 1976).

Inhibition of agglutination. As well as being of use in assays of lectins (or antibodies specific for carbohydrates) agglutination experiments can be used to obtain information about the carbohydrate moieties of glycoproteins or glycopeptides. The ability of a glycoprotein or glycopeptide to inhibit agglutination of cells by a lectin or antibody of known specificity may indicate some of the antigenic structures present. In general this type of information is most valuable when the carbohydrate-binding protein used is highly specific, for example, for a particular blood group determinant. Inhibition by low concentrations of glycoprotein or glycopeptide can then be taken as indicative of the presence of the particular blood-group antigen. The technique has been used to detect the presence of many blood-group antigens in soluble glycoproteins or glycopeptides. Experimentally serial dilutions of the test substance are made in the wells of an agglutination tray. A titre of lectin (or antiserum) rather more than sufficient to produce complete agglutination is added to each of the wells. Then cells are added and agglutination is allowed to proceed. The concentration of test substance required to produce 50% inhibition of agglutination can usually be assessed with an accuracy of ±1 well. In this way the inhibition produced by monosaccharide or other sugar derivatives can be usefully compared. The specificity of agglutination reactions of lectins can be examined using monosaccharides, simple glycosides, disaccharides, glycopeptides or glycoproteins.

7.4. *Lectin affinity chromatography*

This technique exploits the specific binding properties of lectins for the preparative or analytical scale chromatography of glycoproteins and glycopeptides. Lectin coupled covalently to an insoluble support is used tc adsorb glycoconjugates from solution by interaction with their carbohydrate determinants. Elution is achieved by dissociating the glycoprotein-lectin complex using a sugar which competes with glycoprotein for binding to the lectin.

Lectin affinity chromatography is a valuable preparative method which has been applied to the bulk separation of glycoproteins from other proteins (Asperg and Porath, 1970), as a step in the purification of particular glycoproteins (Section 7.4.2; for reviews see Kristiansen, 1975; Dulanay, 1979) and to fractionate glycoproteins into components differing in their carbohydrate moieties (Section 7.4.5; Beeley, 1974; Findlay, 1974; Iwase and Hotta, 1977; Iwase et al., 1981). Membrane-glycoproteins solubilised in detergents can be isolated by this procedure and lectin receptors have been fractionated according to their affinities for a particular lectin or for different lectins (Section 7.4.4).

The method can also be employed in the purification of glycopeptides and structural information may be obtained from the behaviour of glycopeptides on lectin affinity columns (Section 7.4.7 and 6.7.3).

In devising a fractionation procedure for glycopeptides or glycoproteins the following should be considered; (1) choice of a lectin with appropriate affinity; (2) choice of an insoluble support; (3) coupling of the lectin to the support; (4) choice of suitable chromatographic conditions including a method for desorption.

7.4.1. Setting up an affinity chromatography system

(1) Choice of lectin The basic requirements is for the lectin to bind reversibly to the molecules of interest. Whether binding to a particular lectin can occur is determined by the monosaccharide sequence of the oligosaccharide chains of the glycoprotein or glycopeptide. Information about the carbohydrate composition of the glycoprotein may suggest lectins which might be suitable. A much better indication can be obtained by testing the glycoconjugate against a panel of lectins by haemagglutination inhibition (Section

7.3). Kits containing small quantities (1–2 mg) of several lectins suitable for this purpose are available from several suppliers.

A wide range of insolubilised lectins are now available commercially from suppliers including Pharmacia P-L Biochemicals, Vector (B.D.H. in U.K.), Miles, Sigma and Polysciences. The suitability of these lectin derivatives for binding a particular glycoconjugate can readily be determined in small-scale experiments. Of the many lectin derivatives which can now be obtained, insolubilised concanavalin A and lentil lectin have been particularly widely used because of their ready availability and because they react with carbohydrate groups found widely in glycoproteins containing *N*-glycosidically linked oligosaccharides. More selective results may be obtainable with lectins with specificity directed towards less commonly occurring structures.

When a lectin affinity column is loaded with a glycoconjugate and eluted with buffer the sample may pass through the column without binding, it may interact weakly and pass through the column with some retardation or it may be firmly bound and retained on the insolubilised lectins. Either firm binding or retardation can be employed in affinity methods. For glycopeptides it has been shown that firm binding occurs when the association constant for the interaction is 5×10^6 M^{-1} or greater (Baenziger and Fiete, 1979b). The affinity of binding of glycoproteins to insolubilised lectins is influenced not only by the primary structure of the oligosaccharide but also by factors including the number and spacing of carbohydrate units (Baenziger and Fiete, 1979a; Beeley et al., 1983) and the steric accessibility of lectin binding sites. Aggregation of glycoproteins occurring in detergents such as Triton X-100 may also affect the ease with which their interaction with insolubilised lectins can be reversed (Schmidt-Ullrich et al., 1975).

When lectin affinity chromatography is applied to glycoproteins it is often found that only part of the sample binds firmly while some material fails to bind or is only retarded in its elution. This results from carbohydrate heterogeneity and can be exploited in fractionation of molecular species differing in their carbohydrate moieties (Section 7.4.3).

(2) Choice of insoluble support A chemically stable, hydrophilic, open supporting structure which does not absorb lectin or glycoproteins is required. The agarose derivatives Sepharose 2B and 4B have been widely employed. Polyacrylic hydrazide-Sepharose (Lotan et al., 1977) and the polyacrylamide derivative Affi-Gel (Nilsson and Waxdal, 1976; Davey et al., 1976) have also been used.

(3) Coupling of lectin to the support Although a very wide range of procedures for coupling proteins covalently to insoluble supporting media have been described (Lowe, 1979) the most frequently used technique has been the activation of agarose beads with cyanogen bromide (Cuatrecasas, 1970) followed by coupling to lectin (Adair and Kornfeld, 1974; Allan et al., 1972; Davey et al., 1976). Coupling of lectins to polyacrylic hydrazide-Sepharose using glutaraldehyde has been described by Lotan et al. (1977). This procedure is simpler than cyanogen bromide coupling and does not introduce charged groups onto the matrix. Coupling of lectins to Affi-Gel is also a simple technique (Davey et al., 1976; Kornfeld et al., 1981).

Davey et al. (1976) have shown that the conditions used for coupling ConA to cyanogen bromide-activated Sepharose or other support have a profound influence on its chromatographic properties. Coupling at high pH values resulted in lectin which interacted with interferon by largely hydrophobic interactions. However, coupling under conditions likely to lead to single rather than multipoint attachment led to interactions which were carbohydrate-specific. It is known that interaction of ConA with sugar is accompanied by conformational change; multipoint attachment may lead to deformation of the molecule, or prevent the requisite conformational changes taking place.

The density of lectin bound to the matrix can influence the capacity and affinity with which glycoproteins are bound. For a number of lectins coupled to agarose by the CNBr method a density of about 1–2 mg lectin per ml of gel has been employed but considerably higher quantities of some lectins (e.g. 10 mg/ml ConA) have also been used successfully.

(3a) CNBr activation and coupling of agarose to lectins. The method described here has the advantage that continuous titration of the reaction mixture containing the volatile lacrymator cyanogen bromide is not required (March et al., 1974).

One volume of a slurry of washed agarose beads (Sepharose 4B, Pharmacia) consisting of equal volumes of gel and water is added to 1 vol. of 2 M sodium carbonate and mixed by stirring slowly. The rate of stirring is increased and 0.05 vol. of an acetonitrile solution of cyanogen bromide (2 g CNBr/ml) is added rapidly. After stirring vigorously for 1–2 min at 4°C the slurry is poured onto a coarse sintered-glass funnel and washed with 5–10 volumes each of 0.1 M $NaHCO_3$, pH 9.5, water and the buffer to be used in the coupling step. Care should be taken to avoid filtering the beads to a compact cake. A stock solution of CNBr is prepared by adding 12.5 ml dry redistilled acetonitrile to a 25-g bottle of CNBr. The solution is stored at −20°C when not in use. The amount of CNBr solution added can be varied to achieve the desired degree of activation.

The activated Sepharose should, without delay, be resuspended in an equal volume of coupling buffer (0.1 M $NaHCO_3$, pH 8.0, or other non-nucleophilic buffer) containing the lectin at a concentration of 5–10 mg/ml and an appropriate inhibitory monosaccharide or glycoside (0.1 M). Coupling is allowed to proceed for 16–20 h at 4°C with gentle stirring. The few reactive groups remaining can be blocked by further reaction with glycine (1 M). The coupled lectin is then washed extensively with coupling buffer, then with buffered 1 M NaCl, and then with the buffer to be used for affinity chromatography. For lectins with metal ion requirements such as ConA these ions should be included in the washing buffers at the earliest stage where this is feasible in terms of their solubility.

The amount of lectins coupled to the column may be determined by measuring the absorbance at 280 nm of the reaction mixture and washes after coupling. Usually the efficiency of coupling is greater than 90%.

(3b) Coupling of lectins to acrylic hydrazide-Sepharose with glutaral-

dehyde. Polyacrylic hydrazide-Sepharose (commercially available from Miles), 100 ml, is suspended in distilled water (150 ml) and the slurry stirred while 50 ml of 50% (v/v) glutaraldehyde is added. After 4 h at 4°C the gel is washed with cold water until there is no detectable smell of glutaraldehyde. Lectins (wheat germ agglutinin, soybean agglutinin, peanut agglutinin or *Ricinus communis* agglutinin I) are dissolved at a concentration of 5–10 mg/ml in 0.1 M $NaHCO_3$ - 0.15 M NaCl, pH 8.5, containing 0.2 M of the appropriate saccharide inhibitor.

Concanavalin A can be dissolved at the same concentration in 1 M NaCl-0.1 M sodium acetate buffer, pH 6.8, containing 0.2 M methyl α-D-mannopyranoside. The lectin solution is stirred overnight at 4°C with glutaraldehyde-substituted polyacrylic hydrazide-Sepharose at a ratio of 5 mg lectin per ml of packed gel in 2 volumes of buffer. The gels are washed on a sintered glass funnel and the protein concentration in the washings is determined from the absorbance at 280 nm. Lectin-Sepharose conjugates are suspended in 5 mM sodium phosphate-0.1 M NaCl, pH 7.2 (3 vols. per vol. packed gel) and solid $NaBH_4$ is added to give a final concentration of 0.5 mg/ml. After reduction for 3 h at 4°C the gels are washed extensively with the phosphate-0.1 M NaCl buffer. Coupling efficiency is more than 85% with all lectins used (Lotan et al., 1977). Affinity chromatography with these lectin derivatives can be carried out in 0.01 M Tris-HCl, pH 7.2, containing 0.15 M NaCl.

(4) Chromatographic conditions All of the lectin bound to affinity columns is not necessarily available for interaction with glycoproteins. Loss of binding capacity arises from steric restrictions preventing access of glycoproteins to lectin binding sites or through inactivation of some of the coupled lectin. The capacity of affinity adsorbents can be determined by saturating them with a glycoprotein (e.g. fetuin or asialo-fetuin) and measuring the amount of glycoprotein displaced by an inhibitory sugar (Lotan et al., 1977). In all experiments using lectin affinity chromatography different quantities of sample should be applied to columns containing the same quantity

of absorbent to determine whether any unabsorbed material results from exceeding the capacity of the column. The unadsorbed fraction can also be rechromatographed to determine whether it all remains unbound.

Few systematic studies have been carried out on the optimal sample size to use in lectin affinity chromatography. Dulaney (1979) reviewed the application of this technique to the isolation of a variety of glycoprotein enzymes and concluded that there was no correlation between the ratio of protein contained in the sample to the amount of lectin protein bound to the matrix and the extent of purification obtained or the yield of enzyme.

However, the specificity of lectins is relative rather than absolute. When presented with a mixed population of glycoproteins some molecules may be bound with high and others with low affinity. When an excess of such a glycoprotein mixture is applied to a lectin column those molecules binding with high affinity will be adsorbed preferentially. Thus 'overloading' the lectin column will enhance the specificity of the lectin (Adair and Kornfeld, 1974).

Lotan et al. (1977) measured the recovery of glycoprotein bound to polyacrylic hydrazide-Sepharose derivatives of several lectins at different temperatures. Glycoprotein binding at 4°C improved for peanut agglutinin and soybean agglutinin derivatives, remained unchanged for *Ricinus communis* I and wheat germ agglutinin derivatives and decreased (as compared with 23°C) for the insolubilised concanavalin A. Binding at 37°C decreased for all immobilised lectins. Nordren and O'Brien (1976) found that β-galactosidase was not eluted from its complex with concanavalin A-Sepharose by methyl α-mannopyranoside at 2°C but was eluted quantitatively at 22°C. Lectin chromatography should therefore be carried out in the range between 4°C and room temperature, the latter being preferred for ConA-Sepharose provided that the sample is sufficiently stable. Even at elevated temperatures the rates of association and dissociation of the immobilised lectin-glycoprotein complex are often low. It is desirable to run columns at low flow rates and improved binding and reversal of binding may be obtained by stopping the flow through the column for a period of 10–30 min after the sample has been applied and after sugar inhibitor has been added.

The pH range used for lectin affinity chromatography (usually pH 7–8), is limited by the stability and activity of the lectin. Use of low pH conditions with concanavalin A-Sepharose can lead to loss of lectin as not all subunits of the lectin are covalently attached to the matrix. Loss of metal ions from ConA occurs at low pH values and leads to inactivation. Where lectins have a metal ion requirement it is desirable to include the appropriate ions (1 mM) in the buffers used for chromatography.

Inclusion of sodium or potassium chloride in the buffers at concentrations of 0.15–1.0 M improves the recovery of glycoproteins on affinity chromatography (Lotan et al., 1977). The ionic properties of the matrix and ‘coupled’ protein are suppressed at high ionic strength.

Detergents used to solubilise membrane proteins can affect the stability of insolubilised lectins and their interactions with glycoproteins (Kahane et al., 1976; Lotan et al., 1977; Section 7.4.4). Lotan et al. (1977) found that the non-ionic detergents Triton X-100 and Nonidet P-40 were the most suitable for lectin affinity chromatography, having negligible effects on lectins. The zwitterionic detergent *N,N′*-dimethyl-*N*-dodecyl glycine as well as the cationic detergent dodecyl trimethylammonium bromide were suitable for use with immobilised *Ricinus communis* I, peanut agglutinin and wheat germ agglutinin. Deoxycholate produced deleterious effects on most of the lectins. Lectin affinity chromatography can be carried out in the presence of low concentrations (<0.05%) of sodium dodecyl sulphate (Kahane et al., 1976).

(5) Elution of glycoconjugates from lectin affinity adsorbents Insolubilised lectins are commonly used as columns but are also suitable for batch techniques (Nilsson and Waxdal, 1976). Glycoproteins or glycopeptides not bound to the affinity column pass through directly, while weakly bound species may be retarded in their elution. Firmly bound molecules may be eluted specifically by inclusions of an inhibitory sugar (0.1–0.5 M) in the eluting buffer.

Gradient elution with increasing concentrations of inhibitor may give some fractionation of glycoproteins bound with different affinities. Elution can also be achieved in some cases with borate, which forms complexes with *cis*-diols. Change of pH or the addition of denaturants (urea, SDS) may be employed to reverse binding. However, denaturing conditions can dissociate subunits of multimeric lectins and the eluate may contain lectin subunits which were not covalently attached to the matrix.

Hydrophobic interactions between insolubilised lectins, particularly concanavalin A, and glycoproteins may lead to difficulty in reversing binding. Davey et al. (1976) have reported that reversal of binding is improved in the presence of ethylene glycol.

Applications of lectin affinity chromatography

7.4.2. Purification of soluble glycoproteins

Affinity chromatography using insolubilised lectins has been used as a step in the purification of many enzymes and other soluble glycoproteins (Dulaney, 1979). Concanavalin A linked to Sepharose has been particularly widely used because of its ready availability, high capacity and the fact that ConA interacts with many glycoproteins. This does, however, increase the likelihood that ConA chromatography will result in a product containing more than one glycoprotein.

The following procedure can be employed for small-scale preparative chromatography (about 100 mg glycoprotein) on ConA-Sepharose. A slurry of 50 ml concanavalin A-Sepharose (Pharmacia, 10 mg lectin/ml settled gel) is packed in a column 1.5 cm in diameter. The column is pre-washed to remove any loosely bound or degraded ConA with 0.01 M Tris-HCl, pH 7.5, containing 0.5 M NaCl, 1 mM $CaCl_2$ and 1 mM $MnCl_2$ (add $MnCl_2$ after bringing buffer to pH 7.5) followed by a wash with buffer containing the highest concentration of methyl α-mannoside (e.g. 0.5 M) to be used subsequently in eluting glycoprotein. The column is equilibrated in 0.01 M Tris-HCl, pH 7.5, containing 0.15 M NaCl, 1 mM $CaCl_2$ and 1 mM $MnCl_2$ (TCM-saline) and the sample (containing about 100 mg glycoprotein) is applied to the column in the same buffer. Elution with TCM-saline is continued until all unabsorbed material has passed through the column. Bound glycoprotein is eluted first with TCM-saline containing 0.1 M methyl α-D-mannoside and then with TCM-saline containing 0.5 M methyl α-D-mannoside. Eluted protein is detected by its absorbance at 280 nm.

Columns are usually operated at 25°C provided that the sample is stable but can be run at 4°C. Buffers without Mn^{++} and Ca^{++} can be used provided that the buffer pH is above 7. Some dissociations of ConA subunits may occur at lower pH values. For glycoproteins that adsorb strongly to ConA it may be necessary to use higher salt concentrations (e.g. 1 M) in the buffers. ConA-Sepharose columns can be reused many times. When not in use they should be stored in the presence of Mn^{++}, Ca^{++} and an antibacterial agent (e.g. Merthiolate).

The capacity of lectin-affinity columns varies for different glycoproteins. It is advisable to re-chromatograph the breakthrough fraction to ensure that the sample size did not exceed the capacity of the column.

Dulaney (1979) has tabulated the results obtained by a number of workers who have applied ConA chromatography to several enzymes. Recoveries of enzyme activity varied between 43 and 87% and purifications of between 9 and 811-fold were obtained.

Protocols similar to that described above for ConA-Sepharose can be applied to other insolubilised lectins.

7.4.3. Fractionation of glycoprotein species with different carbohydrate groups

Many glycoproteins which are homogeneous in their protein moieties may demonstrate carbohydrate heterogeneity (Section 2.7). Such variations may result in altered affinity for lectins and separation of different species can be achieved by lectin affinity chromatography.

Elution with a gradient of increasing sugar concentration is often valuable.

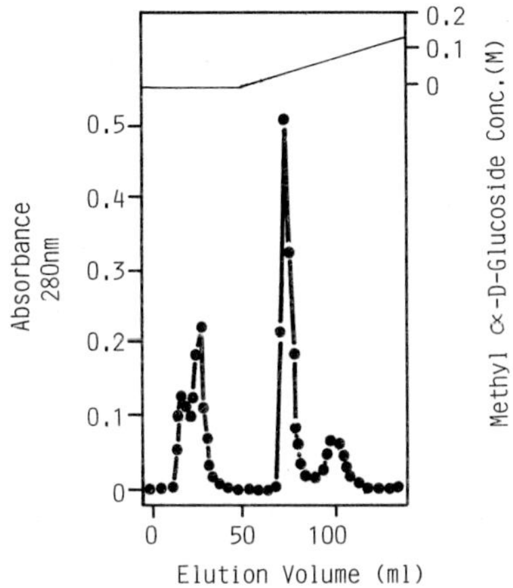

Fig. 7.5. Separation of glycoprotein variants differing in carbohydrate structure by affinity chromatography. Sialic acid-free ovomucoid (20 mg) was applied to a column (1.5 cm × 15 cm) of concanavalin A-Sepharose equilibrated with 0.05 M-Tris buffer, pH 7.5, containing 0.1 M KCl and 0.05 M $CaCl_2$. Bound ovomucoid was eluted with a linear gradient of methyl α-D-glucoside (Beeley, 1974).

Fig. 7.5 shows the separation obtained when hen egg ovomucoid, homogeneous in charge and by immunochemical criteria, was chromatographed on ConA-Sepharose (Beeley, 1974). Similarly, it has been shown by Iwase and Hotta (1977) that ovotransferrin can be resolved into three major fractions by chromatography on ConA-Sepharose and that these components differ in their carbohydrate chains. Iwase et al. (1981) have used the technique to examine ovalbumin microheterogeneity. The existence of fractions of the major human erythrocyte membrane glycoproteins (band 3) differing in their affinity for insolubilised ConA and lentil lectin has also been reported (Findlay, 1974).

7.4.4. *Affinity chromatography of membrane glycoproteins*

Some membrane proteins become soluble in the absence of detergents or chaotropic agents once they have been extracted from the lipid environment of the membrane and can be treated as soluble proteins (Findlay, 1974). Frequently, however, detergent is required for solubilisation of the membrane and to prevent precipitation of the protein. In such cases it is necessary to choose a detergent and use it at a concentration sufficient to solubilise membrane glycoproteins completely yet without affecting the interaction between lectin and glycoprotein. Methods for solubilisation of membrane glycoproteins have been reviewed in Chapter 3 and by Cook (1976).

Allan et al. (1972) showed that lymphocyte membranes could be solubilised in 1% w/v sodium deoxycholate and chromatography on concanavalin A or lentil lectin-Sepharose (Hayman and Crumpton, 1972) could be accomplished in the same solvent. Deoxycholate does cause loss of the activity of several lectins (Lotan et al., 1977) and can be used only in solutions of low ionic strength because of the formation of gels in the presence of salts. Use of divalent ions is restricted by the formation of insoluble salts. Although deoxycholate has been quite widely used the recoveries of glycoproteins obtained have in many cases been low. Adequate extraction of many membrane glycoproteins can be obtained in non-ionic detergents such as Triton X-100 or NP-40. Concentrations of these detergents up to 2.5% v/v produce only slight inhibition of the binding of glycoproteins to insolubilised lectins and non-ionic detergents have been used as solvents for the affinity chromatography of membrane glycoproteins from a variety of sources, including thymocyte plasma membrane (Schmidt-Ullrich et al., 1975), murine lymphocytes (Nilsson and Waxdal, 1976), human erythrocytes (Adair and Kornfeld, 1974), KB cells (Butters and Hughes, 1975) and CHO cells (Gottlieb et al., 1975).

Solubilisation of human erythrocyte membrane glycoproteins can be effected with Triton but after neuraminidase treatment part of the glycoprotein remains unextracted (Pratt and Cook, 1979a). Addition of borate enhances the solubilisation of desialated glycoproteins presumably by providing the negative charge necessary to stabilise the glycoprotein-Triton micelles. Membrane extracts prepared in the presence of Triton and 56 mM borate have been chromatographed on RCA I-Sepharose and WGA-Sepharose (Adair and Kornfeld, 1974). The possibility that borate might interfere with the interaction between some lectins and glycoproteins should, however, be borne in mind.

In general the stability of lectins is increased when they are coupled to an insoluble support. Conditions which lead to rapid inactivation of some lectins may be tolerated by their insolubilised counterparts. Low concentrations of detergents with strong denaturing activity such as SDS may be used with insolubilised lectins (Kahane et al., 1976). A further possibility worth exploring in this field is the use of mixtures of ionic and non-ionic detergents.

Fractionation of the complex mixtures of glycoproteins found in mammalian membranes can be improved by the use of several different lectin affinity absorbents. Gurd and Mahler (1974) used columns of lentil lectin-Sepharose and wheat germ agglutinin-Sepharose connected in sequence to isolate fractions of different binding specificity from extracts of synaptic plasma membrane.

Lectin affinity chromatography on *Ricinus communis* agglutinin (RCA_I and RCA_{II})-Sepharose has been employed to separate cell surface glycoproteins from glycolipids which do not bind (Tsao and Kim, 1981).

7.4.5. Affinity chromatography of glycopeptides

With the increase in availability of glycopeptides of defined structure it has become possible to determine in some detail the structural features responsible for high-affinity binding to lectins. Affinity chromatography will quickly show whether a glycopeptide binds or does not bind to a particular lectin and this may allow deductions about the structure. For columns of ConA- RCA_I- or RCA_{II}-Sepharose it appears that binding to the column will occur only if the glycopeptide binds with an association constant greater than

5×10^6 M^{-1} (Baenziger and Fiete, 1979b). The affinities of different glycopeptides for lectins extend over a wide range and it is possible to distinguish qualitatively between glycopeptides which are (1) eluted directly, (2) retarded in their elution, (3) bound to the column but released in a sharp peak with sugar inhibitor, or (4) eluted as a broad peak (Narasimhan et al., 1979).

Glycopeptides may be isolated from metabolically labelled glycoproteins or they can be labelled by acetylation (with [^{14}C]acetic anhydride) or iodination to facilitate detection of small quantities of material. Lectin affinity chromatography can then be used in conjunction with degradative methods such as the use of glycosidases to determine whether the particular structural features necessary to produce binding are present. Concanavalin A, for example, requires at least two α-linked mannose residues. Removal of one such residue by α-mannosidase digestion will result in loss of binding. Table 7.3 indicates some structural requirements for binding of glycopeptides or oligosaccharides to lectins.

A general scheme for the fractionation of *O*- and *N*-linked oligosaccharide units making use of affinity chromatography of glycopeptides is described in Section 6.7.2. More complete fractionation and characterisation of glycopeptides derived from Asn-linked carbohydrate units can be obtained by sequential chromatography on a series of insolubilised lectins (Cummings and Kornfeld, 1982b). Complex mixtures of glycopeptides can be resolved into fractions suitable for structural analysis. Small amounts of labelled glycopeptides such as are available from cultured cells can be separated and characterised.

Sequential affinity chromatography of Asn-linked glycopeptides (Cummings and Kornfeld, 1982b). Glycoproteins of cultured cells are labelled with 2-[^{3}H]mannose (Section 8.2.3), digested with Pronase and desalted on Sephadex G-25 (Section 6.7.3). Other labelling methods can also be adopted but the use of mannose as precursor has the advantage of labelling principally Asn-linked oligosaccharide units. Initial fractionation of the glycopeptides is carried out on concanavalin A-Sepharose with subsequent further chromatography on other lectins according to the scheme in Fig. 7.6.

TABLE 7.3

Structural requirements for Asn-linked glycopeptide binding to some lectin affinity columns

Concanavalin A (1,2,3,4)

At least two non-substituted or 2-O-substituted α-mannosyl residues are required for retention of glycopeptides on ConA-Sepharose.

Tri- and tetra-antennary complex units are not bound.

Bi-antennary complex structures bind and are eluted with 10 mM methyl α-glucoside.

High-mannose units (and some bi-antennary structures deficient in Gal and/or GlcNAc) and some hybrid structures are eluted with 500 mM methyl α-mannoside.

Pea lectin (5)

In addition to the presence of 2 α-linked Man residues the presence of Fuc attached to the Asn-linked GlcNAc is essential for binding. One α-Man can be substituted at C-6 but not C-4.

Pea lectin-Sepharose can be used to separate glycopeptides containing internal Fuc from the glycopeptide fraction which binds to ConA. In addition a selected population of tri-antennary complex units (not bound by ConA) can bind to pea lectin. Binding to pea lectin is enhanced by exposure of terminal Man residues.

Lentil lectin (5)

Structural requirements for binding are similar to pea lectin except that for lentil lectin exposure of terminal GlcNAc residues enhances binding.

Erythrophytohaemagglutinin (6,7)
(*Phaseolus vulgaris*)

Bi-antennary complex units containing outer Gal on the Man α1–6 branch and an intersecting GlcNAc residue are bound.

Leucophytohaemagglutinin (6)
(*Phaseolus vulgaris*)

Tri- and tetra-antennary complex type units with outer Gal residues and an αMan residue substituted at C-2 and C-6. (Substitution of α-linked Man at C-2 and C-4 prevents binding but the presence of outer NeuAc or Fuc residues does not influence binding.)

Wheat germ agglutinin (7)

High-mannose and some complex units have low affinity. Hybrid structures with an intersecting GalNAc residue bind strongly to WGA-Sepharose. The structure GlcNAcβ1–4Manβ1–4GlcNAcβ1–4GlcNAc-Asn is required for high-affinity binding. Substitution of the Asn-linked GlcNAc with Fuc decreases binding.

1. Baenziger, J. and Fiete, D. (1979) J. Biol. Chem. *254*, 2400–2407; 2. Krusius, T., Finne, J. and Rauvala, H. (1976) FEBS Lett. *71*, 117–120; 3. Ogata, S., Muramatsu, T. and Kobata, A. (1975) J. Biochem. (Tokyo) *78*, 687–696; 4. Narasimhan, S., Wilson, J.R., Martin, E. and Schachter, H. (1979) Can. J. Biochem. *57*, 83–96; 5. Kornfeld, K., Reithman, M.L. and Kornfeld, R. (1981) J. Biol. Chem. *256*, 6633–6640; 6. Cummings, R.D. and Kornfeld, S. (1982) J. Biol. Chem. *257*, 11230–11234; 7. Yamashita, K., Hitoi, A. and Kobata, A. (1983) J. Biol. Chem. *258*, 14753–14755; 8. Yamamoto, K., Tsuji, T., Matsumoto, I. and Osawa, T. (1981) Biochemistry *20*, 5894–5899.

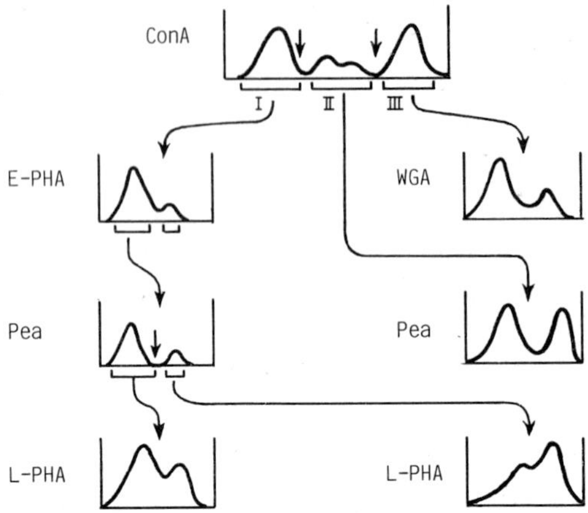

Fig. 7.6. Scheme for the fractionation of glycopeptides by sequential lectin affinity chromatography (Cummings and Kornfeld, 1982b). Columns contain the following lectins attached to agarose: concanavalin A (ConA), *Phaseolus vulgaris* erythrophytohaemagglutinin (E-PHA), pea lectin, *Phaseolus vulgaris* leucophytohaemagglutinin (L-PHA) and wheat germ agglutinin (WGA). The short arrows indicate the application of a new eluting agent.

Labelled glycopeptides are applied to a column (0.7 × 5 cm) containing 2 ml of ConA-Sepharose (Pharmacia) equilibrated with Tris-buffered saline (containing 0.15 M NaCl, 0.01 M Tris adjusted to pH 8.0 with HCl, 1 mM $CaCl_2$ and 1 mM $MnCl_2$). The column is eluted at room temperature with 10 ml TBS, then with 20 ml TBS containing 10 mM methyl α-glucoside and finally with 20 ml TBS containing 0.5 M methyl α-mannoside (or alternatively 0.1 M mannoside at 60°C may be employed). Fractions of 2 ml are collected at a flow rate of about 1 ml/min and aliquots are taken for scintillation counting. Unbound material (fraction I) contains triantennary and tetra-antennary Asn-linked glycopeptides (and Ser/Thr-linked carbohydrate units which should not be labelled). Fraction II, eluted with methyl α-glucoside, contains biantennary complex and some hybrid units, while tightly bound fraction III eluted with methyl α-mannoside contains high-mannose and some hybrid-type structures.

Further chromatography of fractions I–III is carried out according to the scheme shown in Fig. 7.6. Columns of pea and lentil lectin-Sepharose (1 × 7 cm) are equilibrated and operated in the same way as ConA-Sepharose. Preparations (Kornfeld et al., 1981) of *Phaseolus vulgaris* erythrophytohaemagglutinin (EHPA) coupled to Affi-Gel (1.8 mg protein/ml) and leucohaemagglutinin (LPHA)-Affi-Gel (0.6 mg protein/ml) are packed to give columns 0.5 × 30 cm. Commercial preparations of EPHA and LPHA coupled to agarose can also be employed. These columns (and wheat germ agglutinin-agarose) are equilibrated and eluted with PBS/NaN_3 (6.7 mM KH_2PO_4, 0.15 M NaCl, 0.01% NaN_3, pH 7.4) at room temperature at a flow rate of 10 ml/min and fractions (1 ml) are collected.

The structural requirements for binding of glycopeptides to particular lectins are outlined in Table 7.3. EPHA, for example, selects out certain glycopeptide structures containing an intersecting GlcNAc residue, while pea lectin binds only to glycopeptides containing Fuc linked to GlcNAc-Asn. The sequential use of lectin affinity as shown in Fig. 7.6 can result in the fractionation of very complex mixtures into components containing only one or a few oligosaccharide structures. The chromatographic behaviour of several glycopeptides in this system has been tabulated by Cummings and Kornfeld (1982a,b). However, our understanding of the detailed structural requirements for complex oligosaccharides to bind to lectins such as EPHA is continuing to develop (Yamashita et al., 1983) and it is still necessary to support conclusions about carbohydrate structure derived from lectin affinity chromatography with other analytical methods.

Fractionation of glycopeptides differing in galactose content by lectin affinity chromatography. Lectin affinity chromatography can be used to fractionate glycopeptides or oligosaccharides even when the binding affinity is only sufficient to retard elution. This has been elegantly demonstrated by the separation of glycopeptides differing in galactose content by passing them through a column of RCA I-agarose (Kornfeld et al., 1981). For this type of chromatography it is desirable to use a high ratio of insolubilised lectin to glycopeptides and to employ a long, narrow column.

Glycopeptides isolated from IgG were labelled by acetylation with [^{3}H]acetic anhydride and loaded onto a 1 × 50 cm column of RCA I-agarose which was eluted with phosphate-buffered saline pH 7.4. Fractions (2 ml) were collected, counted and the three major peaks in order of elution were shown by structural analysis to contain 0, 1 and 2 residues of galactose (Fig. 7.7).

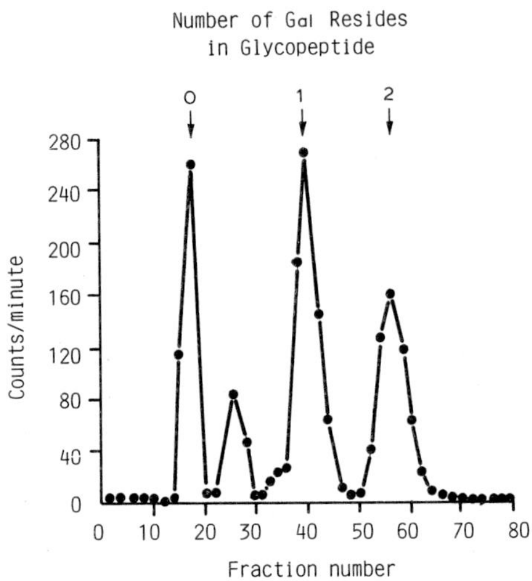

Fig. 7.7. Fractionation of weakly bound glycopeptides by lectin affinity chromatography (Kornfeld et al., 1981). A labelled mixture was passed through a column of RCA I-agarose as described in the text. The structures of the glycopeptides in the three main peaks differed only in their galactose content, the glycopeptide with most galactose being most retarded in elution.

7.5. Lectin staining methods

These techniques make use of the affinity of lectins for specific carbohydrate determinants in the detection of glycoproteins which have been separated by gel electrophoresis or isoelectric focusing. The technique has been used for identification of lectin receptors in membrane preparations. In addition evidence can be obtained about the chemical structure of the carbohydrate units of glycoproteins.

The principle of the method is that glycoproteins are separated on gels (usually in flat beds) and are then precipitated or fixed by a cross-linking agent. The precipitant and fixative are removed and replaced by buffer. Lectin, containing a fluorescent or isotopic label, is added and allowed to bind to the glycoprotein. Unbound lectin is washed from the gel and bands of lectin are detected by fluorescence, autoradiography or fluorography.

An important aspect of the method is that is can be applied to glycoproteins which have been separated under denaturing conditions such as apply in SDS or SDS-urea gel electrophoresis. Although the conformation of the peptide chain may be altered from the native state the carbohydrate moieties of glycoproteins are able to interact with lectins.

Overlay methods using ^{125}I-labelled lectins have been applied to red blood cell membrane glycoproteins (Tanner and Anstee, 1976; Robinson et al., 1975), liver cell membrane fractions (Gurd and Evans, 1976), synaptic plasma membrane fractions (Gurd, 1977), to the total glycoprotein of normal and transformed mouse 3T3 cells, rat and chick fibroblasts (Burridge, 1976), and to hybrid cell membrane glycoproteins (Bramwell and Harris, 1978). Fluorescently labelled lectins have been used to detect glycoproteins from the plasma membrane of *Dictostelium discoideum* (West and McMahon, 1977) and soluble glycoproteins (Furlan et al., 1979).

Lectin staining techniques can also be applied after transfer of glycoproteins from gels to other medium by ‘blotting’ techniques (Towbin et al., 1979).

Staining method of Tanner and Anstee (1976). Although this method was originally applied to a slice obtained from a cylindrical polyacrylamide gel it can be adapted to flat-bed polyacrylamide gels or gel slices 1–2 mm in thickness. This allows comparison between the mobilities of different samples. The following procedure is an adaptation of Tanner and Anstee's method suitable for use with polyacrylamide gels 100 × 200 × 1 mm.

Gels are fixed in 100 ml of 50% v/v methanol for 30 min. Glutaraldehyde, 25% v/v (0.2 ml), is added, mixed, and allowed to react for 1.5 h. This solution is poured off and replaced with 100 ml phosphate-buffered saline, pH 7.4 (PBS) and solid $NaBH_4$ (2 mg) is added. After 1 h the PBS and $NaBH_4$ are replaced with fresh solutions and incubation is continued overnight. The gel is washed twice with PBS, changing the buffer after two hours.

Gels are soaked in PBS (50 ml) containing 0.05% w/v sodium azide, bovine haemoglobin, 0.1 mg/ml, and 1% v/v Tween 20. This solution should be filtered before it is added to the gel. For ConA, Tris-buffered saline containing 1 mM $MnCl_2$ and 1 M $CaCl_2$ replaces PBS. Lectin labelled with ^{125}I (approx. 10^7 dpm) is added and incubated at room temperature for 24 h with occasional gentle mixing. Unbound lectin is removed by washing the gel in PBS containing azide and Tween 20 for 2 days with 2 changes of buffer per day with gentle shaking.

Controls for non-specific binding of lectin can be performed by staining strips of gel on which samples have been run but including a sugar inhibitor (e.g. methyl α-mannoside for concanavalin A) at a concentration 0.1 M in the initial solution to which lectin is added and in all the washes. Another effective internal control is to run standards containing known glycoproteins and non-glycosylated proteins on the gel. The low and high molecular weight calibration kits supplied by Pharmacia are suitable.

The gel can be stained for protein with Coomassie Blue before drying or it may be dried directly. Drying is carried out with the gel on a piece of filter paper placed on a porous plastic bed connected to a vacuum pump with the upper surface of the gel covered with Snap-Wrap. Bands of labelled glycoprotein are detected by autoradiography using Kodirex or other equivalent film. Alternatively the sensitivity of the method can be increased by treating the gel for fluorography (Bonner and Laskey, 1974) prior to drying and exposing to film.

Lectin staining (overlay) method of Burridge (1976, 1978). The gel slices are fixed for 2 h by shaking in a solution containing methanol:water:acetic acid, 5:5:1 by vol. or in a staining solution of the same composition except that it contains 0.1% Coomassie Brilliant Blue. With some lectins (e.g. *Lotus tetragonolobus* agglutinin) the use of Coomassie Blue must be avoided. However, for concanavalin A, *Ricinus communis* agglutinin I, wheat germ agglutinin and red kidney bean agglutinin, prior staining does not affect lectin binding. After shaking with 'fix' or 'stain' solutions these are replaced with a solution of 7.5% methanol, 7.5% acetic acid in water for several hours or until adequate destaining has occurred (this requires several changes of solution). Gel slices are brought back to neutral pH in buffer containing 0.15 M NaCl, 0.1% NaN_3, 50 mM Tris-HCl, pH 7.5.

Gel slices are laid out on a parafilm-covered horizontal glass sheet and excess buffer is gently removed. The iodinated lectin (10^6–2×10^7 cpm/ml) in the Tris-NaCl-NaN_3 buffer containing haemoglobin 2 mg/ml is distributed evenly over the surface of the gel slice. For concanavalin A, *Lens culinans* agglutinin and *Lotus tetragonolobus* lectin Ca^{++} and Mn^{++} (0.5 mM) are added. The gel slices are surrounded by moist filter paper, covered with the lid of a plastic box and allowed to incubate for 1–20 h. After incubation the slices are raised so that the overlay solution runs onto the parafilm from whence it is collected and may be reused at least twice provided no drying out has occurred. Unbound lectin is removed by gently shaking the slices in several changes of Tris-NaCl-NaN_3 buffer over a period of two days. Gels can be stained at this stage if required. Gels are dried down with the radioactive surface of the gel next to the filter paper. Drying and autoradiography or fluorography are carried out as described above for staining by the method of Tanner and Anstee.

Staining gels with fluorescent lectins (West and McMahon, 1977). SDS-polyacrylamide gels are sliced into lanes and fixed with methanol, glutaraldehyde and sodium borohydride as described above (Tanner and Anstee, 1976). For labelling, gel lanes are rocked for three days in a small plastic tray containing 16 ml of a solution composed of 0.1 M NaCl, 0.033 M sodium phosphate, pH 8.0, 0.2% bovine haemoglobin, 0.05% sodium azide and 0.4–0.6 mg lectin, conjugated with fluorescein isothiocyanate, per gel in the presence or absence of inhibitory sugar. Gels are washed for 2 days in two changes (200 ml per lane) of phosphate-buffered saline containing azide and, where appropriate, inhibitory sugar. Gels are illuminated from below with a short-wavelength ultraviolet light box and are photographed using a Wratten No. 65 filter.

Labelling of lectins for glycoprotein staining. Lectins can be labelled with ^{125}I by the chloramine T or lactoperoxidase methods described in Section 7.2.2. Some lectins (e.g. wheat germ agglutinin) can be used without repurification but this step is advisable with concanavalin A and lentil lectins. For lectins requiring metal ions (e.g. Mn^+ and Ca^+ for concanavalin A) these should be included during repurification of the labelled lectin and in all buffers used for binding and washing gels.

FITC-lectins are available commercially from several suppliers, including Vector, Sigma, Pharmacia P-L Biochemicals, Polysciences and Miles, or they can be prepared by standard methods.

Comments on glycoprotein staining methods. The fixation of proteins with glutaraldehyde and $NaBH_4$ is designed to prevent losses of glycoproteins which may occur when gels are soaked in buffer for long periods. It is important to follow the glutaraldehyde fixation procedure closely because an excess of glutaraldehyde leads to loss of Coomassie Blue staining (Wilson V.S., personal communication) and incomplete reduction of aldehyde groups can produce non-specific binding of labelled lectin to the gels. Fig. 7.8 shows the results obtained using this procedure to stain nuclear membrane glycoproteins with ^{125}I-wheat germ agglutinin.

The overlay technique of Burridge (1976, 1978) offers the advantage that small volumes of labelled lectin solution are required and the time taken for binding of lectin may be decreased. Staining with lectin by this method probably occurs only at one surface of the gel rather than at both faces as is the case with immersion of the gels in a solution of labelled lectin (Burridge, 1978).

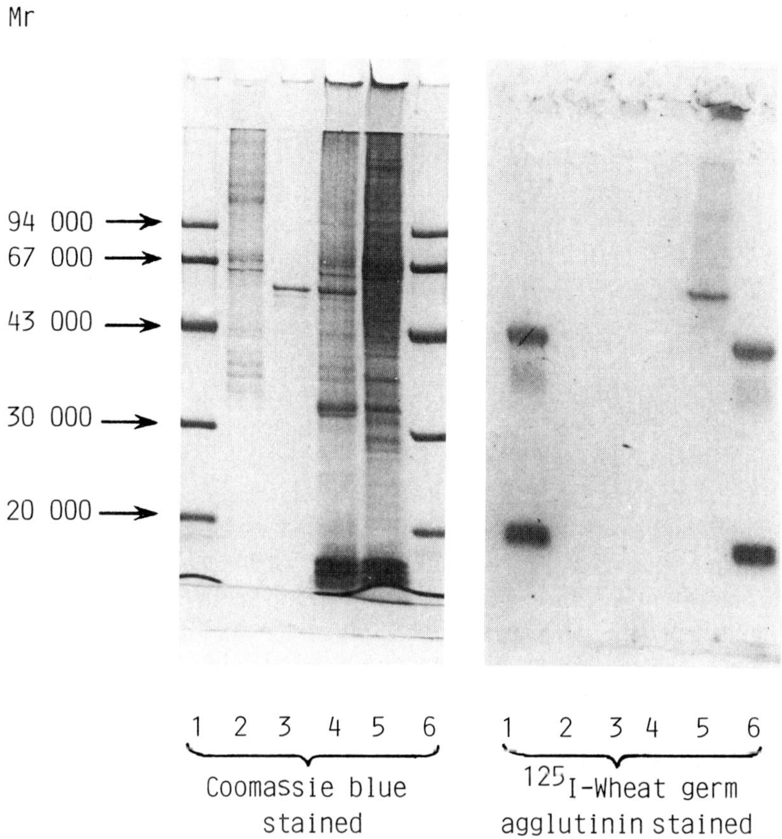

Fig. 7.8. Glycoprotein staining with labelled lectins. Samples containing 10–40 μg protein were separated by SDS-polyacrylamide gel electrophoresis, stained with Coomassie Blue and subsequently were stained with ^{125}I-wheat germ agglutinin and autoradiographed as described in the text (method developed from that of Tanner and Anstee, 1976). Samples were: tracks 1 and 6, Pharmacia molecular weight standards; tracks 2–4, nuclear proteins which do not bind WGA; track 5, rat liver nuclear membrane which contains one major and several minor WGA-binding glycoproteins. The WGA-stained band of M_r 43 000 in tracks 1 and 5 is ovalbumin and degradation products of ovalbumin are also detected.

Although the limits of detection of glycoproteins by lectin staining have not been precisely defined the use of lectins iodinated to high specific activity makes it a technique capable of showing the presence of large numbers of glycoproteins in a variety of biological systems (Burridge, 1976). Many glycoproteins can be detected by lectin staining when they are present in quantities which show very little or no staining with Coomassie Blue. Fluorescent techniques are not capable of such high sensitivity but may be of value in laboratories not equipped for handling ^{125}I.

The use of a single lectin stain will not reveal all of the glycoproteins in a complex mixture such as the components of human erythrocyte membrane (Tanner, 1979). Failure of a protein band to bind labelled concanavalin A therefore does not warrant the conclusion that the band is not a glycoprotein. In searching for glycosylated proteins it is desirable to use several lectins with as wide a range of specificities as possible: suitable kits of different lectins are now available commercially (e.g. from Vector). The activity of labelled lectin preparations should be checked by running glycoproteins known to bind the lectin on the same gel as the sample under examination. Alternatively a preparation of erythrocyte membrane may be used for this purpose.

When a single band on a gel is stained by more than one lectin several interpretations are possible. It may be that there are two different glycoproteins present with the same mobility. Alternatively there may be two different types of carbohydrate groups present within the same protein molecule or a single carbohydrate group may react with different lectins.

Even with high-resolution gels it can be difficult to be certain when examining a complex mixture whether bands stained for protein correspond to these binding a lectin. Staining the gel prior to drying and autoradiography greatly increases the precision with which Coomassie Blue and lectin binding components can be aligned. However, the best test of coincidence of staining is to run two-dimensional gels.

7.6. Lectin immunoprecipitation

This method can be used to distinguish different populations of membrane or other glycoproteins from each other or from non-glycosylated proteins. It also has potential for studying non-covalent interactions between glycoproteins and other molecules. Like the techniques for lectin staining, lectin immunoprecipitation is applicable only on an analytical scale.

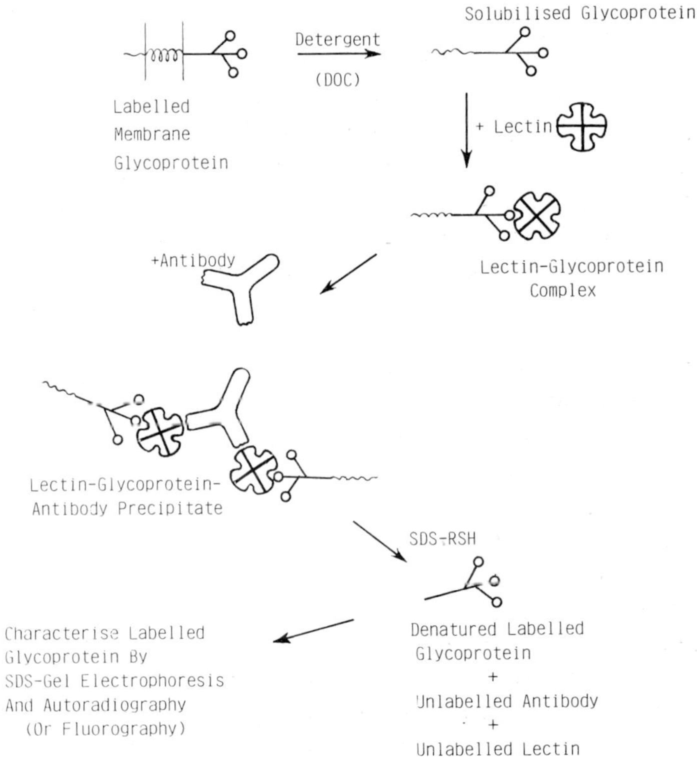

Fig. 7.9. Selective isolation of labelled glycoproteins by lectin immunoprecipitation.

Labelled glycoproteins are allowed to form a complex with a lectin and the complex is then precipitated by addition of anti-lectin antibody. The precipitate, consisting of glycoprotein, lectin and antibody, is dissolved and the glycoproteins can be fractionated under denaturing conditions and detected by virtue of their labelling (Fig. 7.9).

Membrane receptors for particular lectins can be identified by this method provided that they can be solubilised in detergents which do not interfere in the interactions between glycoproteins, lectin and antibody. An interesting aspect of the method is that, if the glycoprotein under investigation were associated with other macromolecules to form a complex, other components of the complex should also appear in the precipitate (Juliano and Li, 1978). As yet, however, the isolation of such glycoprotein associated components has not been experimentally observed by use of this technique.

Glycoprotein 'receptors' for concanavalin A and kidney bean phytohaemagglutinin on lymphocytes and wheat germ agglutinin receptors on CHO cell membranes have been identified (Juliano and Li, 1978) by this method.

Initial labelling of glycoproteins is carried out metabolically by the incorporation of [^{3}H]glucosamine (Juliano and Li, 1978) or by surface labelling with [^{125}I]iodide and lactoperoxidase (Henkart and Fisher, 1975). A high specific activity is desirable because of the small quantities of protein used.

Lectins may be prepared or purchased (Section 7.1.2). Antisera may be prepared as described below or can be purchased from Pharmacia P-L Biochemicals or Vector.

Lectin immunoprecipitation (Juliano and Li, 1978). Plasma membranes isolated from CHO cells are labelled to a specific activity of the order of 1 μCi/mg membrane protein using [^{3}H]glucosamine. A few milligrams of plasma membranes are solubilised in 1% DOC (containing 1 mM phenylmethanesulphonylfluoride to inhibit proteolysis) for 1 h at 4°C and insoluble material is removed by centrifuge.

Anti-lectin sera are raised in rabbits using graded doses of lectins in Freund's adjuvant over a period of 1–2 months. For highly toxic lectins it is necessary to begin with a very low dose (1 μg). Sera obtained from the animals prior to immunisations should be kept for use in control experiments. Specificity of antisera may be tested by immunodiffusion against lectin using plates made up of 1.2% agarose and 100 mM hapten sugar (to inhibit lectin binding to the carbohydrate groups of immunoglobulin) in Tris buffer. The quantity of antiserum giving maximal precipitation of each lectin is determined using ^{125}I-labelled lectins in 10 mM Tris, 0.2% DOC. Generally 50–100 μl of antiserum is sufficient to precipitate 10 μl lectin.

Precipitation of solubilised membrane (about 1–5 μg protein — see below) can be carried out in 0.2% DOC in 10 mM Tris, pH 8.0. Lectin, 10 μg, is added and samples incubated for 1 h at 4°C.

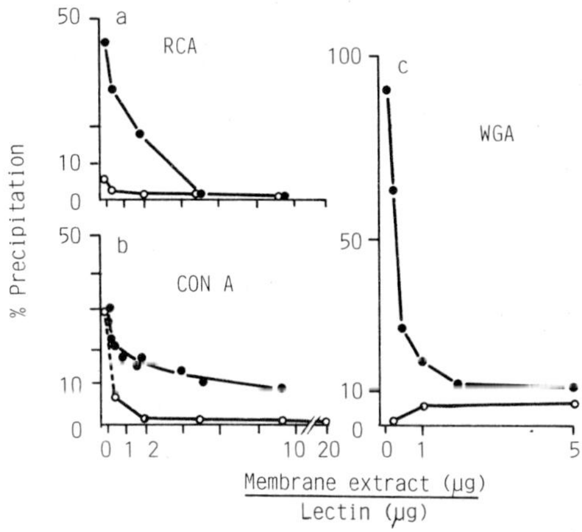

Fig. 7.10. Choice of conditions for lectin immunoprecipitation (Juliano and Li, 1978). Labelled membrane glycoproteins extracted in 1% deoxycholate are diluted in 5 ml of 0.2% DOC in 10 mM Tris-HCl, pH 8. Lectin (5 or 10 μg) is added and the samples are incubated for 1 h at 4°C. Sufficient antibody to precipitate all of the lectin is added and incubation continued for 1 h. The glycoprotein-lectin-antibody complexes are pelleted by centrifugation, washed twice and counted. Incubations are carried out with (○) and without (●) the appropriate inhibitory sugar (0.1 M) for each lectin. The graphs show the percentage of total counts precipitated by *Ricinus communis* agglutinin (a), concanavalin A (b), and wheat germ agglutinin (c), plotted against the ratio of protein in the membrane extract to the quantity of lectin added. Optimal ratios of extract to lectin can be established for each lectin.

Sufficient antiserum to produce maximal precipitation of lectin is then added and incubation is continued for a further hour. Immunoprecipitates are pelleted and washed twice.

The ratio of membrane protein in the extract to the amount of lectin has a very marked effect on the recovery of labelled membrane glycoprotein in the immune precipitate. High recoveries of glycoprotein are obtained only when there is a large excess of lectin (Fig. 7.10). Where the ratio of membrane protein to lectin is high little of the radioactivity may be recovered in the precipitate.

The complex of membrane glycoprotein, lectin and immunoglobulin can be solubilised and dissolved in SDS-mercaptoethanol and separation of the labelled glycoproteins is obtained by SDS gel electrophoresis. A flat-bed gel system is preferable and bands can be detected by fluorography (Bonner and Laskey, 1974). Controls should be performed using sugar inhibitor or serum fromnon-immunised rabbits.

Comments on the method. The quantity of protein added to the sample in the form of lectin and antibody is considerable. This limits the amount of sample which can be applied to gels without producing overloading and makes it necessary to start with membrane labelled to high specific activity.

The ratio of lectin to membrane protein may influence the particular 'spectrum' of glycoprotein molecules which are isolated. The population of molecules obtained at higher ratio of extract to lectin may reflect binding to high-affinity receptors (Juliano and Li, 1978).

It is probable that use of detergents other than DOC or mixed detergent systems could improve the precipitation obtained with some lectins. Lotan et al. (1977) have shown that DOC can have a deleterious effect on the interaction between certain lectins and glycoproteins.

Information about the exposure of membrane glycoproteins can be obtained if labelling is carried out with a reagent which does not penetrate the lipid bilayer. Those glycoproteins which are labelled by such a reagent may be selectively precipitated by a combination of lectin and antibody (Henkart and Fisher, 1975).

7.7. *Lectin precipitation methods*

7.7.1. *Introduction*

Lectins may form precipitates with glycoproteins or polysaccharides in an analogous manner to the precipitin reactions of antibodies. For precipitation to occur the carbohydrate component in the reaction must have more than one binding site for lectin.

Precipitation reactions have been used for the assay of lectins (Goldstein, 1976) and in a range of techniques used to detect glycoproteins. These methods are derived from the antibody techniques of immunoprecipitation, immunodiffusion (Ouchterlony, 1967), immunoelectrophoresis (Grabar and Williams, 1953) and crossed antigen-antibody electrophoresis (Laurell, 1965) with lectins used in place of, or in addition to, antibodies.

Lectin and glycoprotein may be allowed to form precipitates by diffusion in agar gels either with or without prior electrophoresis of the glycoprotein or the glycoproteins may be electrophoresed into a gel containing lectin or through an intermediate gel of immobilised lectin (Bøg-Hansen, 1979). These techniques have been used for the detection of glycoproteins from plasma membrane of *Dictyostelium discoideum* (West and McMahon, 1977), human erythrocyte membrane (Bjerrum and Bøg-Hansen, 1976) and from *Herpes simplex* virus (Vestergaard and Bøg-Hansen, 1975).

7.7.2. *Precipitation of glycoproteins or polysaccharides from solution*

Direct precipitation of glycoproteins or polysaccharides has received limited attention as a preparative or analytical technique. Lectins are not always effective precipitants of glycoproteins. For preparative purposes affinity chromatography (Section 7.4) is a more widely applicable technique or, for small quantities of labelled glycoprotein, the efficiency of precipitation can be greatly improved using the lectin-antibody procedure (Section 7.6). The precipitation of polysaccharides with lectins has been successfully applied to measurement of the activity of concanavalin A (So and Goldstein, 1967; Goldstein, 1976).

7.7.3. Lectin precipitation in gels (lectin immunodiffusion)

If lectin and glycoprotein are allowed to diffuse together from separate wells in agarose gels arcs of precipitation may be obtained as in the Ouchterlony (1967) immunodiffusion technique. Gels containing agarose 1% (w/v) in phosphate buffered saline may be used (except for lectins which bind to agarose). Precipitates may be visible in the gel or may be detected by washing unprecipitated protein out of the gel with 0.75 M NaCl, drying the gel underneath a sheet of filter paper at 55°C anc then staining with Coomassie Blue and destaining.

Control experiments in which a sugar inhibitor (0.1 M) of the lectin is included in the gel should be carried out.

7.7.4. Electrophoresis of glycoprotein into lectin-containing gels

Electrophoresis of glycoproteins into thin layers of lectin-containing gels can be used in a way analogous to rocket immunoelectrophoresis for the identification and quantitation of glycoproteins (Bøg-Hansen et al., 1977). A glycoprotein mixture may first be separated by electrophoresis and then a second electrophoresis at right angles to the first is carried out so that the glycoproteins migrate through an agarose gel containing lectin. Formation of lectin-glycoprotein precipitates occurs when the size of complexes is too great for them to move through the agarose gel. The distance the precipitate extends (i.e. the height of the 'rocket') is dependent on the concentration of glycoprotein. Quantitation of a pure glycoprotein can be made by comparing the size of the rocket produced by direct electrophoresis of a sample of the glycoprotein with standards (Bøg-Hansen et al., 1977).

Bøg-Hansen et al. (1977) carried out electrophoresis in a 1.5 mm layer of 1% w/v agarose using either Veronal (91 mM 5,5-diethyl barbiturate, pH 8.6) or Tris-Veronal (73 mM Tris, 24.5 mM 5,5-

diethyl barbiturate, 0.36 mM calcium lactate and 0.02 mM NaN_3, pH 8.6). For crossed electrophoresis the first dimension was run at 10 V/cm for 30–90 min and the second dimension was electrophoresed overnight at 1.5 V/cm. The lectin concentrations used in the second gel were within the range 13–130 μg/ml. It was observed that the precipitates obtained with lectins had a characteristically 'filled' appearance (Bøg-Hansen et al., 1977) as compared to the arcs of precipitation seen with conventional crossed antibody electrophoresis (Laurell, 1965).

Denaturation of protein does not inhibit the formation of lectin-glycoprotein precipitates in agarose gels (Bøg-Hansen et al., 1977). The identification of glycoproteins after electrophoresis in denaturing conditions, such as SDS-gel electrophoresis, can be achieved using crossed immunoelectrophoresis if the interference of SDS in the lectin-glycoprotein association can be prevented. West and McMahon (1977) have described a method for electrophoresis into a lectin-containing gel of glycoproteins separated by SDS-gel electrophoresis. Their procedure is to place the SDS-polyacrylamide gel strip next to an agarose gel so that the glycoproteins migrate through a layer of gel containing the non-ionic detergent Lubrol before entering the gel containing lectin. It was possible to identify several concanavalin A-binding proteins from the plasma membrane of *Dictyostelium* using this technique (West and McMahon, 1977).

Quantitation of glycoproteins using the rocket technique has also been described (Bøg-Hansen et al., 1977). The limit of detection using Coomassie Blue staining is about 10 ng glycoprotein. Correlation between rocket height and amount of purified protein was not linear.

7.7.5. Electrophoresis through gels containing insolubilised lectin

This technique can be used to identify those antigens in a mixture which bind to a particular lectin. It has also been suggested that the method is useful in evaluating lectin affinity chromatography materials (Bøg-Hansen, 1973).

The basis of the technique is that the sample is electrophoresed through a layer of insolubilised lectin and the components which pass through this barrier are compared with those appearing in the absence of insolubilised lectin. An immobilised lectin barrier has been used prior to the second dimension of crossed immunoelectrophoresis where detection of the components in the sample requires precipitating antibody (Bøg-Hansen, 1973). In principle it should be possible to use this technique in conjunction with other separation and detection methods.

7.8. *Covalent linking of lectins to receptors*

Because of the reversibility of the binding of lectins it is often impossible to establish directly the particular glycoproteins which act as 'receptors'. However, if potentially reactive groups can be attached to the lectin before binding to its receptor and these groups are then activated it should be possible to covalently cross-link the lectin to its receptor (or other molecules in very close proximity) which can subsequently be identified (Jaffe et al., 1979). This experimental technique has been introduced using a heterobifunctional reagent.

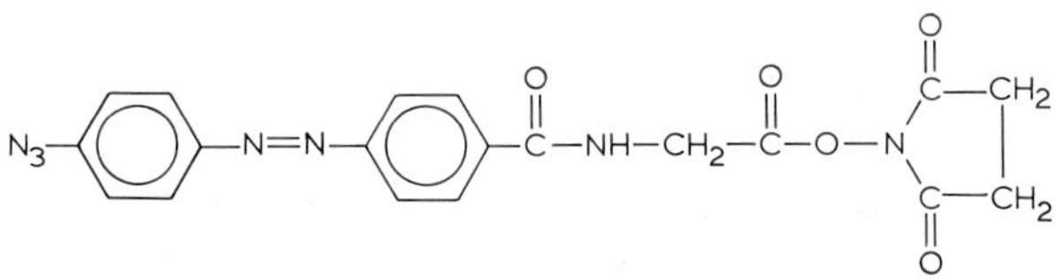

After covalent attachment to lectin and binding to its receptor this reagent can be converted by light to a highly reactive nitrene radical capable of forming covalent bonds with neighbouring residues. Subsequently the cross-link can be cleaved by dithionite and the labelled receptor identified.

Jaffe et al. (1979) have used this technique to identify peanut agglutinin receptors on human erythrocyte membranes.

CHAPTER 8

Radioactive labelling techniques

8.1. Introduction

Isotopic labelling techniques have become of great importance in the detection, characterisation and structural analysis of glycoproteins and proteoglycans as well as in studies of the biosynthesis and functions of these molecules. The high sensitivity with which radioisotopes can be detected and quantitated makes possible the investigation of minute amounts of glycosylated proteins. Selective labelling of particular types of sugar residue or substituents such as sulphate groups can also be of great practical value even when larger quantities of sample are available.

Techniques for labelling glycoproteins and proteoglycans by the biosynthetic incorporation of radioactive precursors into intact cells are described in Sections 8.2.1–5, while in vitro labelling using radioactive nucleotide sugars and glycosyl transferases is considered in Sections 8.2.6–7. Other chemical and enzymatic labelling techniques not involving biosynthetic reactions are given in Section 8.3.

Both in vivo and in vitro labelling can aid the characterisation and structural analysis of glycoproteins and proteoglycans. Labelling can serve to

(1) identify molecules which are glycosylated so aiding their purification (Section 8.2.3);

(2) give information about the types of sugars present (Sections 8.2.2 and 8.3);

(3) act as a marker for the carbohydrate units during structural analysis (Sections 8.2.3 and 8.3);

(4) give information, through the use of specific inhibitors of glycosylation, about the numbers per peptide chain and linkages of glycoprotein carbohydrate groups (Section 8.2.8). In addition, the transfer of labelled sugars to glycoprotein acceptors by purified, highly specific, glycosyl transferases can be used to probe oligosaccharide structure (Sections 8.2.6–7).

Radioactive labelling techniques are of central importance in the investigation of the biosynthesis of glycosylated proteins for detection of intermediates and identification of the subcellular sites of steps in biosynthesis (Sections 8.2.4 and 8.2.5), measuring rates of synthesis (Section 8.2.5), examining the temporal relationship between peptide and oligosaccharide synthesis and processing (Sections 8.2.6–7) and establishing the effects of inhibitors on biosynthesis (Section 8.2.8). In the space of this chapter it will only be possible to mention a few examples of the many possible approaches to studying glycoprotein and proteoglycan synthesis. Some techniques will be considered only in outline and some approaches, of which the author has no experience, will for that reason be omitted. For example, no attempt will be made to discuss the technique of autoradiography for the location of subcellular sites of glycoprotein synthesis.

Some of the techniques employed to introduce radioactive label into glycoproteins are of importance in studies of the function of these molecules. Purified glycosyl transferases can be employed to produce selective modifications of the structure of soluble or membrane-bound glycoproteins (Section 8.2.6). The chemical and enzymatic modifications of sialic acid, galactose and *N*-acetylgalactosamine described in Section 8.3 can also be used to study the functional effects of carbohydrate modification. Glycosylation inhibitors (Section 8.2.8) can produce peptide chains with altered or missing carbohydrate units whose fate can be examined in vivo or in vitro.

The methods discussed in Section 8.3 can be applied to labelling cell surface glycoproteins and the galactose oxidase-NaB^3H_4 procedure has been employed as a vectorial label.

In addition to the specific labelling of the carbohydrate moieties of glycoproteins it is possible to introduce label into their peptide chains either biosynthetically or by modification of amino acid side-chains (e.g. by lactoperoxidase-catalysed iodination). These types of procedures are basically the same for glycoproteins as for other proteins and will not be discussed in this monograph.

8.2. *Biosynthetic labelling*

This section is intended as a short introduction to practical aspects of the isotopic labelling of glycoproteins and proteoglycans, in vivo and in vitro by biosynthetic mechanisms. Comprehensive description of the biosynthesis of these molecules can be found in review articles relating to glycoproteins (Schachter, 1978; Schachter and Roseman, 1980) and proteoglycans (Rodén, 1980).

Mammalian glycoproteins can be labelled in preparations containing intact cells (i.e. whole animals, organ or tissue preparations or cultured cells) by the addition of radioactive monosaccharides or sulphate. Monosaccharides can be converted to sugar nucleotides by the intracellular pathways indicated in Fig. 8.1. The nucleotide sugars (UDP-glucose, UDP-galactose, UDP-*N*-acetylglucosamine, UDP-*N*-acetylgalactosamine, UDP-glucuronic acid, UDP-xylose, GDP-mannose, GDP-fucose and CMP-*N*-acetylneuraminic acid) are important intermediates in glycoprotein and proteoglycan synthesis. Transfer of monosaccharides from nucleotide sugars to the non-reducing terminal positions in carbohydrate acceptors or to the side chains of hydroxyamino acids in polypeptide acceptors catalysed by specific glycosyl transferases is a fundamental mechanism in the assembly of the carbohydrate units of proteoglycans and glycoproteins.

The sugar nucleotides also participate in the interconversion of monosaccharides (e.g. the epimerisation of UDP-glucose toDP-galactose). A monosaccharide may therefore give rise to more than one sugar nucleotide. For example, labelled glucosamine can be converted to UDP-GlcNAc, UDP-GalNAc and CMP-NeuAc and consequently not only labelled GlcNAc but also labelled GalNAc and NeuAc can appear in glycoproteins. The extent of such intercon-

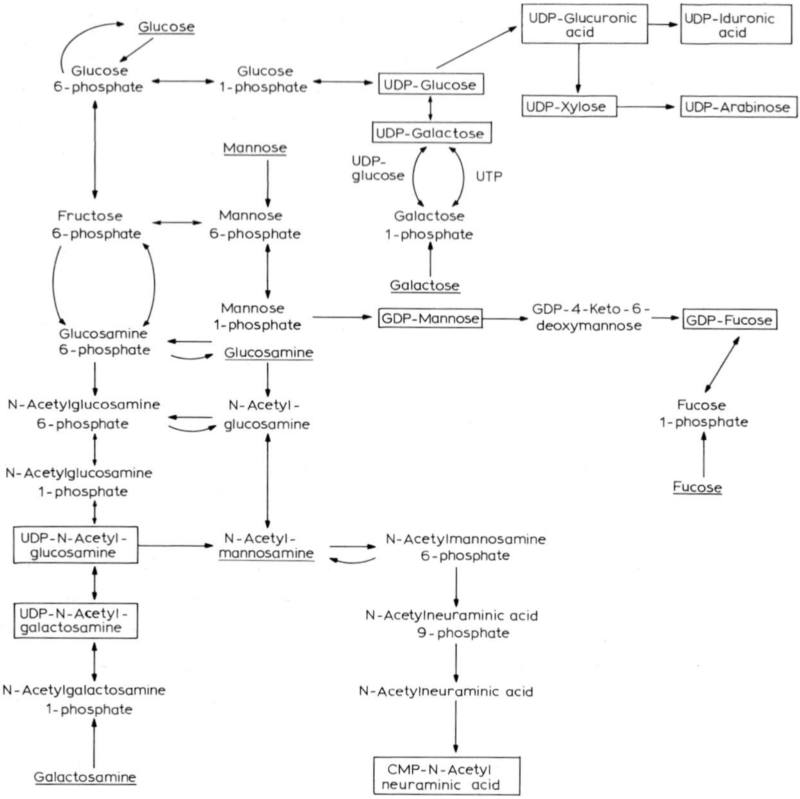

Fig. 8.1. Pathways of biosynthesis of nucleotide sugars (Schachter, 1978). Precursors of glycoprotein or proteoglycan carbohydrate units are underlined.

versions of labelled sugars depends on the monosaccharide employed, the tissue or cell type and experimental factors such as the concentration of glucose in the incubation medium. It is therefore necessary to establish the nature of the labelled monosaccharide actually incorporated into macromolecules under the particular conditions employed in any investigation.

Nucleotide sugars, in contrast to monosaccharides, do not pass readily across the plasma membrane. They are not used for labelling intact cell preparations except in experiments designed to detect extracellular glycosyl transferase activity (see Schur, 1982, for review). However, sugar nucleotides labelled in their monosaccharide components are extensively used for in vitro labelling of glycoproteins and proteoglycans. The activated form of sulphate, 3′-phosphoadenosine 5′-phosphosulphate (PAPS) is employed for in vitro labelling of sulphated glycoconjugates.

The transfer of sugars from nucleotide sugar donors to polypeptide or oligosaccharide acceptors catalysed by highly specific glycosyl transferases probably accounts for the assembly of the mammalian serine- and threonine-linked glycoprotein and proteoglycan oligosaccharide chains and for the addition of peripheral monosaccharide residues to asparagine-linked glycoconjugates. However, the biosynthesis of the 'core' portion of asparagine-linked carbohydrate units present in many glycoproteins and certain proteoglycans (with keratan sulphate type 1 chains) differs in some important ways from this pattern. The mannose- and glucosamine-containing cores of these *N*-linked units are assembled as oligosaccharides attached via a pyrophosphate group to the isoprenoid compound dolichol. This 'lipid intermediate' has the structure shown in Fig. 8.2 and the carbohydrate portion can be transferred en bloc to the side chain of an asparagine residue in the nascent peptide chain of a glycoprotein in a reaction which involves enzymatic recognition of an appropriate acceptor sequence of amino acids. There is usually very close coupling between peptide chain synthesis and glycosylation. The carbohydrate unit is subsequently trimmed by the removal of monosaccharides by exoglycosidases within the endoplasmic reticulum. Further additions of sugars may then occur as a result of the action of glycosyl transferases. The nature of the final product, which may be a high-mannose, hybrid or complex-type of carbohydrate unit, appears to be dependent on the trimming and subsequent processing events which take place. The structures of intermediates and the enzymology of processing have been reviewed (Schachter and Roseman, 1980;

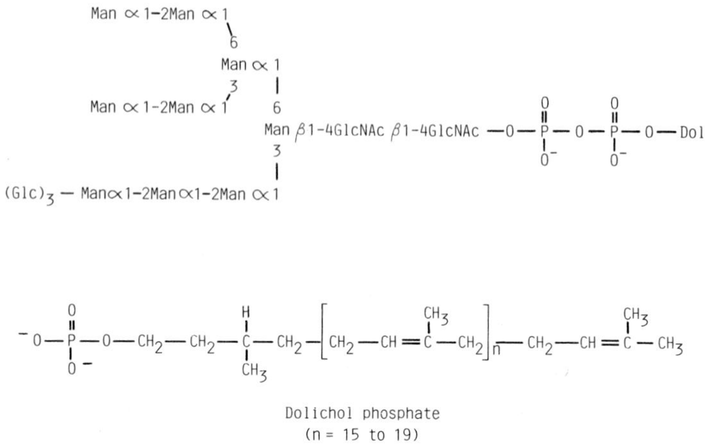

Fig. 8.2. Lipid-linked intermediate in the biosynthesis of asparagine-linked oligosaccharide units. In the upper structure the dolichol moiety of the lipid-linked oligosaccharide ($Glc_3Man_9GlcNAc_2$-PP-Dol) is represented by Dol. The structure of dolichol monophosphate is illustrated in the lower structure.

Kornfeld, 1982; Schwartz and Datema, 1982b). Features of the biosynthesis of asparagine-linked carbohydrate units to bear in mind when designing labelling experiments include the close linkage of core glycosylation to protein synthesis, the requirement for lipid-linked (membrane associated) intermediates and the subsequent processing events which are possible. Advantage may be taken of the specific inhibitors of the assembly of lipid-linked intermediates which are available (Section 8.2.7).

The oligosaccharide chains of proteoglycans can undergo considerable modification between polymer formation and the release of completed molecule from the cell. For example, the L-iduronic acid present in dermatan sulphate, heparin sulphate and heparan sulphate is not incorporated from UDP-iduronate but arises by enzymatic epimerisation of D-glucuronic acid in the polymer chain. A complex sequence of modifications occurs in the biosynthesis of heparin and heparan sulphates (Rodén, 1980). The formation of the *N*-sulphate groups in these molecules involves the initial incorporation of *N*-acetylglucosamine which is then de-*N*-acetylated and *N*-sulphated in the polymer chain. In heparin synthesis these events are followed by epimerisation of glucuronic acid to iduronic acid, 2-*O*-sulphation and 6-*O*-sulphation. The donor for both *N*- and *O*-sulphate groups is PAPS. Complex modifications of oligosaccharide chains often occur in rapid succession in intact cells. In vitro experiments, in which substrate availability can be restricted, are often valuable for defining steps in processing pathways.

Biosynthesis of glycoproteins and proteoglycans is closely linked with the endomembrane systems of animal cells. Some glycoproteins and proteoglycans (forms of heparan sulphate) are intimately associated with the lipid bilayer of cell membranes, while many of these molecules are secretory products of cells. Their carbohydrate units may be subject to modification in different subcellular compartments during their passage through the intracellular membrane system. This has been particularly clearly demonstrated for glycoproteins with asparagine-linked carbohydrate units (for review see Gogstad and Helgeland, 1982) and the type of carbohydrate modification present in glycoproteins may be used as a marker for their cellular destination (Rothman and Fine, 1980). In addition to the processing of the carbohydrate units of glycoproteins the peptide chains of these molecules undergo processing by proteolytic cleavage or other secondary modifications. The effects of these peptide chain modifications on glycosylations are not as yet understood. However, it should be remembered when labelling glycoproteins that differently glycosylated molecules may be present in different intracellular and extracellular compartments (Section 8.2.3).

8.2.1. Labelling intact cells

Glycoproteins and proteoglycans can be radioactively labelled in whole animals, tissue slices or cultured cells by the incorporation of

^{3}H or ^{14}C-labelled monosaccharides or[^{35}S]sulphate. Labelling can be used to identify glycosylated molecules and as an aid in the structural analysis of small amounts of material. Alternatively, slightly different labelling strategies can be used to study biosynthetic and processing pathways.

Precursors which can be employed either singly or in combination for in vivo labelling of glycoproteins and/or proteoglycans are listed in Table 8.1, which indicates the nucleotide sugars formed from them and the predominant monosaccharide residues which are labelled.

TABLE 8.1
Monosaccharide precursors for labelling intact cells

Precursor	Nucleotide sugars	Main sugar residues labelled	Comment
D-Glucosamine	UDP-GlcNAc UDP-GalNAc CMP-NeuAc	GlcNAc GalNAc NeuAc	Good general label for glycoproteins and proteoglycans
L-Fucose	GDP-Fuc	Fuc	Labels some, but not all *N*-linked and *O*-linked glycoprotein units. Little catabolism of fucose occurs. Not usually extensively converted to other sugars.
D-Mannose	GDP-Man + others	Man + others	Label for glycoprotein with *N*-linked units. Specificity of labelling must be checked.
D-Galactose	UDP-Gal + others	Gal + others	Labels many glycoproteins and proteoglycans. Considerable conversion to other sugars and amino acids.
D-Glucose	UDP-Glc + all others	All sugars	Labels glycoproteins and proteoglycans. May be converted to any sugar, but also to amino acids.
D-*N*-Acetylmannosamine	CMP-NeuAc	NeuAc	Specific label in some cell types.

The most useful general precursor for both glycoproteins and proteoglycans is labelled glucosamine. This amino sugar is not converted into hexose because the formation of glucosamine 6-phosphate from fructose 6-phosphate (Fig. 8.1) is essentially irreversible. All types of carbohydrate chains present in mammalian proteoglycans and glycoproteins (except for collagen) are labelled with glucosamine. [^{35}S]sulphate (carrier-free $Na_2^{35}SO_4$) is a good general precursor for proteoglycans (except hyaluronate); some labelling of sulphated glycoproteins can also occur.

Of the precursor monosaccharides which label only a restricted range of glycoproteins L-fucose and D-mannose have been most widely employed. Fucose is a good precursor in the sense that there is relatively little catabolism or conversion to other monosaccharides (Bekesi and Winzler, 1967). This sugar occurs in some, but not all, *O*-linked and *N*-linked carbohydrate units of glycoproteins. The number of residues of fucose per oligosaccharide chain is often low. D-Mannose has been used extensively as a label for *N*-linked carbohydrate units. *N*-Acetylmannosamine appears to be a specific precursor of sialic acid but has not been widely applied. Difficulties arising from the large quantities of glucose present as an energy source, and the conversions of hexoses and their metabolism to amino acids limit the usefulness of glucose as a glycoprotein and proteoglycan precursor.

Monosaccharides labelled with ^{3}H or ^{14}C which are suitable for use as precursors are commercially available (Amersham International; New England Nuclear). Tritiated sugars are now more widely used because of their lower cost and high specific radioactivity which more than offsets the slight disadvantage in liquid scintillation counting efficiency of ^{3}H compared with ^{14}C. Some examples of suitable forms of precursor are D-[6-^{3}H]glucosamine hydrochloride, D-[1-^{14}C]glucosamine hydrochloride, D-[2-^{3}H]mannose, L-[6-^{3}H]fucose, L-[1-^{14}C]fucose. If monosaccharides are obtained from the manufacturer dissolved in solutions containing ethanol they should be evaporated to dryness and redissolving in sterile saline as ethanol may have adverse effects on the growth of cultured cells.

For many purposes it is important to ensure that glycoproteins or

TABLE 8.2

Labelling intact cells for the structural characterisation of glycoproteins and proteoglycans

---

Characterisation of membrane glycoproteins

Plasma membranes of mouse L-929 cells (Hunt and Brown, 1974)

Cultured cells in medium containing 10% calf serum were labelled with D-[^{3}H]glucosamine (4 μCi/ml) and [^{14}C]-amino acids for 48 h.

Glycoproteins from smooth membranes were separated and characterised by SDS-gel electrophoresis. The results were correlated with vectorial galactose-oxidase/NaB^3H_4 labelling of cell surface glycoproteins and with the effects of trypsinising labelled cells.

Synaptic plasma membrane glycoproteins (Gurd and Mahler, 1974)

Labelling was carried out in vivo by intracranial injection of rats with 50 μCi L-[^{3}H]fucose in phosphate-buffered saline. Animals were killed after 16 h.

Synaptic plasma membranes were isolated, glycoproteins were solubilised in 1% sodium deoxycholate and characterised by lectin affinity chromatography.

Labelling asparagine-linked glycoproteins

Glycoproteins from human skin fibroblasts (Tulsiani and Touster, 1983)

Cultured human skin fibroblasts were labelled for 24 h with [^{3}H]mannose (100 μCi/ml) in medium supplemented with dialysed fetal calf serum and 0.1 mg/ml glucose.

After extraction of lipid-linked intermediates with chloroform:methanol:water (1:1:0.3 vol./vol.) the residue was digested with pronase and glycopeptides were purified on Bio-Gel P4. Following endoglycosidase H digestion the carbohydrate units were subjected to exoglycosidase digestion and structural analysis. The effects on carbohydrate structure of the trimming inhibitor swainsonine were reported.

Proteoglycans

Dermatan sulphate and heparan sulphate proteoglycan from cultured rat ovarian granulosa cells (Yanagishita and Hascall, 1983a,b)

Cultured cells in medium containing 10% fetal calf serum were labelled with $Na_2{}^{35}SO_4$ (10–50 μCi/ml, 10–25 mCi/mmol), D-[6-^{3}H]glucosamine HCl (50 μCi/ml, 20–38 Ci/mmol), D-[2-^{3}H]mannose (100 μCi/ml, 16 Ci/mmol) or L-[^{3}H]serine (50 μCi/ml, 2.8 Ci/mmol) at 37°C for 24 h.

Proteoglycans were isolated from the culture medium by gel filtration in 4 M guanidine HCl, DEAE-Sephacel ion-exchange chromatography in 8 M urea containing a zwitterionic detergent and by CsCl centrifugation under dissociative conditions. The proteoglycans were treated with alkaline borohydride and small *O*-linked and *N*-linked oligosaccharides and glycosaminoglycan chains were characterised. The size of the serine-labelled peptide core was estimated and the distribution of oligosaccharide chains on the peptide was examined by isolation of tryptic fragments.

Oligosaccharides in proteoglycans of Swarm rat chondrosarcoma (Lohmander et al., 1980)

Monolayer cultures of chondrocytes were labelled with [^{3}H]glucosamine (50 μCi/ml) or [^{3}H]mannose (100 μCi/ml) for 24 h.

Radioactive proteoglycans were isolated from the culture medium in the presence of unlabelled carrier chondrosarcoma proteoglycan monomer, by CsCl gradient centrifugation under dissociative conditions followed by gel filtration. The proteoglycan monomer was treated with alkaline borohydride and released *O*-linked oligosaccharides and *N*-linked glycopeptides were fractionated by gel filtration on Bio-Gel P10. Oligosaccharide units linked, in the proteoglycan, to serine and threonine side-chains were characterised by primary structure determination.

Isolation of proteoglycans synthesised by cloned bone cells (Hunter et al., 1983)

Cloned bone cells derived from fetal rat calvaria cultured in medium containing 5% dialysed fetal calf serum were labelled with $Na_2{}^{35}SO_4$ (100 μCi/ml) or D-[^{3}H]glucosamine (5 μCi/ml) for 16 h.

Cells and media were extracted with 4 M guanidine HCl containing protease inhibitors, and proteoglycans were purified by gel filtration and ion-exchange chromatography in 8 M urea. Glycosaminoglycan chains were released by Pronase digestion or by alkaline borohydride treatment and were characterised by chemical and enzymic degradation.

Gurd, J.W. and Mahler, H.R. (1974) Biochemistry *13*, 5193–5198; Hunter, G.K., Heersche, J.N.M. and Aubin, J.E. (1983) Biochemistry *22*, 831–837; Hunt, R. and Brown, J.C. (1974) Biochemistry *13*, 22–28; Lohmander, L.S., De Luca, S., Nilsson, B., Hascall, V.C., Caputo, C.B., Kimura, H. and Heinegard, D. (1980) J. Biol. Chem. *255*, 6084–6091; Tulsiani, D.R.P. and Touster, O. (1983) J. Biol. Chem. *258*, 7578–7585; Yanagishita, M. and Hascall, V.C. (1983a) J. Biol. Chem. *288*, 12847–12856; Yanagishita, M. and Hascall, V.C. (1983b) J. Biol. Chem. *288*, 12857–12864.

proteoglycans are labelled to high specific radioactivity. This is, for example, the case when membrane glycoproteins are to be analysed by a separation method such as SDS-gel electrophoresis in which only a limited amount of sample can be applied. In addition to using labelled precursor of high specific activity other experimental factors can be modified. When cultured cells are labelled dilution of the precursor can be minimised by, for example, dialysis of serum present in the growth medium to remove monosaccharides. The carbohydrate energy supply for cells (usually glucose) may compete with precursor for the uptake or dilute intracellular pools of radioactive intermediates. For example, the specific radioactivity of the intracellular pool of UDP-*N*-acetylglucosamine in cells labelled with radioactive glucosamine is regulated (in part) by the relative concentrations of glucose and glucosamine in the medium (Kim and Conrad, 1976; Pouysségur and Pastan, 1977). Decreasing the glucose concentration in the medium or partial substitution with another energy source such as pyruvate can lead to increased incorporation of radioactive glucosamine (Harris and Johnson, 1969). Similarly, the use of galactose as energy supply instead of glucose or mannose increased the incorporation of labelled mannose into glycoproteins of (galactose-adapted) *S. cerevisiae* (Byrd et al., 1982). However, the medium should not be altered in any ways that modify growth or macromolecular synthesis. The length of incubation of cells with labelled precursors can also be adjusted to maximise the specific activity of the glycoconjugates being examined. As well as considering specific radioactivity it is often necessary, for economic reasons, to have in mind the efficiency with which the added label is incorporated. Keeping the volume of medium present during labelling of cells as low as is compatible with their unrestricted growth can increase the efficiency of uptake of label. Some examples of labelling conditions which have been applied in different cells and tissues are given in Table 8.2 and in Section 8.2.3.

Labelling of glycoproteins (or proteoglycans) in experimental animals is carried out by intravenous or intraperitoneal injection of the precursor dissolved in saline. Intraperitoneal administration is technically the simpler procedure to apply to rats or mice but intravenous injection by the caudal vein (Evans, 1978) or the femoral vein (Lawford and Schachter, 1967) probably results in slightly faster distribution. To examine labelling of rat liver glycoproteins after periods as short as 2 min label can be injected directly into the hepatic portal vein (Kreibich et al., 1973) of anaesthetised animals. For labelling cells or tissue slice preparations the isotope dissolved in sterile saline is added to the medium.

After labelling, the nature of the labelled products can be examined either as a group (e.g. hexosamine) or, preferably, by purification of specific molecules (e.g. glucosamine, galactosamine). When cultured cells have been labelled the extracellular material and cells are treated separately; the cells are washed to remove excess unincorporated label. Products (and precursor pools) can be fractionated on the basis of their solubility in trichloroacetic acid and/or lipid solvents (Section

8.2.3). Separation techniques (see Chapter 3) such as gel filtration, ion-exchange chromatography or electrophoresis can be applied to the intact macromolecules or following proteolytic digestion to form glycopeptides or glycosaminoglycans. Specific isolation methods such as immunoprecipitation or affinity chromatography can be employed to select out specific macromolecules.

The macromolecular products should be examined to ascertain the extent to which labelled precursors have been converted to other sugar residues or non-carbohydrate components.

8.2.2. Determination of the extent of interconversion of label

The extents to which labelled precursors are converted to other sugars have been tabulated for several cell types by Yurchenco et al. (1978). However, it is advisable to check the degree of interconversion occurring when setting up any new biosynthetic system. This is done by hydrolysis of the glycoconjugate followed by separation and identification of the radioactive products. Optimal hydrolysis conditions for the release of the labelled sugar should be employed. For glycoproteins labelled with fucose, mannose or galactose hydrolysis can be carried out with Dowex 50 (H^+) at 100°C for 40 h as described in Section 5.8.3. Separation of monosaccharides is described below. After labelling with glucosamine part of the product should be hydrolysed in 4 M HCl for 4 h at 100°C (Section 5.5). The hydrolysate is dried on a rotary evaporator. Free glucosamine and galactosamine can be separated by paper chromatography (see below) or using an amino acid analyser. In the latter case the eluate stream is split so that a known proportion is diverted without addition of ninhydrin to a fraction collector so that the radioactivity in glucosamine and galactosamine can be determined. Another part of the products arising from the incorporation of labelled glucosamine should be examined for the presence of labelled sialic acids by hydrolysis in 0.1 M H_2SO_4 at 80°C for 1 h followed by adsorption on Dowex 50-×2 and elution with 0.3 M formic acid and paper chromatography of the sialic acid (Section 5.7.3). Separation of neutral and amino sugars can be carried out by descending paper chromatography on Whatman No. 1 paper in butanol:pyridine:water (6:4:3, v/v; Muramatsu and Nathenson, 1970; Straus et al., 1970). Spots corresponding to standards are cut out and radioactivity is determined by scintillation counting. Standards are located by staining with $AgNO_3$ and NaOH (Section 6.7.4.3).

8.2.3. Labelling for structural characterisation

Labelling with radioactive sugars or sulphate can be used to (1) show that a macromolecule is glycosylated, (2) indicate what sugars are present, (3) assist in following the purification of glycosylated molecules and (4) be used as a marker in following the degradation of a glycoconjugate by enzymic or chemical methods. Some examples are given in Table 8.2.

To obtain efficient incorporation of precursors into the products of cultured cells, labelling is usually carried out for quite long periods (e.g. 24–48 h). It is essential to analyse the labelled macromolecular products as described in Section 8.2.2 to identify the incorporated monosaccharides.

8.2.4. Labelling to examine biosynthetic pathways

Labelling intact cells with radioactive precursors can be employed to follow the formation of glycoproteins and proteoglycans and to identify intermediates in the biosynthetic process. Labelling experiments can establish the order in which biosynthetic steps take place and their subcellular locations. This section will deal primarily with qualitative studies of biosynthesis of asparagine-linked carbohydrate units. Quantitative measurements of the rates of formation (and breakdown) of proteoglycans and glycoproteins are considered in section 8.2.5.

Several factors can influence the time course and extent of labelling of glycoconjugates (Yurchenco et al., 1978; Schachter, 1978). These include (1) the specific radioactivity and concentration of the precursor, (2) its rates of entry into, and exit from, the cell or tissue, (3) the rates of equilibration of endogenous pools of intermediates, (4) the interconversion of nucleotide-sugar precursors, (5) the formation of lipid-linked intermediates (*N*-linked carbohydrate units only), (6) the temporal relationship between peptide chain synthesis and carbohydrate addition, (7) enzymatic processing of the carbohydrate and peptide components, (8) the movement of macromolecular precursors between intracellular compartments, (9) secretion of extracellular glycoconjugates, (10) degradation of glycoconjugates and reutilisation of label.

The general approach to investigating the biosynthesis of glycoconjugates is to add labelled precursor and then follow the change in specific radioactivity (or total radioactivity) of pools of intermediates and products with time. From the time course of the passage of label between different molecular species it is often possible to deduce precursor-product relationships. For example, if a trace amount of radioactive monosaccharide is injected into an animal the specific activity of the serum pool of that monosaccharide will rise and then fall off with time as the labelled precursor is converted to sugar phosphates, nucleotide phosphates and glycoconjugates within the tissues. The time course of labelling of a product B when its immed-

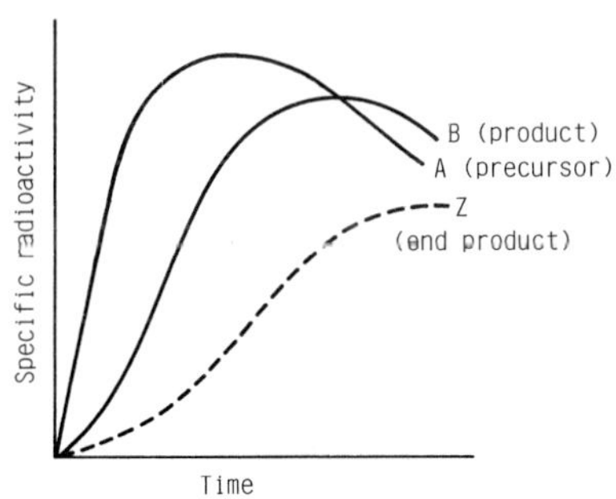

Fig. 8.3. Change in specific activity with time of a labelled biosynthetic precursor and its products. Substance A is the immediate precursor of B and Z is an end product.

iate precursor A is injected as a single dose has been analysed by Zilversmit et al. (1942) and is shown in Fig. 8.3. Criteria indicating that A is the immediate precursor of B include: (1) the specific activity of the immediate precursor A is greater than that of B before the latter reaches its maximum specific activity, (2) after B has reached its maximum specific activity the specific activity of B is greater than that of its precursor and (3) when B has reached its maximum specific activity the specific activity of A equals the specific activity of B.

This analysis of the rate of change of specific activities of A and B involves the assumption that a steady-state exists (i.e. the amounts of substances in the tissues remain constant and the rates of appearance and disappearance of intermediates are constant). As well as establishing precursor-product relationships it is possible to compute the amount of substance turned over by the tissue per unit time (turnover rate), from the curves shown in Fig. 8.3. For a substance which is turning over rapidly the specific activity falls sharply after its maximum is reached. In contrast, the specific activity of an end-product (dotted line in Fig. 8.3) rises in a sigmoid curve until it reaches a plateau.

When labelling cultured cells it is often advantageous to add the labelled precursor to the culture medium for a limited period of time and then to add an excess of the unlabelled precursor. The fate of the pulse label can then be followed, with time, as it is 'chased' through intermediates into final products. When using a labelled monosaccharide as precursor the concentration of unlabelled sugar in the chase medium must be high enough to dilute the radioactivity of the intracellular pools rapidly. However, too high concentrations of some monosaccharides can produce toxic effects (Yurchenco et al., 1978). For example, 0.35 mM D-glucosamine inhibits protein and nucleic acid synthesis in ascites cells (Bekesi et al., 1969).

When working with cultured cells it can often be difficult to obtain sufficient material to determine the specific activity of intermediates. This problem is discussed further in Section 8.2.5. Plotting the time course of the incorporation of counts into different intermediates in pulse-chase experiments can give a good qualitative indication of the pathway of synthesis of glycoconjugates. The measurement of specific activity is to be preferred, when possible, because of the quantitative nature of the relationship between immediate precursor and product and because pool sizes (i.e. the total amounts) of the intermediates are determined. Changes in pool size during the course of an experiment can lead to misinterpretation of the time-course of labelling (see Section 8.2.5).

Before isolating labelled intermediates or products to determine the extent of incorporation the biosynthetic process must be stopped. In many experiments this has been done by the addition of a protein precipitant such as trichloroacetic acid (or a TCA-phosphotungstic acid mixture) to cells. Sugar phosphates and nucleotide sugars can subsequently be isolated from the intracellular TCA-soluble pool and glycoproteins from the TCA-insoluble intracellular and extracellular material (secreted glycoproteins). Precipitation with TCA has the great advantage of simplicity and speed. However, some caution is necessary when interpreting the distribution of label between acid-soluble and -insoluble pools for the following reasons: (1) not all glycoproteins are TCA-precipitated, (2) labelled glycosaminoglycans may occur in the TCA-soluble fractions, (3) labelled glycolipids and lipid-linked precursors of asparagine-linked glycoproteins can occur in the intracellular TCA-insoluble fraction. Thus some characterisation of TCA-soluble and -insoluble fractions is necessary.

A recently developed method for stopping biosynthesis and obtaining preliminary fractionation of lipid intermediates involved in the synthesis of asparagine-linked glycoproteins in cultured cells is to add chloroform:methanol (2:1, vol./vol.) directly to washed cells. This approach allows the isolation of nucleotide sugars, sugar phosphates and glycoprotein products as well as the lipid-linked intermediates.

In experiments designed to examine the compartmentalisation of biosynthetic steps by subcellular fractionation it is only feasible to slow enzymic reactions by cooling the sample until fractionation has been completed. Advantage can be taken of rapid fractionation techniques such as centrifugation on Percoll density gradients to minimise the time taken for isolation. Similarly when specific isolation methods depending on native protein conformation are employed to obtain particular glycoproteins it may not be possible to use protein denaturants to stop biosynthesis.

As there is no single general method for examining glycoprotein biosynthesis some examples of different approaches are given below.

Labelling rat serum and tissue glycoproteins with L-fucose. Bekezi and Winzler (1967) examined the metabolism of L-fucose and its incorporation into glycoproteins. L-[1-^{14}C]fucose (2 μCi, specific activity 0.4 mCi per mmol) was injected intraperitoneally into rats and the labelling of serum glycoproteins and tissues was examined after periods between 5 min and 7 h. Serum glycoproteins were separated electrophoretically and the specific activities of protein bands were determined. Tissue samples (1.5 g wet weight) were homogenised in 10 volumes of 10% trichloroacetic acid. After centrifugation and washing the pellet with TCA the supernatants were combined and utilised for the determination of TCA-soluble radioactivity. The TCA-insoluble fraction was defatted by washing with 95% ethanol and acetone and used for analysis of protein-bound radioactivity. Further fractionation of the TCA-soluble fraction was carried out after ether extraction (to remove TCA) by chromatography on Dowex 1- × 8 using two linear gradients from water to 0.05 M HCl and from 0.05 M to 0.5 M HCl. Two radioactive peaks were obtained. The first to emerge was purified by paper chromatography in butanol:acetic acid:water (4:1:5, vol./vol.) and was shown to be fucose 1-phosphate. The second peak, characterised by paper chromatography in 95% ethanol - 1 M ammonium acetate, pH 7.4 (7.5:3, by vol.) was GDP-fucose. These experiments identified fucose 1-phosphate and GDP-fucose as intermediates in the incorporation of fucose into glycoproteins. Labelled fucose is efficiently incorporated into glycoproteins; in a 6-h period only 1.6% of the radioactivity appeared as CO_2, 30% appeared in urine mainly as free fucose and most of the remainder was present in glycoproteins.

Other examples of methodology for qualitative examination of sugar phosphate and nucleotide sugar intermediates include Harris and Johnston (1969), Hayden et al. (1970) and Yurchenco et al. (1978).

Incorporation in vivo of labelled glucosamine into glycoproteins of rat liver subcellular fractions. The subcellular sites at which glucosamine and sialic acid were incorporated in glycoproteins synthesised in rat liver were examined in an early study by Lawford and Schachter (1966). Fasted rats (160–200 g) were injected intraperitoneally with D-[^{14}C]glucosamine (15 μCi) and were killed after 45 min. Rough and smooth microsomes were isolated by density gradient centrifugation at 0–4°C and were further fractionated into membranes and vesicle contents either by sonication or deoxycholate treatment. Glycoproteins were precipitated from all fractions with 6% TCA 0.5% phosphotungstic acid (final concentration). The specific radioactivities of protein-bound glucosamine and sialic acid were determined after hydrolysis and purification of the monosaccharides (see Section 8.2.2). Protein-bound hexosamine turned over rapidly in the contents fraction of smooth and rough microsomes (but not in the corresponding membranes) as shown by a sharp rise and rapid fall of specific activity. Incorporation into plasma proteins of glucosamine and sialic acid showed the sigmoid curves characteristic of the formation of final product. Sialic acid labelling (Fig. 8.4) showed rapid turnover

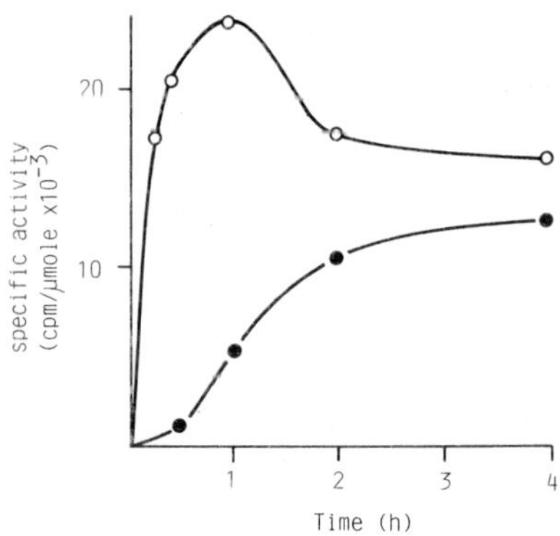

Fig. 8.4. Incorporation of labelled glucosamine into protein bound sialic acid in rat liver microsomes and plasma proteins. Closed circles (●) represent specific activity of plasma protein bound sialic acid and open circles (○) the specific activity of smooth microsome contents (Lawford and Schachter, 1966).

in the smooth microsome contents but not the rough membrane contents. The results suggested that glucosamine was incorporated into glycoprotein in both the rough and smooth endoplasmic reticulum but sialic acid incorporation is localised in the smooth membranes.

In these experiments the total glycoprotein produced by the liver was examined. The results obtained represent the average pattern of labelling of many different glycoproteins containing predominantly asparagine-linked oligosaccharides. Other studies have examined the subcellular sites of labelling of specific glycoproteins and have shown a similar pattern (see Schachter, 1978 for review).

The use of TCA and phosphotungstic acid described above is a useful general procedure because some glycoproteins which are soluble in TCA are precipitated by the mixture.

Kreibich and Sabatini (1974) have developed further the methodology for labelling and separating microsomal contents and membranes. They showed that rapidly synthesised proteins and glycoproteins, destined for secretion, occurred in the microsomal contents. Some proteins and glycoproteins associated with the microsomal membranes were synthesised rapidly and some were synthesised slowly. Labelling of rapidly synthesised proteins was carried out by injection of 4 μCi L-[4,5-^{3}H]leucine (40 Ci per mmol) directly into the portal vein of the anaesthetised rat, which was killed after 2 min. In other experiments labelled leucine (1 μCi) or D-[^{3}H]glucosamine (3.5 μCi, 1.9 Ci per mol) was injected intraperitoneally and the animals were killed after 30 min. After the isolation of subcellular fractions and the separation of endoplasmic reticulum membranes and contents, labelled glycoproteins and proteins were fractionated by SDS-gel electrophoresis.

The technique described below was used to separate vesicle contents from membranes of rough or smooth microsomes of rat liver (Kreibich and Sabatini, 1974). Microsomes were suspended in 50 mM Tris-HCl, pH 7.5, 50 mM KCl, 5 mM $MgCl_2$ at a protein concentration of 3–6 mg/ml. Sodium deoxycholate (0.49%) was added with vortex mixing to give a final detergent concentration of 0.049%. After incubation for 30 min at 0°C the soluble (cisternal contents) and sedimentable (membrane) subfractions were separated by centrifugation for 1 h at 100 000 *g*. Little solubilisation of membrane lipid occurs under these conditions. The efficiency of solubilisation is sensitive to changes in ionic strength and detergent-protein ratio (Kreibich et al., 1973). Extraction of non-glycosylated cisternal contents, such as rat serum albumin, is more complete than that of glycoproteins (Kreibich and Sabatini, 1974); the serum glycoproteins en route for secretion interact reversibly with the endoplasmic reticulum membrane (Gogstad and Helgeland, 1978).

Labelling lipid-linked intermediates in cultured CHO cells. Lipid-linked intermediates of two types have been identified in the biosynthesis of *N*-linked carbohydrate units of glycoproteins. These are dolichol monophosphate hexoses and derivatives of dolichol pyrophosphate containing from one to fourteen monosaccharide units. There is evidence for the occurrence of a common dolichol pyrophosphate-linked oligosaccharide intermediate in the synthesis of *N*-linked glycoproteins (Robbins et al., 1977). The formation of lipid-linked oligosaccharides has been examined in tissue slices (Spiro et al., 1976) and in several cultured cell lines (Robbins, 1977; Tabas et al., 1978; Hunt, 1980). Several studies have employed virally infected cells with the advantage that a defined viral component is the main glycoprotein synthesised by the cell. The following methodology (Rosner et al., 1982) can be applied to labelling lipid-linked intermediates in virally infected or non-infected cultured Chinese hamster ovary (CHO) cells.

Prior to labelling, cells are grown adhering to glass 9-cm Petri dishes to a density of about 2×10^6 cells/dish in growth medium (5 ml per dish) consisting of alpha minimal essential medium supplemented with 10% fetal calf serum and buffered with bicarbonate equilibrated with an atmosphere of 5% CO_2. Labelling of the mannose, *N*-acetylglucosamine and glucose residues of the oligosaccharide-lipid intermediate can be carried out with D-[2-^{3}H]mannose, D-[6-^{3}H]glucosamine or D-[1-^{3}H]galactose respectively. Labelled mannose is most commonly used as precursor because mannose is not present in glycolipids. Glucosamine is also a useful label and similar amounts of radioactivity are employed to those described below for mannose. The incorporation of radioactivity from galactose into glucose residues in lipid-linked oligosaccharides involves the interconversion of UDP-galactose to UDP-glucose. For labelling with galactose large amounts of radioactivity (about 4 mCi/ml) are required and a medium with low glucose concentration (see below) should be employed. In attempting to label glucose residues in lipid intermediates the use of clone 13 CHO cells may be advantageous. These cells, selected for their resistance to wheat germ agglutinin, are deficient in cell membrane sialic acid and galactose. Unlike wild-type cells they incorporate radiolabelled galactose into membranes only as glucose (Li et al., 1978).

Immediately prior to labelling the cells are washed three times with 5-ml portions of warm glucose-free Eagle's minimal essential medium (MEM). Labelling is carried out in 0.7 ml of a low-glucose medium comprising labelled sugar (e.g. D-[2-^{3}H]mannose 100–200 μCi/ml, 10–20 Ci/mmol) in glucose-free MEM supplemented with 0.1 mg/ml glucose, 5 mM sodium pyruvate, 10% dialysed fetal calf serum and enough bicarbonate to maintain the pH in a 5% CO_2 atmosphere.

Labelling times of 10–30 min give optimal incorporation of [^{3}H]mannose. If labelling is carried out for periods shorter than five minutes 0.5–1 mCi/ml of [^{3}H]mannose is necessary. Label can be 'chased' by removing the radioactive medium, washing the cells three times and continuing incubation with complete medium (Rosner et al., 1982).

Extraction of lipid-linked intermediates. Procedures for the isolation of lipid-linked intermediates in glycoprotein biosynthesis are based on selective extraction with solvent mixtures. Dolichol monophosphate-monosaccharides, small dolichol pyrophosphate-linked oligosaccharides (less than $Man_5GlcNAc_2$; Datema et al., 1980) are, like glycolipids, soluble in chloroform:methanol (2:1, v/v). However, the larger dolichol pyrophosphate oligosaccharides are insoluble in both chloroform:methanol (2:1) and in water but they can be extracted with chloroform:methanol:water (1:1:0.3 by vol.; Parodi et al. 1972). Sequential extractions of cells with solvents can therefore separate fractions containing different classes of biosynthetic intermediates (Fig. 8.5).

Lipid intermediates can be isolated from labelled cultured cells after removal of label, washing the cells in cold salts solution and harvesting the cells in the usual way (Waechter et al., 1973; Robbins et al., 1977; Li et al., 1978). However, less degradation of the lipid-linked oligosaccharides occurs using the following method (Rosner et al., 1982). After rapidly removing the medium, pre-cooled (0°C) chloroform:methanol (2:1 v/v, 5 ml) is added immediately to the glass Petri dishes containing the cells. The cells are removed with a Teflon scraper and transferred, with 1 ml of wash (same solvent) to a conical glass centrifuge tube. Samples can be stored at −20°C at this stage before extraction.

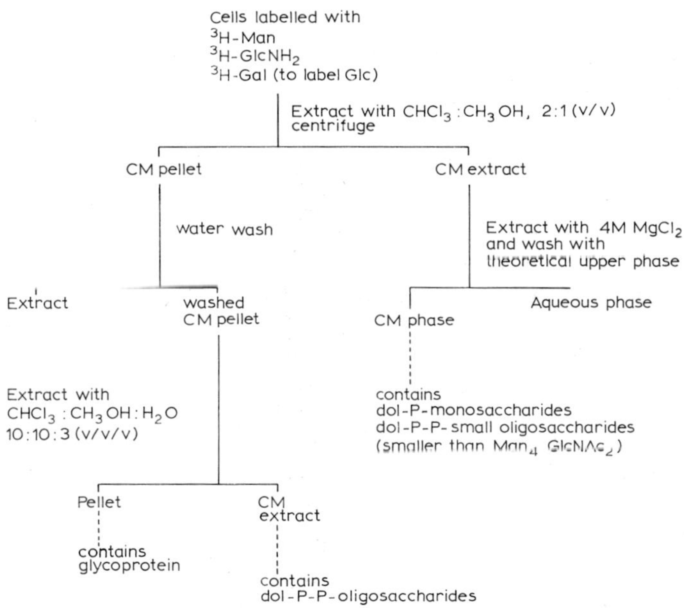

Fig. 8.5. Isolation of labelled lipid-linked intermediates in the synthesis of *N*-linked carbohydrate units of glycoproteins.

The suspension of cells in chloroform:methanol (2:1 v/v) is allowed to stand at room temperature for 15 min with periodic vortex mixing and it is then centrifuged at 1000 *g* for 10 min. After two further extractions each with 5 ml of chloroform:methanol (2:1 v/v), the pellet is dried under a stream of nitrogen. Water (1 ml) is added and a suspension produced by brief sonication. The sample is centrifuged and extraction with 1-ml portions of water is repeated until no further water-soluble counts are extracted (this may require 5–7 extractions). Lipid-linked oligosaccharides are extracted from the pellet with three portions (4 ml, 3 ml, 3 ml) of chloroform:methanol:water (10:10:3, v/v/v). These extracts are pooled and stored at −20°C. They are used for characterisation of the oligosaccharide lipids. The pellet remaining after the extraction contains labelled denatured cell glycoproteins. This material can be solubilised (by heating in 1% SDS at 100°C for 15 min) and counted, or it can be solubilised in sample buffer for SDS-polyacrylamide gel electrophoresis and the labelled peptides separated and characterised.

The chloroform:methanol (2:1 v/v) washes normally contain mainly lipid-linked mono- and di-saccharides but lipid-pyrophosphate oligosaccharides containing up to six sugar residues may be present. Before further examination of this fraction removal of water-soluble radioactive contaminants is necessary. This is done by washing the chloroform:methanol extract first with 1 ml of 4 M $MgCl_2$. The tubes are then rinsed twice with 2 ml of an artificial upper phase (methanol:water:1 M $MgCl_2$:chloroform, 96:94:0.8:6 by volume) without mixing and the upper phase is discarded. A third portion (1 ml) of artificial upper phase is added and the contents of the tube are mixed. After centrifuging the upper phase is removed and the lower phase is retained for analysis. The phosphorylated products can be examined by DEAE-cellulose and thin-layer chromatography (Sharma et al., 1974) or oligosaccharide units can be fractionated and characterised as described below.

Characterisation of lipid-linked intermediates. Lipid-linked oligosaccharides appearing in the chloroform:methanol or chloroform: methanol:water extracts can be cleaved by mild acid hydrolysis and their labelled oligosaccharide units analysed by chromatography on Bio-Gel P4 (Rosner et al., 1982) or by paper chromatography (Tabas and Kornfeld, 1982). Hydrolysis is carried out by dissolving the lipid oligosaccharide in *n*-propanol (0.5 ml), adding 0.1 M HCl (1 ml) and incubating in a capped tube at 100°C for 20 min. After evaporating

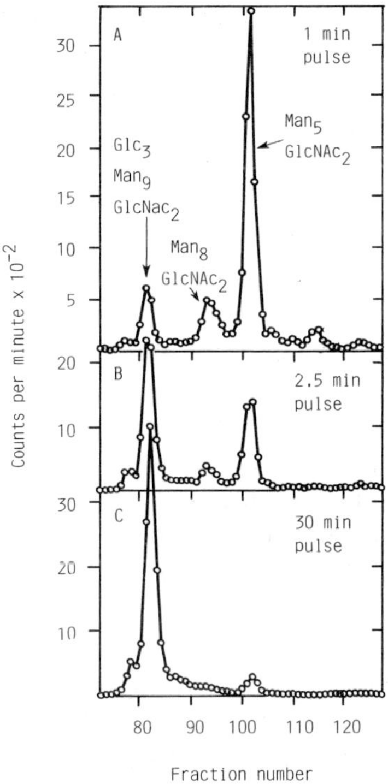

Fig. 8.6. Pulse-labelled oligosaccharide units released from lipid-linked intermediates by mild acid hydrolysis. CHO cells were labelled with [^{3}H]mannose, lipid-linked intermediates were isolated and hydrolysed and then chromatographed on Bio-Gel P-4 as described in the text. Peaks corresponding to the oligosaccharides $Glc_3Man_9GlcNAc_2$, $Man_8GlcNAc_2$ and $Man_5GlcNAc_2$ are identified. Redrawn from Hubbard and Robbins (1980).

the hydrolysate to dryness at 60–70°C the residue is redissolved in water (0.5 ml). Oligosaccharides can be separated and characterised by chromatography on Bio-Gel P4 using the methodology described in Section 6.9.4.1 (also see 6.7.4.2). Alternatively, more rapid separations can be carried out on Bio-Gel P4 (400 mesh) columns 1 × 210 cm or 1 × 115 cm eluting with 0.1 M pyridine acetate buffer, pH 5.0, and collecting 0.5-ml samples for scintillation counting. Samples (200 μl) are applied containing internal markers for the void volume (200 μg bovine serum albumin) and of the included volume ([^{14}C]mannose, 1000 cpm). The separation of a series of structurally characterised oligosaccharides ranging from $Glc_3Man_9GlcNAc_2$ to $Man_1GlcNAc_2$ can be obtained and a calibration curve relating log molecular weight to composition of oligosaccharide is given by Hubbard and Robbins (1980). An example of the use of this type of analysis to follow the formation of lipid linked oligosaccharides is shown in Fig. 8.6.

Dolichyl pyrophosphoryl oligosaccharides can also be resolved, without hydrolysis, by high-performance liquid chromatography (Wells et al., 1982).

Labelling processing intermediates in cultured cells. Oligosaccharides transferred from lipid-linked intermediates to protein rapidly undergo processing as a result of the action of specific exoglycosidases. The protein-bound oligosaccharides labelled in their glucose, mannose or *N*-acetylglucosamine residues can be isolated from the pellet fraction remaining after extraction of lipid-linked intermediates. The pellets are digested exhaustively with Pronase (Section 6.4.2) and solubilised glycopeptides are desalted by gel filtration. After digestion with endo-β-*N*-acetylglucosaminidase H or C_{II} the released oligosaccharides can be fractionated by gel filtration on Bio-Gel P4 or by paper chromatography in ethyl acetate-pyridine-acetic acid-water (5:5:1:3 by vol.). The type of oligosaccharide which can be obtained in this way (e.g. $Glc_1Man_9GlcNAc$) contains a single GlcNAc residue as a result of cleavage of the GlcNAcβ(1–4)GlcNAc linkage by the endoglycosidase treatment. This type of oligosaccharide with a single reducing terminal GlcNAc residue is released by

Endo H digestion of dolichol pyrophosphate-linked oligosaccharides as well as from protein-bound oligosaccharides. Analysis of oligosaccharides released by Endo H digestion can therefore be used to compare lipid-linked oligosaccharides with protein-linked high-mannose type oligosaccharides. Separation of this series of oligosaccharides ($Glc_3Man_9GlcNAc_1$ to $Man_1GlcNAc$) by gel filtration on Bio-Gel P4 columns has been documented by Hunt (1980).

Detailed methodology for the isolation of labelled processing intermediate derivatives $Glc_3Man_9GlcNAc$, $Glc_2Man_9GlcNAc$, $Glc_1Man_9GlcNAc$, $Man_{8\text{-}9}GlcNAc$ and $Man_5GlcNAc$ from CHO cells is described by Tabas and Kornfeld (1982).

8.2.5. Measuring rates of synthesis

Most studies of glycoprotein and proteoglycan biosynthesis have been concerned with establishing qualitatively the pathways by which these molecules are formed in living cells and tissues. Isotopic labelling techniques can also yield quantitative measurements of the flow of intermediates along biosynthetic pathways and the absolute rates of synthesis of end products.

To determine the rate of synthesis of a macromolecule it is necessary to know the rate of incorporation of labelled precursor into the macromolecule (in dpm per nmole per min) and the specific radioactivity of the immediate metabolic precursor (in dpm per nmole). Measurements of absolute rates of synthesis of glycoproteins and proteoglycans have been made in animal tissues in vivo and in cultured cells. The latter type of system has the advantage that it provides a homogeneous population of cells although there are limits to the number of cells which can be obtained conveniently. The concentration of nucleotide sugars in cells is low (usually 0.5–8 nmoles per 10^6 cells; Yurchenco et al., 1978), necessitating the use of micromethods for their direct quantitation. Incorporation of $^{35}SO_4$ into sulphated glycosaminoglycans in cultured chondrocytes gives a direct measure of the rates of synthesis of these molecules (Kim and Conrad, 1976) and can be used, in conjunction with glucosamine labelling, indirectly to estimate the specific radioactivity of the UDP-*N*-acetylglucosamine pool. Some examples of the measurement of rates of synthesis of macromolecules are outlined below.

Rates of synthesis of proteoglycans and glycoproteins in neonatal rat skin (Hardingham and Phelps, 1968). In this pioneering study the rates of biosynthesis of proteoglycans and glycoproteins were measured by following the turnover of labelled precursors. Neonatal rats were injected intraperitoneally with [U-^{14}C]glucose (10^6 cpm, 2 nmoles). After labelling for different times up to 2 h the animals were decapitated and the skin was freeze-clamped at −40°C, powdered and extracted with 0.4 M $HClO_4$. The neutralised extract was fractionated on Dowex 1 to separate hexose monophosphates, *N*-acetylglucosamine 6-phosphate and UDP-derivatives of *N*-acetylglucosamine, *N*-acetylgalactosamine, glucose, galactose and glucuronic acid. For each of the intermediates the tissue content (nmoles/g wet wt. of tissue), intracellular concentration (mM, obtained by measurement of the cell water), turnover time (min) and turnover rate (nmoles/g wet wt. per min) could be calculated (Zilversmit et al., 1942). The rate of synthesis of UDP-glucuronic acid was assumed to represent total proteoglycan synthesis, while the formation of UDP-hexosamine was assumed to give the total synthesis of proteoglycans and glycoproteins.

This approach can give some overall assessment of the synthesis of proteoglycans and glycoproteins averaged throughout the different cell types present in the tissue but does not provide information about specific macromolecules.

Rates of synthesis of proteoglycans and glycoproteins by chondrocytes (Kim and Conrad, 1976). The kinetics of incorporation of $^{35}SO_4^{2-}$ and [^{14}C]glucosamine into glycoproteins and defined glycosaminoglycans were measured in a primary culture of chick embryo chondrocytes. Glucose and 10% fetal calf serum were present in the growth medium. The glycoconjugates were digested with thermolysin and then incubated with a mixture of chondroitinases ABC and AC-II. Digests were fractionated by chromatography on paper and DEAE-paper to resolve the unsaturated disaccharides arising from chondroitin-6-SO_4, chondroitin-4-SO_4, chondroitin and hyaluronic acid (also see Section 6.9.4.1).

Incorporation of label from $^{35}SO_4^{2-}$ into sulphated proteoglycans occurred at linear rates until the glucose concentration in the medium was below 1 μmol/ml and would no longer support growth. Labelled sulphate from the medium was activated (as PAPS) and transformed to sulphated proteoglycans without significant dilution. The rates of synthesis of sulphated glycosaminoglycan chains could therefore be calculated directly from the counts per min recovered in sulphated proteoglycan divided by the specific activity (cpm per nmole) of free SO_4^{2-} in the medium. This latter figure is obtained from the cpm of labelled sulphate added to the medium divided by the total sulphate present in the medium (including the serum and antibiotics).

In this system the incorporation of labelled glucosamine is non-linear with time as a result of a progressive increase in the specific activity of the UDP-*N*-acetylhexosamine pool as glucose in the culture medium is depleted (Fig. 8.7). Thus the absolute rate of synthesis of polymers containing *N*-acetylglucosamine cannot be obtained directly

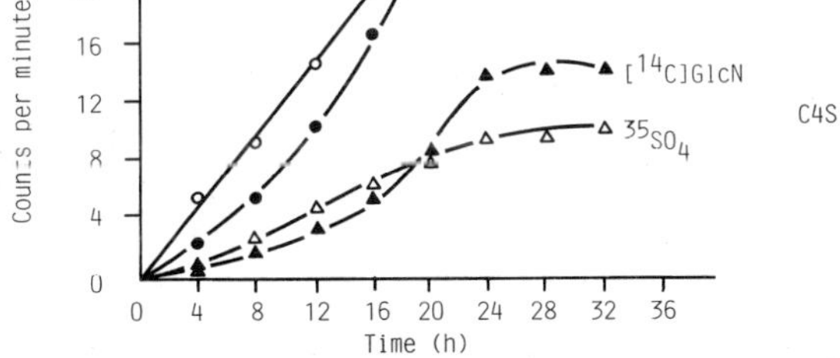

Fig. 8.7. Incorporation of $^{35}SO_4^{2-}$ and [^{14}C]glucosamine into chondroitin sulphate in chick embryo chondrocytes (Kim and Conrad, 1976). C6S is chondroitin-6-SO_4 and C4S is chondroitin-4-SO_4.

from the [^{14}C]glucosamine incorporation. However, it is possible to calculate the specific radioactivity of the UDP-*N*-acetylhexosamine pool at each stage in the labelling period from the relative rates of appearance of counts from $^{35}SO_4^{2-}$ and [^{14}C]glucosamine in chondroitin-4-SO_4 and chondroitin-6-SO_4.

Rates of synthesis of fucosyl glycoproteins in cultured cells (Yurchenco and Atkinson, 1977). Radioactive fucose is a precursor which labels glycoproteins in animal cells. Little interconversion of fucose to other monosaccharides present in complex carbohydrates has been detected. Direct measurement of pool sizes and changes in specific radioactivity of fucose in intermediates and macromolecular products can be performed in cultured cells using a micromethod for the quantitation of fucose. From these measurements the rate at which fucose flows through precursor pools into end-products such as membrane and secreted glycoproteins can be estimated.

After labelling cells with [^{3}H]fucose intermediates and products can be isolated by methodology described in detail by Yurchenco et al. (1978). Free fucose is released from GDP-fucose or macromolecules by hydrolysis. Quantitation of the [^{3}H]fucose (after purification from other sugars) is performed by a micromethod which employs [γ-^{32}P]ATP of known specific activity together with the enzyme fucose kinase, which catalyses the reaction

$$[^3\text{H}]\text{fucose} + [\gamma\text{-}^{32}\text{P}]\text{ATP} \rightarrow [^3\text{H}]\text{fucose 1-}[^{32}\text{P}]\text{phosphate} + \text{ADP}.$$

The quantity of fucose phosphorylated can be determined and the specific radioactivity of the [^{3}H]fucose in the GDP-fucose and macromolecular fucose pools can be calculated. This method requires about 2×10^8 cells (HeLa) for each estimation of both cellular and plasma membrane-bound glycoproteins.

As fucose is present only in certain of the complex type of *N*-linked carbohydrate units of glycoproteins the method only reflects the (averaged) metabolism of part of the glycoprotein population. Methods for quantitating the biosynthesis of other sugars in glycoproteins are reviewed by Yurchenco et al. (1978).

8.2.6. In vitro (cell-free) labelling of glycoproteins

The carbohydrate or peptide chains of glycoproteins can be labelled in the absence of intact cells to examine (1) the relationship between peptide chain synthesis and glycosylation, (2) the biosynthesis of carbohydrate units and (3) to investigate carbohydrate structure and its relation to function. Cell-free systems range in complexity from crude subcellular fractions, capable of many biosynthetic reactions, to purified enzymes catalysing a single type of reaction between substrates of defined structure. Studies of glycoprotein synthesis have generally progressed towards better-defined systems and it is now possible to assemble the carbohydrate units of some glycoproteins (with serine- or threonine-linked oligosaccharides) entirely by the use of a succession of purified enzymes and sugar nucleotides (Beyer and Hill, 1982). However, the intimate association between the synthesis of asparagine-linked carbohydrate units and subcellular membranes has necessitated more complex systems for studying the synthesis of glycoproteins containing GlcNAc-Asn linkages.

In vitro protein synthesis and glycosylation. In vitro synthesis of the peptide chains of glycoproteins can, in the presence of microsomal membranes, be coupled with the addition of *N*-linked carbohydrate groups. The temporal and spatial relationship between glycosylation and peptide chain synthesis can be examined in this type of system (for review see Bergman and Kuehl, 1982). The study of the cell-free synthesis of vesicular stomatitis virus (VSV) glycoprotein (G) by Rothman and Lodish (1977) is a good example of this approach. VSV glycoprotein contains two asparagine-linked carbohydrate units attached to a single peptide chain. After synthesis part of the peptide chain remains embedded in the lipid bilayer of the endoplasmic reticulum. In vivo the glycoprotein migrates through the endoplasmic reticulum to the cell surface. The components required for cell-free synthesis include purified messenger RNA for VSV glycoprotein, a crude wheat germ lysate, an energy source, amino acids (including labelled methionine) and canine pancreatic microsomes. Canine microsomes are used because they are free of RNAase. Synthesis can be detected by SDS-polyacrylamide gel electrophoresis and autoradiography of the labelled peptide chains. Glycosylated chains are distinguished from unglycosylated molecules by small increments in molecular weight detected by SDS-polyacrylamide gel electrophoresis. Synthesis of carbohydrate-free peptide chains was found to occur if microsomes were omitted from the system (Katz et al., 1977). In the presence of microsomes the peptide chains were glycosylated. All but a short section of the peptide was inserted into the microsomes as judged by the insensitivity of most of the molecule to cleavage by trypsin.

To examine the temporal relationship between protein synthesis and glycosylation it is necessary to synchronise the synthesis of peptide chains. This can be done by adding an initiation inhibitor (7-methylguanosine 5′-phosphate, mGp) a short time after peptide synthesis has begun. The time taken for the synthesis of a complete peptide chain can therefore be determined. The relationship between peptide chain synthesis and membrane insertion was examined by the addition of microsomes to the synchronised protein-synthesis system at different times and determining whether membrane insertion, conferring trypsin insensitivity to the peptide chain, had taken place. The timing of the addition of carbohydrate groups was investigated by the addition of Triton X-100 to microsomes at intervals after VSV glycoprotein synthesis was initiated and examining the peptide chains formed by SDS-gel electrophoresis. Triton X-100 was found to inhibit glycosylation without altering the rate of peptide chain synthesis. Labelled products with discrete M_r values corresponding to 0, 1 or 2 carbohydrate groups were detectable at different times after commencing peptide chain synthesis (Fig. 8.8). Assuming the rate of peptide synthesis to be constant it was possible to correlate the times at which carbohydrate units were added with the proportion of the peptide chain synthesised. The results show that glycosylation is closely coupled to the passage of the nascent peptide chain through the endoplasmic reticulum membrane.

The system employed contained the following components:

1. VSV mRNA isolated from infected BHK 21 cells (1×10^8 cells)

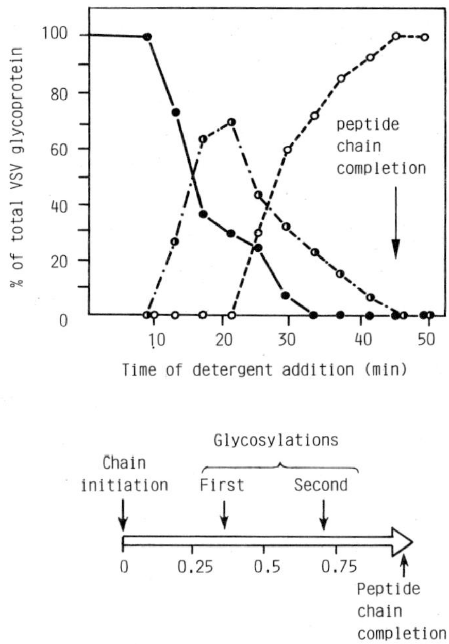

Fig. 8.8. Glycosylation of VSV glycoprotein synthesised in vitro (Rothman and Lodish, 1977). The upper figure shows the amounts of VSVG-specific translation products containing no oligosaccharide (●), one oligosaccharide unit (◐), and two oligosaccharide units (○) detected in the final product by SDS-gel electrophoresis. Triton X-100 was added at different times after the synchronised initiation of peptide chain synthesis (0 min) to inhibit subsequent glycosylations (but not peptide synthesis). The relationship between peptide-chain synthesis and glycosylation is indicated in the lower figure.

purified by binding and elution from oligo(dT)-cellulose and then suspended in 0.5 ml water and stored at −70°C (Katz et al. 1977).
2. Canine pancreatic microsomes prepared as described by Katz et al. (1977) and stored at −20°C.
3. Wheat germ extract prepared by the method of Roberts and Patterson (1973) as modified by Katz et al. (1977) and stored at −70°C.

The cell-free protein synthesis reaction mixture contained, in 0.5 ml, wheat germ extract (0.15 ml), VSV mRNA (20 μl) and the following at the final concentrations indicated: HEPES buffer, pH 7.2 (27 mM), KCl (120 mM), magnesium acetate (1.5 mM), dithiothreitol (5.6 mM), ATP (2.6 mM), GTP (0.26 mM), creatine phosphate (16 mM), rabbit skeletal muscle creatine phosphokinase (42 μg/ml, Calbiochem), spermidine (0.73 mM), 19 amino acids except methionine (130 μM each) and [^{35}S]methionine (0.5 mCi/ml, specific activity 600 Ci/mmol). This mixture (without mRNA) was preincubated before addition of mRNA at zero time. After 2 min 25 μl of 20 mM 7-methylguanosine 5′-phosphate was added to prevent subsequent initiations.

Membrane addition: 50-μl samples of the protein synthesis cocktail were mixed with 0.5 μl pancreatic microsomes (stock suspension giving A_{280} of 50 in 1% SDS) and incubated at 23°C.

Inhibition of glycosylation with Triton X-100: to samples (25 μl) containing protein-synthesis reaction mixture and membranes Triton X-100 (2.5 μl, 10% w/v) was added. After incubation samples were precipitated with boiling TCA (50%) and prepared for SDS-gel electrophoresis.

Glycosylation of asparagine residues is not the only structural modification which can affect the molecular weight of peptides on SDS-polyacrylamide gels. Confirmation that higher molecular weight forms of a protein (as seen in VSV glycoprotein) are due to differences in glycosylation can often be obtained by studies with inhibitors such as tunicamycin (Section 8.2.8) or by the use of endoglycosidases.

In vitro transfer of lipid-linked oligosaccharides to proteins or synthetic peptides. It is possible to label lipid-linked oligosaccharide precursors of asparagine-linked glycoproteins and then follow the transfer of the oligosaccharides to peptide chains or to synthetic peptides. Structural requirements of acceptors can be examined in this way. Labelling is usually carried out with radioactive mannose or glucosamine. Detailed methodology has been published (Elting and Lennarz, 1982).

Labelling glycoproteins with radioactive sugar nucleotides. In cell-free systems it is possible to study the transfer of label from radioactive nucleotide sugars to endogenous or exogenous acceptors.

Several of the glycosyl transferases which catalyse this type of reaction have been purified, some to homogeneity. Examination of the subcellular location and the specificities of these enzymes has led to significant advances in understanding of the biosynthesis of the *O*-linked carbohydrate units and the peripheral (non-core) portions of *N*-linked units (Beyer and Hill, 1982; Schachter, 1978).

Nucleotide sugars used for cell-free labelling of glycoproteins include UDP-*N*-acetyl-D-glucosamine, UDP-*N*-acetyl-D-galactosamine, UDP-D-galactose, GDP-L-fucose, GDP-mannose and CMP-sialic acid. Labelled (^{14}C or ^{3}H) nucleotide sugars are commercially available from Amersham International and from New England Nuclear. The appreciable cost of these compounds is a necessary consideration when designing experiments.

In many early experimental investigations of glycoprotein synthesis (for review see Schachter, 1978) labelled nucleotide sugars were added to membrane fractions and incorporation of label into endogenous precursors was examined. For example, Wagner and Cynkin (1969a,b) measured the transfer of [^{14}C]glucosamine from UDP-*N*-acetyl[^{14}C]glucosamine into endogenous acceptors in microsomal and Golgi fractions of rat liver. Reaction mixtures contained, in 100 μl, membrane suspension (0.3–2 mg protein), 0.8 nmoles UDP-*N*-acetyl[^{14}C]glucosamine (15 000 cpm), 0.9 μmoles $MnCl_2$, 0.5 μmoles EDTA and 5 μmoles Tris buffer, pH 7.0. After incubation at 37°C for 5–60 min the radioactivity incorporated into TCA-insoluble material was measured. The endogenous substrates were found to be mainly compartmentalised within the membranes and could be released by detergent treatment. Considerably higher incorporation was observed in the Golgi apparatus fractions than in either rough or smooth endoplasmic reticulum. D'Amico and Kern (1970) demonstrated incorporation of sialic acid from CMP-sialic acid (2 μM in reaction mixture) into microsomal fractions of lymph node cells. The product was characterised as immunoglobulin G by immunoprecipitation. Exogenous desialylated immunoglobulin G would also act as an acceptor for labelled sialic acid and activity was stimulated by addition of 0.1% Triton X-100.

The nature of labelled endogenous products is often difficult to define. Whenever possible it is advisable to use an exogenous substrate of known structure. Measurement of the incorporation of label can then be used as a glycosyl transferase assay.

Controls for endogenous incorporation are necessary unless the enzyme has been purified. Other enzyme activities present in crude extracts (e.g. nucleotide pyrophosphatase, CMP-sialic acid hydrolase, glycosidases and phosphatases) can lead to formation of radioactive side-reaction products including sugar phosphates and free sugars. In assaying impure preparations it is desirable to establish the extent to which substrate is degraded by these reactions as well as measuring the formation of product. This can be achieved by separation of components of the reaction mixture by paper electrophoresis in borate buffer as described below (Roseman et al., 1966; Schachter and Roseman, 1980). Other, more rapid techniques for the measurement of labelled products are also considered later in the section.

The substrate employed can be a glycoprotein treated with glycosidases (Section 6.9.4.1) or by Smith degradation (Section 6.9.4.4) to remove terminal sugars. Acceptors of this type may resemble the natural substrates of glycosyl transferases but because of their complex and heterogeneous structures there can be some uncertainty about the group which is acting as substrate for the glycosyl transferase. From this point of view small oligosaccharides or glycopeptides of defined structure are preferable as substrates. For example the disaccharide lactose can be used as substrate in some sialyl transferase assays.

Optimal conditions of acceptor concentration must be established for each glycosyl transferase as an excess can produce inhibition of activity. The concentration of nucleotide sugar must be maintained at saturating levels (although this can prove to be expensive) if valid quantitative results are to be obtained. In addition many of the mammalian glycosyl transferases have a divalent metal ion requirement and they may also require detergent for maximal activity.

Separation of products by paper electrophoresis. This electrophoretic technique can be adapted for the assay of a variety of glycosyl transferases. The details given below relate to the assay of rat mammary gland sialyl transferase (Carlson and Jourdian, 1966).

The assay mixture contains 10 μl of 5 mM CMP-*N*-acetylneuraminic acid (^{14}C-labelled, 1.3×10^6 cpm/mole) and 5 μl 1.0 M potassium phosphate buffer, pH 6.9. The reaction is started by addition of 25 μl of enzyme preparation. Controls with heat-inactivated enzyme and an 'endogenous substrate' control in which no lactose is present are prepared. After incubation at 37°C for 30 min the reaction mixture is cooled rapidly and aliquots are applied to Whatman 3MM paper. The paper is moistened with 0.05 M sodium tetraborate, pH 9.0, and electrophoresis is carried out for 40 min at 3.5 kV (for a paper strip 52 cm in length between electrode compartments) with the paper immersed in white spirit. (Appropriate fire precautions must be taken when setting up this type of apparatus.) After electrophoresis the dried paper is cut into strips and counted. Labelled CMP-*N*-acetylneuraminic acid migrates rapidly towards the anode and is separated from neuraminic acid and labelled sialyl lactose.

The products of other glycosyl transferases and other types of receptor can also be separated in this system (Schachter and Roseman, 1980). Sugar phosphates usually migrate rapidly towards the anode and can often be resolved from nucleotide phosphate. Large molecular weight acceptors usually remain near the origin although they may become spread over an area extending several cm from the origin. Endogenous acceptors also appear in this area. Although free hexose, fucose and sialic acid have sufficient negative charge, in the presence of borate, to produce appreciable movement towards the anode the *N*-acetylhexosamines remain close to the origin. These sugars can be removed by descending chromatography in 80% ethanol.

Other glycosyltransferase assays. The substrates in glycosyltransferase assays are often partially deglycosylated glycoproteins. Labelled macromolecular products of glycosyltransferases can be separated from labelled nucleotide sugars (and their degradation products) by gel filtration. Schwytzer and Hill (1977a) have described an assay for an α(1–3) *N*-acetylgalactosaminyl transferase based on this principle. Following incubation of UDP-[^{14}C]GalNAc with asialo porcine submaxillary mucin and glycosyltransferase the reaction was stopped by the addition of EDTA to chelate calcium. The reaction mixture (60 μl) was then applied to a column (0.8 × 10 cm) of Sephadex G-50 (fine) equilibrated and eluted with 0.01 M Tris-HCl, pH 7.5, 0.2 M NaCl and 0.01% (w/v) sodium azide. Fractions of the eluate containing the radioactive macromolecular products emerged between 1.5 ml and 3.5 ml. This portion of the eluate was retained for scintillation counting. Columns could be reused after washing to remove nucleotide sugars. Similar assays have been employed for several other glycosyl transferases (Sadler et al., 1982).

A more rapid and convenient (but slightly less precise) assay involves the precipitation of macromolecular products and their separation from labelled substrate by filtration. In the procedure described by Letts et al. (1974) partially deglycosylated derivatives of α_1-acid glycoprotein were employed as substrates for galactosyl and *N*-acetylglucosaminyl transferases. Assays (in a total volume of 50 μl) were terminated by the addition of 50 μl of 10% trichloroacetic acid -2% phosphotungstic acid (PTA). The resulting precipitate was collected by vacuum filtration on a Whatman glass-fibre filter. After washing with 5% TCA-1% PTA (30 ml) and chloroform:methanol:water (2:1:1, by vol.) the filters were prepared for scintillation counting. Some modifications of the concentration of TCA and PTA may be required for other substrates; Schwytzer and Hill (1977a) employed 15% TCA-5% PTA for precipitation of the products obtained when asialo porcine submaxillary mucin was used as a glycosyltransferase substrate.

Different assay methods are required when the substrate is of low molecular weight. An assay which can be used for oligosaccharide substrates relies on the adsorption of untreated labelled substrate on Dowex 1 while the labelled products are not retained by the column (Beyer et al., 1980; Schwytzer and Hill, 1977b). The reaction mixture containing the low molecular weight acceptor (e.g. lactose) in 50 μl is diluted with water at 0°C and applied to a 0.5-ml column of Dowex 1-×8 (Cl^- 100–200 mesh) previously washed with a solution of lactose. The column is rinsed with water (1 ml) and the eluate is taken for scintillation counting. The column is discarded after use. As hydrolysis products of the substrate can appear in the eluate appropriate controls must be employed. This assay is suitable only for purified glycosyltransferases.

Purified glycosyltransferases. These enzymes have been studied extensively in relation to their role in the biosynthesis of the carbohydrate units of glycoproteins and proteoglycans (reviewed by Schachter, 1978; Rodén, 1980; Beyer et al., 1981). The formation of each type of glycosidic linkage requires a different transferase and the specificities of these enzymes towards the acceptor to which each sugar residue is transferred is often very exacting. Technical difficulties in the purification of glycosyl transferases in quantity have to a considerable extent been overcome through the introduction of affinity chromatography techniques (Beyer et al., 1981; Sadler et al., 1982). Purification of these enzymes opens the way for them to be used as highly specific tools in the analysis of glycoprotein structure and function.

Glycosyl transferases which have been obtained in highly purified form are listed in Table 8.3. These enzymes can be assayed by the methods described previously. The difficulties in the purification of these enzymes have arisen because of their low concentration in tissues and because many of the enzymes are membrane-bound and labile. A detailed discussion of approaches to the purification of these enzymes can be found in the review by Beyer et al. (1981). The example given below indicates the potential use of these purified enzymes in functional and structural studies of glycoproteins (see also Beyer et al., 1981).

Application of glycosyltransferases to investigation of glycoprotein structure and function (Paulson et al., 1977). The purified rabbit liver membrane glycoprotein that binds galactosyl residues of asialo-plasma glycoproteins is known to lose binding activity after enzymic removal of sialic acid from its oligosaccharide prosthetic group. It is possible to replace sialic acid residues by incubating the purified binding protein with labelled CMP-sialic acid and a sialyl transferase. Replacement of sialic acid can be correlated with restoration of

TABLE 8.3
Some highly purified glycosyltransferases

	Sugar donor	Typical product	Comment	Source and purification
Fucosyl transferases				
β-Galactoside α1–2-fucosyltransferase	GDP-Fuc	Fucα1–2Galβ1–X	Broad specificity for residue X. Can be used to quantitate exposed Gal. Product has group H activity	Porcine submaxillary gland (1)
N-Acetylglucosaminyl α1–3-fucosyltransferase	GDP-Fuc	Fucα1–3GlcNAc	Both activities co-purify. GlcNAc can be replaced by Glc as substrate	Human milk (1)
N-Acetylglucosaminyl α1–4-fucosyltransferase	GDP-Fuc	Fucα1–4GlcNAc		
Sialytransferases				
β-Galactoside α2–3-sialyltransferase	CMP-NeuAc	NeuAcα2–3Galβ1–X	Galβ1–4GlcNAc is not a substrate while Galβ1–3GalNAc is good substrate	Porcine submaxillary gland (1)
α-*N*-Acetylgalactosaminide α2–6-sialyltransferase	CMP-NeuAc	NeuAcα2–6 GalNAcα1–Thr/Ser	Specific for *O*-linked carbohydrate units; substitution of GalNAc with Gal at 3 position does not prevent sialylation	Porcine submaxillary gland (1)
β-Galactoside α2–6-sialyltransferase	CMP-NeuAc	NeuAcα2–6Galβ1–4	Galβ1–4GlcNAc units (of complex *N*-linked oligosaccharide) are good substrates	Bovine colostrum (1)
Galactosyltransferases				
β-*N*-Acetylglucosaminide β1–4-galactosyltransferase	UDP-Gal	Galβ1–4GlcNAc	Active with GlcNAc and glycoprotein substrates; substrate specificity can be modified by	Bovine milk (2) Bovine colostrum (3)

(Fucosyl α1–2)Galactoside α1–3-galactosyltransferase	UDP-Gal	Fucα1–2, Galα1–3 → Galα1–X	Blood group B active product	Human serum (4)
N-Acetylgalactosaminyl transferase				
(Fucosyl α1–2)Galactoside α1–3-*N*-acetylgalactosaminyl transferase	UDP-GalNAc	Fucα1–2, GalNAcα1–3 → Galα1–X	Product is blood group A active. Can be used to detect Fucα1–2Gal linkage in glycoproteins	Porcine submaxillary gland (1)
N-Acetylglucosaminyl transferase				
α-Mannoside β1–2-*N*-acetylglucosaminyl transferase	UDP-GlcNAc	GlcNAcβ1–2Manα1–3	*N*-Acetylglucosaminyl transferase I involved in synthesis of complex *N*-linked oligosaccharide units. Best oligosaccharide substrate is $Man_5GlcNAc_2$	Rabbit liver (1)
Glucosyltransferase				
Galactoside β1–2-glucosyltransferase	UDP-Glc	Glcα1–2Galβ1–O-Hyl	Acts on denatured collagens containing Gal–Hyl linkages	Chick embryo (5)
Xylosyltransferase				
Chondroitin sulphate core protein-β-xylosyltransferase	UDP-Xyl	Xylβ1–O-Ser	Acts on chondroitin sulphate core protein in proteoglycan synthesis	Rat chondrosarcoma (6)

1. Sadler, J.E., Beyer, T.A., Oppenheimer, C.L., Paulson, J.C., Prieels, J.-P., Rearick, J.I. and Hill, R.L. (1982) Methods Enzymol. *83*, 458–514; 2. Trayer, I.P. and Hill, R.L. (1971) J. Biol. Chem. *246*, 6666–6675; 3. Powell, J.T. and Brew, K. (1974) Eur. J. Biochem. *48*, 217–228; 4. Nagai, M., Dave, V., Muensch, H. and Yoshida, A. (1978) J. Biol. Chem. *253*, 380–381; 5. Myllylä, R., Anttinen, H., Risteli, L. and Kivirikko, K.I. (1977) Biochim. Biophys. Acta *480*, 113–121; 6. Schwartz, N.B. and Dorfman, A. (1975) Arch. Biochem. Biophys. *171*, 136–144.

binding activity. The structure of the carbohydrate units present in the desialylated binding protein which act as acceptors for labelled sialic acid can also be inferred from the specificity of the purified sialyl transferase. Replacement of sialic acid residues was carried out in incubation mixtures (50 μl) containing 5 μmol of sodium phosphate, pH 6.8, 50 μg bovine serum albumin (to stabilise the enzyme), 34 nmol of CMP-[4-^{14}C]NeuAc (1 mCi/mmol), 40 milliunits of sialyl transferase (purified from bovine colostrum; 1 unit of enzyme produces 1 μmol product/min) and desialylated asialo-glycoprotein binding protein (27 μg). Samples were incubated at 37°C and were assayed for the incorporation of [^{14}C]NeuAc and the level of binding activity.

It was shown that incubation with β-D-galactoside α(2–6) sialyl transferase (EC 2.4.99.1) and CMP-NeuAc could restore 95% of the sialic acid removed by neuraminidase and 80% of the binding activity was recovered. The existence of a terminal sequence at the non-reducing end of the carbohydrate chains of NeuAcα2–6Galβ1– was also established on the basis of the specificity of the transferase.

8.2.7. In vitro (cell-free) labelling of proteoglycans

In vitro labelling methods have played an important part in elucidation of the biosynthesis of proteoglycans. These studies have, to date, largely been confined to the carbohydrate moieties of these macromolecules. Little information is as yet available about the cell-free biosynthesis of the peptide cores of proteoglycans. The transfer of labelled sugars from sugar nucleotides onto endogenous and exogenous acceptors has been studied in several systems (Table 8.4). These investigations have led to at least partial characterisation of several of the steps involved in the assembly of the proteoglycans (see Rodén, 1980) and have given strong support to the 'one enzyme one linkage' hypothesis (Hagopian et al., 1968). Modifications of oligosaccharide units which occur after oligosaccharide chain assembly such as the epimerisation of D-glucuronate residues to L-iduronate and the de-*N*-acetylation and subsequent *N*-sulphation of hexosamine in heparin can also be examined. Reaction sequences

TABLE 8.4
Some cell-free systems used to examine reactions of proteoglycan synthesis

Heparin	Mouse mastocytoma (Dunn and Potter, 1957) (microsomal fraction)	Nucleotide sugar incorporation, deactylation and sulphation (1,2)
	Mouse mastocytoma (Furth et al., 1957)	(3)
Chondroitin sulphate	Chick embryo epiphyseal cartilage (microsomal fraction)	Nucleotide sugar incorporation (4)
	Rat chondrosarcoma	(5) Xylosyl transferase
Dermatan sulphate	Human embryonic skin fibroblasts (particulate fraction)	Incorporation of nucleotide sugars, sulphation, glucuronate epimerisation (6)
Hyaluronic acid	Group A *Streptococcus*	(7)
	Fetal rat skin (particulate fraction)	(8)

1. Silbert, J.E. (1967) J. Biol. Chem. *242*, 5146–5152; 2. Silbert, J.E. (1967) J. Biol. Chem. *242*, 5153–5157; 3. Helting, T. (1972) J. Biol. Chem. *247*, 4327–4332; 4. Silbert, J.E. (1964) J. Biol. Chem. *239*, 1310–1315; 5. Schwartz, N.B. and Dorfman, A. (1975) Arch. Biochem. Biophys. *171*, 136–144; 6. Malmstrom, A., Fransson, L., Höök, M. and Lindahl, U. (1975) J. Biol. Chem. *250*, 3419–3425; 7. Stoolmiller, A.C. and Dorfman, A. (1969) J. Biol. Chem. *244*, 236–246; 8. Schiller, S., Slover, G.A. and Dorfman, A. (1961) Biochem. Biophys. Res. Commun. *5*, 344–348.

which occur rapidly in vivo can often be separated into individual steps in vitro by limitation of the supply of substrate. For example, the relationship between de-*N*-acetylation of heparin and *N*-sulphation can be examined in vitro by restricting the supply of PAPS (Silbert, 1967; Rodén, 1980).

Labelled precursors which are of value in studies of proteoglycan synthesis in vitro include UDP-D-xylose, UDP-D-glucuronic acid, UDP-D-galactose, UDP-*N*-acetyl-D-glucosamine and UDP-*N*-acetyl-D-galactosamine which can be obtained as ^{14}C or ^{3}H derivatives. Acetyl-labelled UDP-GlcNAc has been used in studies of de-*N*-acetylation. UDP-iduronic acid is not a proteoglycan precursor; iduronate in proteoglycans arises from enzymically catalysed epimerisation at C-5 of glucuronate residues. Consequently when UDP-[5-^{3}H]glucuronic acid is used as precursor for the synthesis of proteoglycans in which glucuronate is epimerised to iduronate the epimerisation occurring at C-5 can be followed by the release of ^{3}H (as tritiated water) in this reaction (Malmström et al., 1980). [^{35}S]PAPS is also a useful precursor for sulphated proteoglycans. Keratan sulphate type 1 has a protein-carbohydrate linkage region involving mannose residues and GDP-Man would be a suitable precursor for studying the synthesis of this part of the molecule.

In vitro synthesis can be shown to occur if a radioactive sugar is transferred from the appropriate nucleotide sugar to an endogenous acceptor having the properties of the proteoglycan in question. For

TABLE 8.5

Chondroitin sulfate glycosyltransferases and some of their exogenous substrates

Enzyme	Exogenous acceptors
Galactosyltransferase I	D-Xylose *O*-β-D-Xylosyl-L-serine Methyl, ethyl, butyl and octyl β-D-xylopyranoside Benzyl β-D-xylopyranoside *p*-Nitrophenyl β-D-xylopyranoside
Galactosyltransferase II	4-*O*-β-D-Galactosyl-D-xylose 4-*O*-β-D-Galactosyl-*O*-β-D-xylosyl-L-serine
Glucuronosyltransferase I	3-*O*-β-D-Galactosyl-D-galactose
N-Acetylgalactosaminyltransferase	*O*-β-D-Gal-(1–3)-β-D-Gal-(1–4)-D-Xyl-GlcA-GalNAc-GlcA-GalNAc-GlcA-GalNAc (chondroitin hexasaccharide) GlcA-GalNAc-4S-GlcA-GalNAc-4S-GlcA-GalNAc-4S (chondroitin 4-sulphate hexasaccharide)
Glucuronosyltransferase II	GalNAc-GlcA-GalNAc-GlcA-GalNAc (chondroitin pentasaccharide) GalNAc-6S-GlcA-GalNAc-6S-GlcA-GalNAc-6S (chondroitin 6-sulphate hexasaccharide)
Xylosyltransferase	Smith-degraded cartilage proteoglycan HF deglycosylated cartilage proteoglycan [1] Silk fibroin [2] L-Serylglycylglycine

From Rodén and Schwartz (1975) except for [1] Coudron et al. (1980) and [2] Campbell et al. (1984).

further characterisation of the glycosyl transferases (and other enzymes) involved in biosynthesis it is highly desirable to be able to employ exogenous substrates of defined structure. Two routes are available in searching for suitable exogenous substrates; one is partial degradation of the normal end product of the biosynthetic pathway by enzymatic or chemical methods (e.g. Smith degradation); the other is the use of synthetic acceptors. Both of these approaches have been successfully applied in the search for acceptors for the glycosyl transferases and sulphotransferases involved in chondroitin sulphate biosynthesis (Table 8.5). Of particular interest is the finding that D-xylose or simple xylosides will act as acceptors for the galactosyl transferase I. Addition of xylosides to a cell-free system leads to the formation of oligosaccharide chains which are not attached to core protein. This permits the separate study of oligosaccharides and core protein synthesis.

The availability of exogenous acceptors allows examination of the kinetics and substrate specificities of the biosynthetic enzymes. Simple assay procedures facilitate enzyme purification. As with the glycosyl transferases of glycoprotein synthesis the enzymes involved in proteoglycan synthesis have proved difficult to purify. Most of the glycosyl transferases are membrane-associated and there is some evidence for the association of transferases into multi-enzyme complexes (Schwartz et al., 1974b; Schwartz, 1975). Solubilisation of several of the enzymes of chondroitin sulphate biosynthesis has been achieved by the use of the non-ionic detergent Nonidet P-40 in 0.5 M KCl (Schwartz and Rodén, 1975). The xylosyl transferase has been purified to homogeneity (Schwartz and Dorfman, 1975) by a procedure which includes affinity chromatography on insolubilised substrate (core protein deglycosylated by Smith degradation).

While this approach has successfully shown the pathway of synthesis of chondroitin sulphates and most of the steps in heparin synthesis have been worked out in outline, the search for exogenous precursors of hyaluronate has been less productive. Alternative possibilities, such as the assembly of carbohydrate on a lipid precursor, have been examined without, to date, revealing their involvement. In the case of keratan sulphate type 1, biosynthesis is inhibited by tunicamycin, indicating the involvement of a lipid precursor in the biosynthetic pathway, as would be anticipated from the glycosylamine nature of the protein-carbohydrate linkage (Hart and Lennarz, 1978).

8.2.8. Inhibitors of glycoprotein and proteoglycan biosynthesis

The biosynthesis of glycoproteins and proteoglycans depends on the formation of peptide chains to which monosaccharides or oligosaccharides are added. Inhibitors of peptide chain synthesis such as cycloheximide and puromycin will consequently also inhibit the incorporation of labelled sugars into the carbohydrate units of glycoproteins and proteoglycans. However, there can be wide variations in the time taken for this effect to be observed. The incorporation of labelled mannose into the core of asparagine-linked glycoproteins is rapidly inhibited following the cessation of protein synthesis. However, glycosylation of peripheral residues is much less closely coupled to translation and may continue for a long time in the absence of protein synthesis. For example, cycloheximide (100 μg/ml) completely inhibits the incorporation of labelled amino acids into the glycoprotein ovomucoid but labelling with [^{14}C]glucosamine was not significantly decreased after seven hours incubation of tissue slices with the inhibitor (McCairns and Beeley, 1972). In this and many other cases it appears that the total intracellular pool of glycosylated protein is quite large. Both the size and the rate of turnover of pools of acceptors and donors for monosaccharide precursors are important factors in labelling experiments carried out in the presence of inhibitors.

Protein-synthesis inhibitors can be employed to examine the relationship between translation and glycosylation of glycoproteins. The general approach is to isolate nascent peptide chains after short periods of labelling, in vivo, with glucosamine or mannose (Uhr, 1970; Bergman and Kuehl, 1977; Kiely et al., 1976; Lawford and Schachter, 1967). Technical aspects of these experiments and the results obtained have been reviewed (Bergman and Kuehl, 1982) and will not be considered further here.

Since the introduction of tunicamycin and related antibiotics it has become possible to inhibit specifically the synthesis of *N*-linked oligosaccharide units in glycoproteins (and corneal keratan sulphate). The availability of specific inhibitors, as well as aiding biosynthetic studies, provides a means for producing 'naked', carbohydrate-deficient, peptide chains. Studying these underglycosylated molecules can give information about the function of carbohydrate groups and about their structure (number and type of carbohydrate unit per peptide chain). In addition to inhibition by tunicamycin it is now possible to block selectively several stages in the assembly and processing of *N*-linked carbohydrate units of glycoproteins. Much less progress has been made in finding effective inhibitors for the glycosylation of serine- and threonine-linked carbohydrate units. The β-D-xylosides have proved valuable in studying the biosynthesis of proteoglycans.

This section describes conditions for the use of some of the more selective inhibitors of glycosylation of glycoprotein and proteoglycan.

Inhibitors of glycoprotein glycosylation. Several inhibitors of the glycosylation of glycoproteins containing asparagine-linked carbohydrate units have been described (Table 8.6). The mechanisms of action of some of these inhibitors have been established in detail; examination of the effects of inhibitors on viral glycoprotein synthesis has been particularly valuable in this regard (Schwartz and Datema, 1982a).

Of the inhibitors affecting *N*-linked oligosaccharide units the antibiotic tunicamycin has been the most valuable. Tunicamycin inhibits the first step in the assembly of dolichol diphosphate-bound oligosaccharides (the transfer of GlcNAc from UDP-GlcNAc to dolichol monophosphate producing GlcNAc-PP-dolichol). This antibiotic can be used with intact cells or in cell-free systems to produce peptide chains free of *N*-linked carbohydrate units. Concentrations in the range 0.1–10 μg/ml of the inhibitor are used in both types of system.

TABLE 8.6
Inhibitors of glycosylation and trimming of asparagine-linked oligosaccharides

Protein glycosylation inhibitors	Reactions inhibited	Comments on use	References
Antibiotics			
Tunicamycin	Formation of GlcNAc-PP-Dol from UDP-GlcNAc + P-Dol	Can be used with intact cells (0.1–10 μg/ml*) or cell-free systems (0.5–5 μg/ml) (Mycospociodins, Streptovirudins and A2401 are related antibiotics)	(1)
Amphomycin	Formation of dolichol-linked monosaccharides	Cell-free systems, 500 μg/ml	(1)
Bacitracin	Formation of P-Dol from PP-Dol	Inhibitor probably does not cross cell membranes	(2)
Showdomycin	Formation of Glc-P-Dol	Cell-free system, 5–10 μg/ml	(1)
Diumycin	1. Formation of $GlcNAc_2$-PP-Dol 2. Formation of Man-P-Dol and GlcNAc-PP-Dol	Cell-free system, 50 μg/ml	(1)
Sugar analogues			
2-Deoxy-D-glucose	1. Formation of dolichol-linked monosaccharides by trapping P-Dol as dGlc-P-Dol 2. Assembly of lipid-linked oligosaccharides by incorporation of dGlc instead of Glc or Man	Intact cells 1–50 μM, or 10–120 μM GDP-dGlc or UDP-dGlc in cell-free system	(1)
2-Deoxy-2-fluoro-D-glucose	Formation of Man-P-Dol and Glc-P-Dol is inhibited; UDP-Glc pool size decreased	Intact cells 0.2–0.3 mM	(1)

2-Deoxy-2-fluoro-D-mannose	Formation of lipid-linked oligosaccharides; mechanism not known	Intact cells 0.2–10 mM	(1)
D-Glucosamine	Formation of lipid-linked oligosaccharides; mechanism not known	Intact cells 4–40 mM, inactive in cell-free systems	(1)
Oligosaccharide trimming			
1-Deoxynojirimycin	Inhibits release by glucosidase I of outer glucose residues from Glc_3Man_9-$GlcNAc_2$. Also inhibits synthesis of $Glc_3Man_9GlcNAc_2$-PP-Dol	Intact cells (5 mM) or cell-free (*Saccharomyces cerevisiae*) 1–50 μM (nojirimycin and *N*-methyl-l-deoxynojirimycin are related inhibitors)	(3-5)
Castanospermine	Inhibits release by glucosidase I of outer glucose residues from $Glc_3Man_9GlcNAc_2$	Intact cells, 10–50 μg/ml	(6)
Bromochonuritol	Inhibits release by glucosidase II of outer glucose residue from $Glc_3Man_9GlcNAc_2$		(7)
Swainsonine	Inhibits α-mannosidase II leading to formation of hybrid-type oligosaccharide units with high-mannose and complex-type branches	Intact cells 0.1–2 μg/ml; cell-free, 0.1–10 μM	(8-11)

* The concentrations of inhibitors used in certain experiments should be treated only as a guide for obtaining optimal inhibitor concentrations

1. Schwartz, R.T. and Datema, R. (1982) Methods Enzymol. *83*, 432–443; 2. Wedgewood, J.F. and Strominger, J.L. (1980) J. Biol. Chem. *255*, 1120–1123; 3. Saunier, B., Kilker, R.D., Tkacz, J.S., Quaroni, A. and Herskovics, A. (1982) J. Biol. Chem. *257*, 14155–14161; 4. Gross, V., Andus, T., Tran-Tri, T.A., Schwartz, R.T., Decker, K. and Heinrich, P.C. (1983) J. Biol. Chem. *258*, 12203–11209; 5. Truscheit, E., Frommer, W., Junge, B., Müller, L., Schmidt, D.D. and Wingender, W. (1981) Angew. Chemie. *93*, 738–755; 6. Pan, Y.T., Hori, H., Saul, R., Sanford, B.A., Molyneux, R.J. and Elbein, A.D. (1983) Biochemistry *22*, 3975–3984; 7. Datema, R., Romero, P.A., Legler, G. and Schwartz, R.T. (1982) Proc. Natl. Acad. Sci. U.S.A. *79*, 6787–6791; 8. Arumughan, R.G. and Tanzer, M.L. (1983) J. Biol. Chem. *258*, 11883–11889; 9. Elbein, A.D., Dorling, P.R., Vosbeck, K. and Horisberger, M. (1982) J. Biol. Chem. *257*, 1573–1576; 10. Tulsiani, D.R.P., Harris, T.M. and Touster, O. (1983) J. Biol. Chem. *257*, 7936–7939; 11. Gross, V., Tran-Tri, T.A., Vosbeck, K. and Heinrich, P.C. (1983) J. Biol. Chem. *258*, 4032–4036.

Tunicamycin, in some systems, produces partial inhibition of protein synthesis. The optimal concentration of the inhibitor to use in experiments with cultured cells can be determined by measuring the incorporation of labelled mannose and leucine into glycoprotein after preincubation (1 h is usually long enough to produce maximal inhibition) with different concentrations of tunicamycin. A concentration of inhibitor producing >90% inhibition of mannose incorporation with only a small (<10%) inhibition of protein synthesis can usually be found. It is also possible to measure the inhibition of incorporation of labelled mannose into lipid-linked intermediates (see Section 8.2.4 and Schwartz and Datema, 1982a). Tunicamycin can be dissolved in dilute alkali (25 mM NaOH) or dimethyl sulphoxide; thelatter solvent can itself sometimes affect cell growth and/or protein synthesis. As the inhibitor binds strongly to cell membranes its effects cannot readily be reversed by washing cells to which it has been added. Commercial preparations of the antibiotic usually contain mixtures of a group of molecules with related structures but which show some differences in activity (Mahoney and Duskin, 1979; Eckhardt et al., 1980).

The effect of tunicamycin on specific glycoproteins can often be detected as a decrease in M_r of the polypeptide on SDS-gel electrophoresis. This change in size can be used to estimate the total molecular weight of asparagine-linked carbohydrate units present in the untreated glycoprotein (Polonoff et al., 1982). Such estimates are, of course, subject to considerable errors because of the small changes in mobilities usually observed and because of the inaccuracies inherent in measuring glycoprotein molecular weights by SDS-gel electrophoresis (Section 4.3.2). Tunicamycin inhibition can also be used to determine the number of asparagine-linked carbohydrate units present in immunoglobulins (Cushley et al., 1983). Labelled immunoglobulin μ-chains synthesised in the absence of inhibitor or in the presence of different concentrations of tunicamycin were isolated by immunoprecipitation and separated by SDS-gel electrophoresis. When glycosylation was only partially inhibited bands corresponding to different partially glycosylated molecules were detected after prolonged fluorography. Counting the number of glycosylated species gave the number of asparagine-linked carbohydrate units. The fully glycosylated and completely unglycosylated chains were, however, the predominant molecular species. The biosynthetic approach to the characterisation of glycoprotein structure has been extended by the use of the threonine analogue β-hydroxynorvaline to prevent transfer of *N*-linked units to Asn-X-Thr sequences (Hortin and Boime, 1980; Polonoff et al., 1982). In the presence of this inhibitor (1–20 mM) in the culture medium of NRK cells infected with murine leukaemia virus only Asn-X-Ser sequences become glycosylated; analysis of the partially glycosylated products gave an estimate of the number of carbohydrate groups attached to Asn-X-Thr as opposed to Asn-X-Ser sequences (Polonoff et al., 1982). Another amino acid analogue which interferes with *N*-glycosylation in vitro is DL-*threo*-fluoroasparagine (Hortin et al., 1983). This asparagine analogue becomes incorporated in newly formed peptide chains and prevents *N*-glycosylation. It is effective at a concentration of 1 mM.

In addition to its uses in analysing glycoprotein structure tunicamycin has been extensively employed in studies of the functional effects of the absence of *N*-glycosylation on soluble and membrane-bound glycoproteins and on cellular differentiation (for reviews see Schwartz and Datema, 1982b; Olden et al., 1982).

While tunicamycin has been the most widely applied inhibitor of *N*-glycosylation of glycoproteins several of the other inhibitors listed in Table 8.6 deserve consideration. For example, bacitracin, which inhibits the formation of dolichol monophosphate, apparently does not enter intact cells (Wedgewood and Strominger, 1980) and can be used to distinguish between lipid-dependent oligosaccharide synthesis occurring at cell surfaces (in plants and fungi) and biosynthesis at intracellular sites.

Of the sugars and sugar analogues listed as inhibitors in Table 8.6 2-deoxyglucose has been most widely used. In addition to the in vitro effects of 2-deoxyglucose the UDP- and GDP-derivatives are inhibitory in vitro. The inhibitory action of 2-deoxyglucose is complex; four steps in the biosynthesis of glycoprotein oligosaccharide units can be affected (Schwartz and Datema, 1982b). As with other sugar analogues 2-deoxyglucose is employed in the growth medium of cultured cells, at a concentration (5–10 mM) similar to that of glucose. Lower concentrations (0.2–2 mM) can be used for cells grown on pyruvate as sole carbon source. Side-effects of the sugar analogue are more likely in the latter case (Schwartz and Datema, 1982a). Because of the complexity of inhibition and side-effects of the inhibitor, for most purposes the use of tunicamycin is more satisfactory.

The involvement of lipid-linked intermediates in the pathway of synthesis of *N*-linked carbohydrate units has the result that a variety of compounds inhibiting the biosynthesis of dolichol monophosphate, such as 25-hydroxycholesterol, 7-ketocholesterol, 3-hydroxy-3-methylglutaric acid and compactin, can influence glycosylation (Schwartz and Datema, 1982b). Other inhibitors acting at the level of intracellular transport processes (e.g. cytochalasin B) can also affect glycosylation.

There has been considerable recent interest in the discovery of inhibitors of the processing of oligosaccharides following the transfer of $Glc_3Man_9GlcNAc_2$ from Dol-PP to protein (Schwartz and Datema, 1984). The indozilidine alkaloid, swainsonine, which has been isolated from *Swainsona canescens* (Colegate et al., 1979) or from *Astragalus lentiginosus* (Molyneux and James, 1982), is an inhibitor in vitro of α-mannosidase II (Tulsiani et al., 1982). Addition of swainsonine (0.1–2 μg/ml) to cultured cells has been shown to lead to the production of glycoproteins with altered carbohydrate units; instead of complex-type oligosaccharide chains human cellular fibronectin and rat hepatocyte α_1-antitrypsin produce oligosaccharides of the 'hybrid type' (Gross et al., 1983; Arumughan and Tanzer, 1983). This structural change can most readily be detected experimentally as an increase in the oligosaccharide chains of glycopeptides which are susceptible to hydrolysis by endoglycosidase H (Section 6.9.4.1).

The antibiotic 1-deoxynojirimycin is an inhibitor, in vitro, of processing glucosidases. When added to cultured intestinal epithelial cells 1-deoxynojirimycin (5 mM) decreased the proportion of radioactive mannose incorporated into complex oligosaccharides and rendered the high-mannose oligosaccharides less susceptible to α-mannosidase, indicating that removal of terminal glucose residues is also inhibited in vivo (Saunier et al., 1982).

Inhibitors of processing have great potential value as tools for examining the role of specific types of carbohydrate structure in glycoprotein function.

Proteoglycan biosynthesis inhibitors. The synthesis of keratan sulphate (type 1, *N*-linked) in embryonic chick cornea can be inhibited by 1–2.5 μg/ml tunicamycin (Hart and Lennarz, 1978). Intact corneas were preincubated with tunicamycin for up to 2 h prior to labelling with [^{3}H]glucosamine and $^{35}SO_4^{2-}$. In addition to its specific effect on the lipid-linked intermediates involved in keratan sulphate synthesis the antibiotic decreases total proteoglycan synthesis as a result of lower availability of core proteins due to non-specific inhibition of protein synthesis.

Biosynthesis of the carbohydrate chains of proteoglycans containing the Xyl-Ser protein-carbohydrate linkage (e.g. chondroitin sulphate proteoglycan) can be inhibited by β-D-xylosides. The xylosides act as exogenous acceptors for the synthesis of sulphated glycosaminoglycan chains of similar size to the carbohydrate chains normally attached to core protein and have been shown to be active in several cell types (Schwartz et al., 1974b). Addition of xylosides to cells capable of producing proteoglycans usually results in stimulation of the rate of formation of glycosaminoglycan chains as a consequence of increasing the concentration of acceptors for the first galactosyl transferase involved in formation of the 'linkage region' of the oligosaccharide chain. Xylosides have been shown to act as inhibitors of the glycosylation of native polypeptide acceptors (Schwartz, 1977). Rather than being employed as inhibitors of proteoglycan synthesis, β-xylosides have been used to assess the capacity of cells to synthesise sulphated glycosaminoglycan chains independently of the production of core protein.

The xylosides *p*-nitrophenyl-β-D-xylopyranoside and 4-methylumbelliferyl-β-D-xyloside are effective at low concentrations (1 mM); the free sugar D-xylose is also inhibitory but at higher concentration (40 mM). Schwartz (1977) incubated chick embryo cartilage cells in culture medium containing *p*-nitrophenyl-β-D-xyloside (1 mM) for 48 h prior to labelling glycosaminoglycans. Hart and Lennarz (1978) employed the 4-methylumbelliferyl derivative (1 mg/ml) during labelling (for 20.5 h) of embryonic chick cornea.

8.3. Labelling terminal sialic acid, galactose and N-acetylgalactosamine

Two methods are available for introducing a tritium label specifically into the carbohydrate units of glycoproteins which contain non-reducing terminal residues of galactose, *N*-acetylgalactosamine or sialic acid. Both techniques involve the initial oxidation of a monosaccharide residue to produce an aldehyde which is then labelled by reduction with NaB^3H_4. Sialic acid residues are selectively oxidised with low concentrations of periodate, while galactose and *N*-acetylgalactosamine can be oxidised enzymatically by galactose oxidase (Fig. 8.9). Both methods were originally developed for the modification of soluble glycoproteins but have been adapted for application to membrane-bound glycoconjugates. The galactose oxidase-borotritide procedure has been extensively used for side-specific labelling of cell surface glycoproteins (and glycolipids).

8.3.1. Sialic acid labelling with $NaIO_4$ and NaB^3H_4

Sialic acid is rapidly oxidised by $NaIO_4$ provided that the hydroxyls at positions C-7–C-9 are unsubstituted. The C-7 aldehyde (Fig. 8.9) is the major product. Selective modification requires mild conditions with a low concentration (2–5 mM) and a small molar excess (5–10-fold) of periodate (Van Lenten and Ashwell, 1972; Liao et al., 1973).

The following procedure for the labelling of soluble glycoproteins is a slight modification of the method described by Van Lenten and Ashwell (1972). The glycoprotein is dissolved in 0.1 M sodium ace-

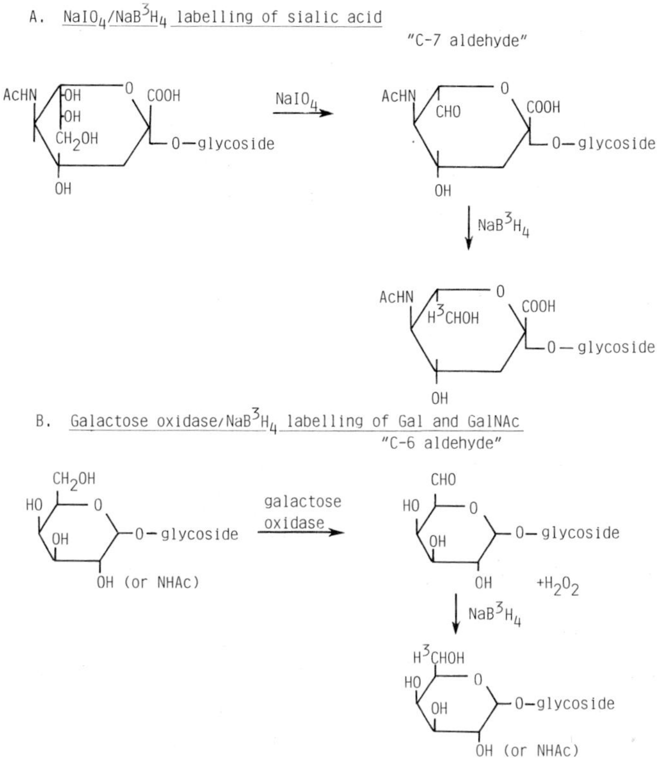

Fig. 8.9. Labelling of glycoproteins containing sialic acid, galactose or *N*-acetylgalactosamine with ^3H.

tate, pH 5.6, containing 0.15 M NaCl to give a solution having 1 μmole of sialic acid residues per ml. After cooling on ice, an equal volume of 10 mM $NaIO_4$ is added and the mixture is kept at 0°C for 10 min in the dark. The reaction is stopped by addition of an excess of ethylene glycol and the solution is dialysed overnight at 4°C against 0.05 M sodium phosphate, pH 7.5, containing 0.15 M NaCl. Labelling is carried out with sodium or potassium borohydride (200–300 mCi/mmole) which is made up as a solution (100 mM) in 0.01 M NaOH. Labelled borohydride, although rapidly decomposed under even mildly acidic conditions, is relatively stable in alkali. Stock solutions in 0.01 M alkali can be stored in aliquots at −70°C for several months.

For every μmole of sialic acid in the original sample 15 μl (1.5 μmoles) of borohydride solution are added and the mixture is allowed to react for 30 min at room temperature. To ensure complete reduction a further addition (3 μmoles/μmole of sialic acid) is made and the solution allowed to react for 30 minutes. Excess borohydride can be removed by dialysis.

Labelling of glycoproteins to specific activities of about 10^7 dpm/mg can be obtained by this procedure; much higher specific radioactivities can be achieved by scaling down the quantity of glycoprotein and using tritiated borohydride of higher specific activity. However, some radioactivity is incorporated into glycoproteins treated with KB^3H_4 without periodate oxidation. This non-specific labelling can amount to 4–5% of the specific incorporation associated with carbohydrate (Van Lenten and Ashwell, 1972). Side-reactions which can be encountered as a result of borohydride treatment of proteins include disulphide-bond reduction and peptide-bond cleavage. The integrity of the primary structure and conformation of glycoproteins labelled by this method should be examined before functional studies are carried out.

A similar procedure can be applied to label sialic acid residues in cell membranes (Blumenfeld et al., 1972; Liao et al., 1973). The following development of this method has been used for labelling the sialic acid residues of glycoproteins of human erythrocyte membranes (Gahmberg et al., 1976b; Gahmberg and Andersson, 1977).

To 0.5 ml of washed, packed cells add 1 ml of phosphate-buffered saline (5 mM sodium phosphate, 0.15 M NaCl, pH 7.4) and sodium metaperiodate to a final concentration of 2 mM. After incubating at room temperature in the dark for 10 min excess periodate is removed by washing three times with phosphate-buffered saline. The cells are resuspended in 0.5 ml phosphate-buffered saline and reduced with 50 μl (0.5 mCi) NaB^3H_4 (8 μCi/mmol dissolved, 10 mCi/ml, in 0.01 M NaOH and stored at −70°C) for 30 min at room temperature. After washing three times in phosphate-buffered saline the labelled membranes are isolated (Fairbanks et al., 1971). The membrane glycoproteins are separated by SDS-polyacrylamide gel electrophoresis and labelled bands are detected by fluorography (or, with considerably lower resolution, by slicing the gel and counting slices).

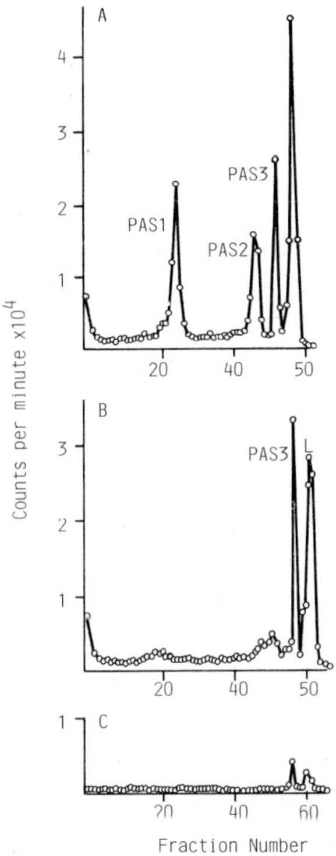

Fig. 8.10. Labelling erythrocyte membrane glycoproteins by treatment with periodate and NaB^3H_4 (Gahmberg et al., 1976b). Normal and blood group En(a−) erythrocytes were treated with $NaIO_4$ and labelled with NaB^3H_4. The membrane components were separated by SDS-polyacrylamide gel electrophoresis and the gels were sliced and counted. A, normal cells; B, En(a−) cells, and C, normal cells treated with NaB^3H_4 without periodate oxidation. The results demonstrate theabsence of periodic acid-Schiff staining sialoglycoproteins (PAS 1 and 2) in En(a−) cells.

Fig. 8.10 shows the results obtained when this technique was applied to normal human erythrocytes and those from an individual with blood group En(a−) erythrocytes. It can be seen that the En(a−) membranes lack the major glycoprotein bands PAS1 and PAS2 (Gahmberg et al., 1976b). Little labelling of the membrane occurs when periodate oxidation is omitted but this control should always be included.

As can be seen in Fig. 8.10, treatment with periodate and NaB^3H_4 results in appreciable labelling of lipid as well as glycoprotein.

The modified seven-carbon sialic acid derivative (5-acetamido-3,5-dideoxy-L-arabino-2-heptulosonic acid) is a substrate for *Clostridium perfringens* neuraminidase but the rate of hydrolysis is decreased (Liao et al., 1973). 'Seven-carbon sialic acid' reacts with the thiobarbituric acid and other reagents used to detect sialic acid but with different colour yield.

8.3.2. Galactose and N-acetylgalactosamine labelling

The reactions by which an atom of tritium can be incorporated at C-6 of galactose and *N*-acetylgalactosamine residues in glycoconjugates are outlined in Fig. 8.9. In contrast to the periodate/borotritide labelling of sialic acid, the modification of Gal/GalNAc regenerates the original sugar. The reversibility of this modification has been exploited in studies of the functional role of galactose residues in glycoproteins. Because galactose oxidase does not modify galactose residues in glycoproteins when substituted with sialic acid it is usually necessary to remove sialic acid by treatment with neuraminidase to obtain high levels of galactose modification. Labelling with galactose oxidase/borotritide can be applied to soluble or membrane-bound glycoproteins (and glycolipids). Because galactose oxidase does not penetrate the cell membrane it can be employed for the specific modification and labelling of surface carbohydrates.

Galactose oxidase from *Dactylium dendroides* can be obtained from several commercial sources. Preparations should be tested for the presence of proteolytic or neuraminidase activities by incubation with the chromogenic substrate azoalbumin (Sigma) and submaxillary mucin. If necessary galactose oxidase can readily be purified by affinity chromatography on Sepharose (Gahmberg and Hakomori, 1975).

Modification of soluble glycoproteins. This procedure developed by Morell and Ashwell (1972) requires some modification for particular glycoproteins to optimise the enzymic oxidation of galactose residues and to allow separation of the labelled glycoprotein from the enzymes required for the reaction. This separation can often be achieved in a single step at the end of the procedure but this must be investigated for each glycoprotein. In the following example the glycoprotein ceruloplasmin was repurified by DEAE-cellulose chromatography after each step.

Ceruloplasmin is incubated with neuraminidase (*Vibrio cholerae*) to remove sialic acid and asialoceruloplasmin is isolated by ion-exchange chromatography. (Alternatively neuraminidase can be selectively adsorbed by passing through a column of insolubilised anti-neuraminidase antibody.) Asialoceruloplasmin (0.54% w/v), galactose oxidase (12.7 units/ml) and horseradish peroxidase (18 units/ml) are incubated at 25°C for 66 h at pH 7.0 in a buffer consisting of 0.02 M sodium phosphate, 0.45 M sodium acetate, 0.15 M NaCl and 1% toluene. The oxidised protein is isolated by DEAE-cellulose chromatography. Oxidised asialoceruloplasmin is diluted to 0.2% protein with 0.05 M NaCl. For each mg of protein 1.6 μmoles of NaB^3H_4 is added. After 10 min at room temperature the pH is lowered to 5.6 to decompose borohydride and the labelled glycoprotein is repurified on a column of DEAE-cellulose. Labelling of glycopeptides can also be carried out; purification of the product can be achieved in a single gel filtration step.

Modification of cell surface glycoproteins and glycolipids. Procedures for labelling galactose and *N*-acetylgalactosamine residues exposed at the surface of cells have been described for erythrocytes (Gahmberg and Hakomori, 1973; Steck and Dawson, 1974; Gahmberg et al., 1976b; Gahmberg, 1976), fibroblasts (Gahmberg and Hakomori, 1975) and lymphoid cells (Gahmberg et al., 1976a). The cells are incubated in isotonic buffer either with galactose oxidase alone or with a mixture of galactose oxidase and neuraminidase. The modified cells are then labelled with NaB^3H_4. As a control unmodified cells should also be subjected to labelling. Membrane glycoproteins can then be separated by SDS-gel electrophoresis and labelled bands located by fluorography.

The following protocol has been standardised for the comparison of several types of lymphoid cell (Gahmberg et al., 1976a) but can also be applied to fibroblasts (Gahmberg, 1978). Requirements include protease- and neuraminidase-free *Dactylium dendroides* galactose oxidase (see above), protease-free *Vibrio cholerae* neuraminidase (suitable enzyme is obtainable from Behringwerke) and NaB^3H_4 of high specific activity (approx. 8 mCi/mmol, Amersham International).

Cells are washed in Dulbecco's phosphate-buffered saline (containing NaCl 8.00 g, KCl 0.20 g, $Na_2HPO_4 \cdot 2H_2O$ 1.15 g, KH_2PO_4 0.2 g, $CaCl_2$ 0.1 g and $MgCl_2 \cdot 6H_2O$ 0.1 g per litre; the calcium and magnesium are dissolved separately and added to a solution of the other salts). Three tubes, each containing 20×10^6 cells made up to 1 ml with Dulbecco's phosphate-buffered saline (PBS), are prepared. To the first tube add galactose oxidase alone (5 units dissolved in 25 μl PBS); to a second tube both galactose oxidase (5 units) and neuraminidase (12.5 units) are added. No enzyme is added to the third, control, tube. After incubation at room temperature for 30 min with gentle shaking the cells are washed three times in PBS by centrifugation. The cells are resuspended in 0.5 ml PBS and labelled with NaB^3H_4 (0.5 mCi in 25 μl 0.01 M NaOH) for 30 min at room temperature. Then the cells are washed four times in PBS and an aliquot is counted for radioactivity.

Because DNA interferes in the fractionation of proteins by SDS-gel electrophoresis it can be advantageous to remove cells' nuclei before preparing samples for running on gels (Gahmberg et al., 1978). This can be done by treating cells at 0°C with 1% Triton X-100 in PBS containing 2 mM phenylmethylsulphonyl fluoride and centrifuging. Supernatants are then prepared for electrophoresis by treating with

SDS, a thiol-reducing agent and boiling. After separation by SDS-gel electrophoresis protein bands are located by staining with Coomassie Blue and labelled glycoproteins are detected by fluorography (Section 7.5).

It is generally found that labelling is greatly increased in samples treated with neuraminidase as well as galactose oxidase. The difference in labelling can give some indication of the extent to which glycoproteins are sialylated. Correlation between bands which are labelled strongly only after neuraminidase treatment and bands labelled by the $NaIO_4/NaB^3H_4$ method are often observed. Glycolipids as well as glycoproteins are labelled by the galactose oxidase/NaB^3H_4 method.

The method described above allows the identification of glycoproteins by labelling them to high specific activity under conditions which do not lead to peptide-chain cleavage. However, only a fraction of the terminal non-reducing galactose and *N*-acetylgalactosamine residues are labelled. Thus the procedure is qualitative rather than quantitative. The quantitative modification of surface glycoconjugates of human erythrocytes using higher concentrations of NaB^3H_4 has been reported (Gattegno et al., 1981). There is, however, some danger of peptide-chain degradation occurring if the conditions employed for reduction are too severe.

Appendix

Suppliers of equipment and reagents for glycoprotein and proteoglycan techniques

A. Equipment

Manufacturer	Equipment referred to in text
Beckman Instruments Inc. Spinco Division 1117 California Avenue Palo Alto CA 94304 U.S.A.	Analytical ultracentrifuge, Microfuge, microzone electrophoresis apparatus
Beckman-RIIC Limited Progress Road Sands Industrial Estate High Wycombe HP12 4JL Buckinghamshire England	
Japan Electron Optics Laboratory Co. Ltd. 1418 Nakagami Akishima Tokyo Japan	Carbohydrate analyser
LKB-Producter AB Box 305 S-161 26 Bromma Sweden	LKB 4400 amino acid analyser, isoelectric focusing (and chromatography) apparatus

LKB Instruments, Inc. 9319 Gaither Road Gaitherburg MD 20877 U.S.A.	
LKB Instruments, Ltd. 232 Addington Road S. Croydon Surrey CR2 8YD England	
Pierce Chemical Co. Inc. Box 117 Rockford Illinois 61105 U.S.A.	Tubes with Teflon-coated septa for GLC samples and Reacti-vials
Pierce and Warriner (UK) Ltd. 44 Upper Northgate Street Chester Cheshire CH1 4EF England	
Pye-Unicam, Ltd. York Street Cambridge, CB 2PX England	Equipment for GLC (and HPLC)
Pharmacia Fine Chemicals AB Box 175 S-75104 Uppsala 1 Sweden	Equipment for isoelectric focusing (and chromatography)
Technicon Instrument Corp. 511 Benedict Avenue Tarrytown, NY 10591 U.S.A.	Carbohydrate and amino acid analysers

Technicon Instrument Corp.
Evans House
Hamilton Close
Basingstoke
Hants
England

B. Chromatographic media, proteins and reagents

Manufacturer	Materials referred to in text
Aldrich Chemical Co. P.O. Box 355 Milwaukee Wisconsin 53201 U.S.A.	Wide range of chemicals, including trifluoromethane sulphonic acid
Aldrich Chemical Co. The Old Brickyard New Road Gillingham Dorset SP8 4JL England	
Amersham International plc. White Lion Road Buckinghamshire, HP 79LL England	Radiochemicals
Applied Science Laboratories Inc. P.O. Box 440 State College Pennsylvania 16801 U.S.A.	Gas chromatography stationary phases and reagents
(in U.K. **Pierce and Warriner** address as given above)	

Bio-Rad Laboratories Inc. 2200 Wright Avenue Richmond California 94804 U.S.A.	Gel filtration (Bio-Gel) media, ion-exchange resins and Enzymobeads
Bio-Rad Laboratories Ltd. Greenhill Crescent Holywell Industrial Estate Watford Hertfordshire England	
Boehringer Mannheim GmbH Biochemica Postfach 310120 D-6800 Mannheim F.R.G.	Carbohydrates, lectins, endo- and exo-glycosidases
BCL — The Boehringer Corporation (London) Ltd. Bell Lane Lewes East Sussex, BN 71 LG England	
BDH Chemicals Ltd. Broom Road Poole Dorset BH 124 NN England	Lectins (Vector) and lectin-affinity media. Ion-exchange resins
Calbiochem-Behring Corp. 10933 No. Torrey Pines Road La Jolla California 92037 U.S.A. (in U.K. **Cambridge Bioscience** 149-153 St. Neots Road Hardwick Cambridge CB3 7QJ)	Pronase, neuraminidase and lectins

Eastman Kodak Co.
343 State Street
Rochestester
NY 14650
U.S.A.

X-ray film for autoradiography and fluorography. Cyanogen bromide

(in U.K. **Kodak Ltd.**
P.O. Box 10
Dallimore Road
Manchester M23 9NJ)

Fluka AG Chemische Fabrik
CH-9470 Buchs
Switzerland

Carbohydrates and lectins

(in U.K. **Fluorochem Ltd.**
Peakdale Road
Glossop
Derbeyshire, SK 139 XE
England)

New England Nuclear
549 Albany Street
Boston
Ma 02118
U.S.A.

Radiochemicals

Pharmacia P-L Biochemicals
Box 175
S-75104 Uppsala 1
Sweden

Gel filtration, ion-exchange and lectin-affinity media, lectins and carbohydrates

Pharmacia Ltd.
Pharmacia House
Midsummer Boulevard
Milton Keynes MK 9 3HP
England

Polysciences Inc.
400 Valley Road
Warrington
PA 18976
U.S.A.

Lectins

Polysciences Ltd. 34 Tenter Road Moulton Park Northampton NN3 1P2 England	
Sigma Chemical Co. P.O. Box 14508 St. Louis MO 63178 U.S.A.	Lectins, lectin-affinity media, exo- and endo-glycosidases, tunicamycin, carbohydrates
Sigma Chemical Co. Ltd. Fancy Road Poole Dorset BH 177NH England	
Miles Laboratories Inc. P.O. Box 2000 Elkhart IN 46515 U.S.A.	Lectins, lectin-affinity media, exo- and endo-glycosidases
Scientific Division of Miles Laboratories Ltd. P.O. Box 37 Stoke Court Stoke Poges Slough, SL2 4LY England	
Vector Laboratories, Inc. Burlinghame California U.S.A.	Lectins and lectin-affinity media

(in U.K. **BDH Chemicals Ltd.**
see above)

Waters Associates, Inc. Milford Massachussetts 01757 U.S.A.	HPLC separation media (and instruments)
Waters Associates (Inst.) Ltd. 324 Chester Road Hartford Northwich Cheshire CW8 2AH England	
Whatman Ltd. Springfield Mill Maidstone Kent England	Ion-exchange cellulose materials, chromatography paper
Whatman Inc. 9 Bridewell Place Clifton, NJ 07014 U.S.A.	

References

ACKERS, G.K. (1975) in The Proteins, 3rd Edn. Vol. 1 (Neurath, H. and Hill, R.L., eds), pp. 1–94, Academic Press, New York.

ADAIR, W.L. and KORNFELD, S. (1974) J. Biol. Chem. *249*, 4696–4704.

AKREM, A.A., AVVAKUMOV, G.V. and STREL'CHYONOK, O.A. (1979) J. Chromatog. *176*, 207–216.

ALLAN, D., AUGER, J. and CRUMPTON, M.J. (1972) Nature (New Biol.) *236*, 23–25.

ALLEN, A. (1981) in Physiology of the Gastrointestinal Tract, Vol. 1 (Johnson, L.R., ed), pp. 617–639, Raven Press, New York.

ALLEN, A. (1983) Trends Biochem Sci. *8*, 169–173.

ALLEN, A., PAIN, R. and ROBSON, T.R. (1976) Nature *264*, 88–89.

ALLEN, A.K. and NEUBERGER, A. (1978) Biochem. J. *171*, 665–674.

ALLEN, A.K. and NEUBERGER, A. (1973) Biochem. J. *135*, 307–314.

ALLEN, A.K. and NEUBERGER, A. (1975) FEBS Lett. *50*, 362–364.

ALLEN, A.K., DESAI, N.N., NEUBERGER, A. and CREETH, J.M. (1978) Biochem. J. *171*, 665–674.

ALLEN, A.K., NEUBERGER, A. and SHARON, N. (1973) Biochem. J. *131*, 155–162.

ALLEN, G. (1980) in Sequencing of Proteins and Peptides, Laboratory Techniques in Molecular Biology (Work, T.S. and Burdon, R.H., eds), North-Holland, Amsterdam.

AMINOFF, D. (1959) Biochem. J. *81*, 384–392.

AMINOFF, D., GATHMANN, W.D., MCLEAN, C.M. and YADOMAC, T. (1980) Anal. Biochem. *101*, 44–53.

AMOS, W.B. (1976) Anal. Biochem. *70*, 612–615.

ANDERSON, B., HOFFMAN, P. and MEYER, K. (1965) J. Biol. Chem. *240*, 156–167.

ANDREWS, P. (1965) Biochem. J. *96*, 595–606.

ANTONOPOULOS, C.A., AXELSON, C.A., HEINEGARD, D. and GARDELL, S. (1974) Biochim. Biophys. Acta *338*, 108–119.

ARNOTT, S., GUSS, J.M., HUKINS, D.W.L., DEA, I.C.M. and REES, D.A. (1974) J. Mol. Biol. *88*, 175–184.

ARNOTT, S., GUSS, J.M., HUKINS, D.W.L. and MATTHEWS, M.B. (1973) Science *180*, 743–745.

ARUMUGHAN, R.G. and TANZER, M.L. (1983) J. Biol. Chem. *258*, 11883–11889.

ASHWELL, G. and HARFORD, J. (1982) Annu. Rev. Biochem. *51*, 531–554.

ASPBERG, K. and PORATH, J (1970) Acta Chem. Scand. *24*, 1839–1841.

ASPINALL, G.O. and FERRIER, R.J. (1957) Chem. and Ind. 1216.

ATKINS, E.D.T. and ISAAC, D.H. (1973) J. Mol. Biol. *80*, 773–779.

ATKINS, E.D.T. and MACKIE, W. (1972) Biopolymers *11*, 1685–1691.
ATKINS, E.D.T. and SHEEHAN, J.K. (1973) Science *179*, 562–564.
AUNE, K.C. (1978) Methods Enzymol. *48*, 163–185.
AVIGARD, G. (1969) Carbohydr. Res. *11*, 119–123.
AVIGARD, G. (1975) Methods Enzymol. *41*, 27–29.
AWDEH, Z.L., WILLIAMSON, A.R. and ASKONAS, B.A. (1968) Nature *219*, 66–67.
AXELSSON, O. and LINDAHL, U. (1971) J. Biol. Chem. *246*, 74–82.
BAENZIGER, J.U. and FIETE, D. (1979a) J. Biol. Chem. *254*, 2400–2407.
BAENZIGER, J.U. and FIETE, D. (1979b) J. Biol. Chem. *254*, 9795–9799.
BAENZIGER, J.U. and NATOWICZ, M. (1981) Anal. Biochem. *112*, 357–361.
BAKER, R.M., HIRSCHBERG, C.B., O'BRIEN, W.A., AWEBUCH, T.E. and WATSON, D. (1982) Methods Enzymol. *83*, 444–458.
BARONDES, S.H. (1981) Annu. Rev. Biochem. *50*, 207–231.
BAXTER, A. (1974) Ph. D. Thesis, University of Glasgow.
BAYARD, B. and FOURNET, B. (1976) Carbohydr. Res. *46*, 75–86.
BAYARD, B. and MONTREUIL, J. (1974) Collq. Int. Cent. Natl. Rech. Sci. *220*, 209–218.
BAYARD, B. and MONTREUIL, J. (1972) Carbohydr. Res. *24*, 427–443.
BAYARD, B., FOURNET, B., BOUQUELET, S., STRECKER, G., SPIK, G. and MONTREUIL, J. (1972) Carbohydr. Res. *24*, 445–456.
BAYARD, B., KERKAERT, J.P., ROUX, D. and STRECKER, G. (1979) Protides Biol. Fluids *27*, 153–156.
BAYER, E.A., SKUTELSKY, E. and WILCHECK, M. (1982) Methods Enzymol. *83*, 195–215.
BEMILLER, J.N. (1967) Advanc. Carbohydr. Chem. *22*, 25–108.
BECKER, J.W., REEKE, G.N., WANG, J.L., CUNNINGHAM, B.A. and EDELMAN, G.M. (1975) J. Biol. Chem. *250*, 1513–1524.
BEELEY, J.A., STEVENSON, S.M. and BEELEY, J.G. (1972) Biochim. Biophys. Acta *285*, 293–300.
BEELEY, J.G. (1971) Biochem. J. *123*, 399–405.
BEELEY, J.G. (1974) Biochem. Soc. Symp. *40*, 27–36.
BEELEY, J.G. (1976a) Biochem. J. *155*, 345–351.
BEELEY, J.G. (1976b) Biochem. J. *159*, 335–345.
BEELEY, J.G. and JEVONS, F.R. (1963) Biochem. J. *88*, 15–19.
BEELEY, J.G. and MCCAIRNS, E. (1972) Biochim. Biophys. Acta *271*, 204–213.
BEELEY, J.G., ALTOUMAH, B. and JARDINE, J.J. (1983) Biochem. Soc. Trans. *11*, 103–104.
BEKESI, J.G. and WINZLER, R.J. (1967) J. Biol. Chem. *242*, 3873–3879.
BEKESI, J.G., BEKESI, E. and WINZLER, R.J. (1969) J. Biol. Chem. *244*, 3766–3772.
BELLA, A.M. and KIM, Y.S. (1970) J. Chromatog. *51*, 314–315.
BERGMAN, L.W. and KUEHL, W.M. (1982) The Glycoconjugates, Vol. 3 (Horowitz, M.I., ed), pp. 81–98, Academic Press, New York.
BERGMAN, L.W. and KUEHL, W.M. (1977) Biochemistry *16*, 4490–4497.
BETTELEHIM, F.A. (1977) in The Glycoconjugates (Horowitz, M.I., ed), Vol. 1, pp. 111–128, Academic Press, New York.
BEYER, T.A. and HILL, R.T. (1982) in The Glycoconjugates (Horowitz, M., ed),

Vol. 3, pp. 25–46, Academic Press, New York.
BEYER, T.A., SADLER, J.E., REARICK, J.I., PAULSON, J.C. and HILL, R.L. (1981) Advanc. Enzymol. *52*, 23–175.
BEYER, T.A., SADLER, J.E. and HILL, R.L. (1980) J. Biol. Chem. *255*, 5364–5372.
BHATTI, T. and CLAMP, J.R. (1968) Biochim. Biophys. Acta *170*, 206–208.
BHAVANANDAN, V.P. and KATLIC, A.W. (1979) J. Biol. Chem. *254*, 4000–4008.
BHUSHANA RAO, K.S.P., VAN ROOST, E., MASSON, P.L., HEREMANS, J.F. and ANDRE, F. (1973) Biochim. Biophys. Acta *317*, 286–302.
BIANCHI, P., NADER, H.B., TAKAHASHI, H.K., OSIMA, M., STRANO, A.H. and DIETRICH, C.P. (1980) J. Chromatog. *196*, 455–462.
BIRNIE, G.D. and RICKWOOD, D. (1978) Centrifugal Separations in Molecular and Cellular Biology, Butterworths, London.
BITTER, T. and MUIR, H.M. (1962) Anal. Biochem. *4*, 330–334.
BITTIGER, H. and SCHNEBLI, H.P. (eds) (1976) Concanavalin A as a Tool; Wiley, London.
BJERRUM, O.J. and BØG-HANSEN, T.C. (1976) in Biochemical Analysis of Membranes (Maddy, A.H., ed), pp. 378–426, Chapman and Hall, London.
BJÖRLING, H. (1976) in Plasma Proteins (Blombäck, B. and Hanson, L.A., eds), pp. 29–37, Wiley, Chichester.
BJORNDAHL, H., HELLERQUIST, C.G., LINDBERG, B. and SVENSSON, S. (1970) Angew. Chem. Int. Ed. Engl. *9*, 610–619.
BLOCK, R. and BURGER, M.M. (1974) Biochem. Biophys. Res. Commun. *58*, 13–19.
BLUMENFELD, O.O. and ZVILICHOVSKY, B. (1973) Methods Enzymol. *28*, 245–252.
BLUMENFELD, O.O., GALLOP, P.M. and LIAO, T.H. (1972) Biochem. Biophys. Res. Commun. *48*, 242–251.
BOAS, N.F. (1953) J. Biol. Chem. *204*, 553–563.
BØG-HANSEN, T.C. (1979) in Proceedings, 3rd International Symposium on Affinity Chromatography and Molecular Interactions (Egly, J.M., ed), pp. 399–416, Inserm Symposia Series, Paris.
BØG-HANSEN, T.C., BJERRUM, O.J. and BROGREN, C.H. (1977) Anal. Biochem. *81*, 78–87.
BØG-HANSEN, T.C. (1973) Anal. Biochem. *56*, 480–488.
BOLTON, A.E. and HUNTER, W.M. (1973) Biochem. J. *133*, 529–538.
BONNER, W.M. and LASKEY, R.A. (1974) Eur. J. Biochem. *46*, 83–88.
BOUCHARD, P., MOROUX, Y., TIXIER, R., PRIVAT, J.P. and MONSIGNY, M. (1976) Biochimie *58*, 1247–1253.
BOUQUELET, S., STRECKER, G., MONTREUIL, J. and SPIK, G. (1980) Biochimie *62*, 43–49.
BRAMWELL, M.E. and HARRIS, H. (1978) Proc. Roy. Soc. Lond. B. *201*, 87–106.
BRANDT, K.D. and MUIR, H. (1971) Biochem. J. *121*, 261–270.
BREEN, M., WEINSTEIN, H.G., BLACK, L.J., BORCHERDING, M.S. and SITTIG, R.A. (1976) Methods Carbohydr. Chem. *7*, 101–115.
BREIMER, M.E., HANSSON, G.C., KARLSSON, K.A., LEFFLER, H., PILMOTT, W. and SAMUELSSON, B.E. (1979) FEBS Lett. *89*, 42–46.
BREWER, C.F. (1979) Biochem. Biophys. Res. Commun. *90*, 117–122.
BREWER, C.F. and KAISER, H. (1975) Proc. Natl. Acad. Sci. USA *72*, 3421–3423.

BRIDES, E.B., LI, E. and KORNFELD, S. (1977) J. Biol. Chem. *252*, 1107–1116.
BRISSON, J.-R. and CARVER, J.P. (1983) Biochemistry *22*, 3680–3686.
BROWN, J.C. and HUNT, R.C. (1978) Int. Rev. Cytol. *52*, 277–349.
BRUNETTI, P., SWANSON, A. and ROSEMAN, S. (1963) Methods Enzymol. *6*, 465–473.
BRUNGRABER, E.G., BROWN, B.D. and HOF, H. (1971) Clin. Chim. Acta *32*, 159–170.
BURRIDGE, K. (1976) Proc. Natl. Acad. Sci. USA *73*, 4457–4461.
BURRIDGE, K. (1978) Methods Enzymol. *50*, 54–63.
BUTLER, W.T. (1963) J. Immunol. *90*, 663–671.
BUTTERS, T.D. and HUGHES, R.C. (1974) Biochem. J. *140*, 469–478.
BUTTERS, T.D. and HUGHES, R.C. (1975) Biochem. J. *150*, 59–69.
BYRD, J.G., TARENTINO, A.L., MALEY, F., ATKINSON, P.H. and TRIMBLE, R.B. (1982) J. Biol. Chem. *259*, 14657–14666.
CAMPBELL, A.M. (1984) Monoclonal Antibody Technology, Techniques in Biochemistry and Molecular Biology (Burdon, R.H. and van Knippenberg, P.H., eds), Elsevier, Amsterdam.
CAMPBELL, P., JACOBSSON, I., BENZING-PURDIE, L., RODÉN, L. and FESSLER, J.H. (1984) Anal. Biochem. *137*, 505–516.
CAPPELLETTI, R., DEL ROSA, M. and CHLARUGI, V.P. (1979) Anal. Biochem. *99*, 311–315.
CARLSON, D.M. (1968) J. Biol. Chem. *243*, 616–626.
CARLSON, D.M. and JOURDIAN, G.W. (1966) Methods Enzymol. *8*, 358–361.
CASASSA, E.F. and EISENBERG, H. (1964) Advanc. Protein Chem. *19*, 287–395.
CESSI, C. and PILEGO, F. (1960) Biochem. J. *77*, 508–510.
CHAMBERS, R., CLAMP, J.R., BAYARD, B. and MONTREUIL, J. (1973) Biochimie *55*, 1195–1198.
CHANDRASEKARAN, E.V. and BEMILLER, J.N. (1980) Methods Carbohydr. Chem. *8*, 89–96.
CHANG, K. and CUATRECASAS, P. (1976) in Concanavalin A as a Tool (Bittiger, H. and Schnebli, H.P., eds), pp. 173–180, Wiley, London.
CIFONELLI, J.A. (1968) Carbohydr. Res. *8*, 233–242.
CIFONELLI, J.A. and KING, J. (1972) Carbohydr. Res. *21*, 173–186.
CIFONELLI, J.A. and KING, J. (1973) Biochim. Biophys. Acta *320*, 331–340.
CLAMP, J. (1974) Biochem. Soc. Symp. *40*, 3–16.
CLAMP, J.R., BHATTI, T. and CHAMBERS, R.E. (1971) Methods Biochem. Anal. *19*, 229–334.
CLAMP, J.R., BHATTI, T. and CHAMBERS, R.E. (1972) in Glycoproteins, 2nd Ed., Part A (Gottschalk, A., ed), pp. 300–321, Elsevier, Amsterdam.
CLELLAND, R.L. (1977) Arch. Biochem. Biophys. *180*, 57–68.
CLINE, M.J. and LIVINGSTONE, D.C. (1971) Nature New Biol. *232*, 155–156.
CODINGTON, J.F., LINSLEY, K.B. and SILBER, C. (1976) Methods Carbohydr. Chem. *7*, 226–232.
COLEGATE, S.M., DORLING, P.R. and HUXSTABLE, C.R. (1979) Austral. J. Chem. *32*, 2257–2264.
CONCHIE, J. (1976) Methods Carbohydr. Chem. *7*, 195–199.
CONRAD, H.E. (1977) Methods Carbohydr. Chem. *7*, 71–75.
CONRAD, H.E., VARBONCOUER, E. and JAMES, M.E. (1973) Anal. Biochem. *51*,

486–500.

COOK, G.M.W. (1976) in Biochemical Analysis of Membranes (Maddy, A.H., ed), pp. 283–351, Chapman and Hall, London.

COREY, E.J. and CHAYKOVSKY, M. (1965) J. Amer. Chem. Soc. *87*, 1345–1353.

COUDRON, C., ELLIS, K., PHILIPSON, L. and SCHWARZ, N.B. (1980) Biochem. Biophys. Res. Comm. *92*, 618–623.

COXTON, B. (1972) Methods Carbohydr. Chem. *6*, 513–539.

CREETH, J.M. (1978) Brit. Med. Bull. *34*, 17–24.

CREETH, J.M. and KNIGHT, C.G. (1968) in Solution Properties of Natural Polymers, p. 303, The Chemical Society, London.

CREETH, J.M. and PAIN, R.H. (1967) Progr. Biophys. Mol. Biol. *17*, 217.

CREETH, J.M., BHASKAR, K.R., DONALD, A.S.R. and MORGAN, W.T.J. (1974) Biochem. J. *143*, 159–170.

CUATRECASAS, P. (1970) J. Biol. Chem. *245*, 3059–3065.

CUATRECASAS, P. (1973) Biochemistry *12*, 1312–1323.

CUATRECASAS, P. and HOLLENBERG, M.D. (1974) Biochem. Biophys. Res. Commun. *62*, 31–41.

CUMMINGS, R.D. and KORNFELD, S. (1982a) J. Biol. Chem. *257*, 11230–11234.

CUMMINGS, R.D. and KORNFELD, S. (1982b) J. Biol. Chem. *257*, 11235–11240.

CUNNINGHAM, L., FORD, J.D. and RAINEY, J.M. (1965) Biochim. Biophys. Acta *101*, 233–235.

CURWEN, K.D. and SMITH, S.C. (1977) Anal. Biochem. *79*, 291–301.

CUSHLEY, W., SINGER, H.H. and WILLIAMSON, A.R. (1983) Biosci. Reports *3*, 331–336.

D'AMICO, R.P. and KERN, M. (1970) Biochim. Biophys. Acta *215*, 78–87.

DARKE, A., FINER, E.G., MOORHOUSE, R. and REES, D.A. (1975) J. Mol. Biol. *99*, 477–486.

DATEMA, R., SCHWARZ, R.T. and JANKOWSKI, A.W. (1980) Eur. J. Biochem. *109*, 331–341.

DAVEY, M.W., SULKOWSKI, E. and CARTER, W.A. (1976) Biochemistry *15*, 704–713.

DAVIDSON, E.A. (1966) Methods Enzymol. *8*, 52–59.

DAVIDSON, E.A. and MEYER, K. (1954) J. Amer. Chem. Soc. *76*, 5686–5689.

DEBRAY, J., PIERCE-CRETAL, A., SPIK, G. and MONTREUIL, J. (1983) Lectins, Vol. 3 (Bøg-Hansen, T.C. and Spengler, G.A., eds), pp.335–350, Walter de Gruyter, Berlin.

DEISENHOFER, J. (1981) Biochemistry *20*, 2361–2370.

DELL, A., MORRIS, H.R., EGGE, H., VON NICOLAI, H. and STRECKER, G. (1983) Carbohydr. Res. *115*, 41–52.

DIETRICH, C.P., MCDUFFIE, W.M. and SAMPIO, L.O. (1977) J. Chromatog. *130*, 299–304.

DIJKSTRA, B.W., VLIEGENTHART, F.G., STRECKER, G. and MONTREUIL, J. (1983) Eur. J. Biochem. *130*, 111–115.

DILL, K. and ALLERHAND, A. (1979) J. Biol. Chem. *254*, 4524–4531.

DISCHE, Z. (1947) J. Biol. Chem. *167*, 189–198.

DISCHE, Z. and BORENFREUND, E. (1950) J. Biol. Chem. *184*, 517–522.

DISCHE, Z. and Shettles, L.B. (1948) J. Biol. Chem. *175*, 595–603.

DIXON, H.F.B. (1981) Nature *292*, 192.

DODGSON, K.S. and PRICE, R.G. (1961) Biochem. J. *84*, 106–110.

DONALD, A.S.R. (1968) J. Chromatog. *35*, 106–107.

DORLAND, L. (1979) Structure Determination of the Complex Carbohydrate Chains of Glycoproteins by 360 MH ^{1}H-NMR Spectroscopy, Ph.D. Thesis, Rijksuniversiteit of Utrecht.

DORLAND, L., SCHUT, B.L., VLIEGENTHART, J.F.G., STRECKER, G., FOURNET, B., SPIK, G. and MONTREUIL, J. (1977a) Eur. J. Biochem. *73*, 93–97.

DORLAND, L., HAVERKAMP, J., SCHUT, B.L., VLIEGENTHART, J.F.G., SPIK, G., STRECKER, G., FOURNET, B. and MONTREUIL, J. (1977b) FEBS Lett. *77*, 15–20.

DOWNS, F. and PIGMAN, W. (1976a) Methods Carbohydr. Chem. *7*, 200–204.

DOWNS, F. and PIGMAN, W. (1976b) Methods Carbohydr. Chem. *7*, 244–248.

DOWNS, F., HERP, A., MOSCHERA, J. and PIGMAN, W. (1973) Biochim. Biophys. Acta *328*, 182–192.

DRICKAMER, L.K. (1977) J. Biol. Chem. *252*, 6909–6917.

DUBOIS, M., GILLES, K.A., HAMILTON, J.K., REBERS, P.A. and SMITH, F. (1956) Anal. Chem. *28*, 350–356.

DULANEY, J.T. (1979) Mol. Cell Biochem. *21*, 43–63.

DURANT, G.J., HENDRICKSON, H.R. and MONTGOMERY, R. (1962) Arch. Biochem. Biophys. *99*, 418–425.

DYER, J.T. (1956) Methods Biochem. Anal. *3*, 112–152.

EASTOE, J. (1972) in Glycoproteins, 2nd Ed., Part A (Gottschalk, A., ed), pp. 158–207, Elsevier, Amsterdam.

ECKHART, A.E., HAYES, C.E. and GOLDSTEIN, I.J. (1976) Anal. Biochem. *73*, 192–197.

ECKHARDT, K., WETSTEIN, H., THRUM, H. and IHN, W. (1980) J. Antibiot. *33*, 908–910.

EDELMAN, G.M. and WANG, J.L. (1978) J. Biol. Chem. *253*, 3016–3022.

EDELSTEIN, S.J. and SCHACHMAN, H.K. (1967) J. Biol. Chem. *242*, 306–311.

EDGE, A.S.B., FALTYNEK, C.R., HOF, L., REICHERT, L.E. and WEBER, P. (1981) Anal. Biochem. *118*, 131–137.

EGMOND, M.R., WILLIAMS, R.J.P., WELSH, E.J. and REES, D.A. (1979) Eur. J. Biochem. *97*, 73–83.

ELDER, J.H. and ALEXANDER, S. (1982) Proc. Natl. Acad. Sci. USA *79*, 4540–4544.

ELDER, J.P. (1979) Methods Enzymol. *61*, 12–25.

ELTING, J.J. and LENNARZ, W.J. (1982) Methods Enzymol. *83*, 408–418.

EMES, J.H. and PIERCE, R.H. (1975) Biochem. J. *145*, 549–556.

ENDO, Y., YAMASHITA, K., HAN, Y.N., IWANAGA, S. and KOBATA, A. (1977) J. Biochem. (Tokyo) *82*, 545–550.

ENDO, Y., YAMASHITA, K., TACHIBANA, Y., TOJO, S. and KOBATA, A. (1979) J. Biochem. (Tokyo) *85*, 669–679.

ESHDAT, Y. and SHARON, N. (1982) Methods Enzymol. *83*, 386–391.

EVANS, W.H. (1978) in Preparation and Characterisation of Plasma Membranes, Laboratory Techniques in Biochemistry and Molecular Biology (Work, T. and Work, E., eds), North-Holland, Amsterdam.

EYLAR, E.H., MADOFF, M.A., BRODY, O.V. and ONCLEY, J.Y. (1962) J. Biol. Chem.

237, 1992–2000.
FAIRBANKS, G., STECK, T.L. and WALLACH, D.H.F. (1971) Biochemistry *10*, 2606–2617.
FANGER, M.W. and SMYTH, D.G. (1970) Anal. Biochem. *34*, 494–499.
FARWELL, D.C. and DION, A.S. (1981) Anal. Biochem. *113*, 423–437.
FEIZI, T. (1981) Immunol. Common. *10*, 127–156.
FELLER, M., BEHNKE, D. and GRUENSTEIN, E. (1979) Biochim. Biophys. Acta *586*, 315–329.
FINCH, P.R., YUEN, R., SCHACHTER, H. and MOSCARELLO, M.A. (1969) Anal. Biochem. *31*, 296–305.
FINCHER, G.B., SAWYER, W.H. and STONE, B.A. (1974) Biochem. J. *139*, 535–545.
FINDLAY, J.B.C. (1974) J. Biol. Chem. *249*, 4398–4403.
FINNE, J. and KRUSIUS, T. (1982) Methods Enzymol. *83*, 269–277.
FISCHER, L. (1969) in Laboratory Techniques in Biochemistry and Molecular Biology Vol. 1 (Work, T.S. and Work, E., eds), pp. 157–396, North-Holland, Amsterdam.
FISCHER, F.G. and NEBEL, G.G. (1955) Z. Physiol. Chem. *302*, 10–18.
FLOWERS, H.M. and SHARON, M. (1979) Advanc. Enzymol. *48*, 29–95.
FORIERS, A., DE NEVE, R., KANAREK, L. and STROSBERG, A.D. (1978) Proc. Natl. Acad. Sci. USA *75*, 1136–1139.
FOURNET, B., DHALLUIN, J., STRECKER, G., MONTREUIL, J., BOSSO, G. and DEFAYE, J., (1980) Anal. Biochem. *108*, 35–56.
FOURNET, B., MONTREUIL, J., STRECKER, G., DORLAND, L., HAVERKAMP, J., VLIEGENTHART, J.F.G., BINETTE, J.P. and SCMID, K. (1978) Biochemistry *17*, 5206–5214.
FOURNET, B., STRECKER, G., LEROY, Y. and MONTREUIL, J. (1981) Anal. Biochem. *116*, 489–502.
FRANÇOIS-GERARD, C., GERDAY, C. and BEELEY, J.G. (1979) Biochem. J. *177*, 679–685.
FRANKS, F. and MORRIS, E.R. (1978) Biochim. Biophys. Acta *540*, 346–356.
FRANSSON, L. (1974) Carbohydr. Res. *36*, 339–348.
FRANSSON, L. and CARLSTEDT, I. (1974) Carbohydr. Res. *36*, 349–358.
FRANZEN, L., SVENSSON, S. and LARM, O. (1980) J. Biol. Chem. *255*, 5090–5093.
FUKUDA, M., DELL, A. and FUKUDA, M.N. (1984) J. Biol. Chem. *259*, 4782–4791.
FUKUDA, M.N. and MATSUMURA, G. (1976) J. Biol. Chem. *251*, 6218–6225.
FUNDERBURGH, J.L. and CHANDLER, J.W. (1978) Anal. Biochem. *91*, 464–472.
FURLAIN, M., PERRET, B.A. and BECK, E.A. (1979) Anal. Biochem. *96*, 208–214.
GAHMBERG, C.G. (1976) J. Biol. Chem. *251*, 510–515.
GAHMBERG, C.G. (1978) Methods Enzymol. *50*, 204–206.
GAHMBERG, C.G. and ANDERSSON, L.C. (1977) J. Biol. Chem. *252*, 5888–5894.
GAHMBERG, C.G. and HAKOMORI, S. (1973) J. Biol. Chem. *248*, 4311–4317.
GAHMBERG, C.G. and HAKOMORI, S.-I. (1975) J. Biol. Chem. *250*, 2438–2448.
GAHMBERG, C.G., HÄYRA, P. and ANDERSSON, L.C. (1976a) J. Cell Biol. *68*, 642–653.
GAHMBERG, C.G., MYLLYLA, G., LEIKOLA, J., PIRKOLA, A. and NORDLING, S. (1976b) J. Biol. Chem. *251*, 6108–6116.
GALLACHER, J.T., KENT, P.W., PHILLS, R. and RICHARDSON, P. (1977) in Mucus

in Health and Disease (Elstein, M. and Parke, D.V., eds), pp. 91–102, Plenum Press, New York and London.

GATTEGNO, L., PERRET, G., FABIA, F., BLADIER, D. and CORNILLOT, P. (1981) Carbohydr. Res. *95*, 283–290.

GIBBONS, R.A. (1972) in Glycoproteins (Gottschalk, A., ed), Part A, pp. 31–40, Elsevier, Amsterdam.

GIBBONS, R.A. and SELWOOD, R. (1973) in The Biology of the Cervix (Blandau, R.J. and Moghiss, K., eds), pp. 251–265, University of Chicago Press, Chicago.

GLAZER, A.N., DELANGE, R.J. and SIGMAN, D.S. (1975) in Chemical Modification of Proteins, Laboratory Techniques in Biochemistry and Molecular Biology (Work, T.S. and Work, E., eds), Elsevier, Amsterdam.

GLOSSMAN, H. and NEVILLE, D.M. (1971) J. Biol. Chem. *246*, 6339–6346.

GOGSTAD, G.O. and HELGELAND, L. (1978) Biochim. Biophys. Acta *508*, 551–564.

GOGSTAD, G.O. and HELGELAND, L. (1982) The Glycoconjugates, Vol. 3 (Horowitz, M.I., ed), pp. 99–112, Academic Press, New York.

GOLDSTEIN, I.J. (1976) in Concanavalin A as a Tool (Bittiger, H. and Schnebli, H.P., eds), pp. 161–167, Wiley, London.

GOLDSTEIN, I.J. and HAYES, C.E. (1978) Advanc. Carbohydr. Chem. Biochem. *35*, 127–340.

GOLDSTEIN, I.J., HAMMARSTRÖM, S. and SUNDBLAD, G. (1975) Biochim. Biophys. Acta *405*, 53–61.

GOLDSTEIN, I.J., HAY, G.W., LEWIS, B.A. and SMITH, F. (1965) Methods Carbohydr. Chem. *5*, 361–377.

GOLDSTEIN, I.J. HUGHES, R.C., MONSIGNY, M., OSAWA, T. and SHARON, N. (1980) Nature *285*, 66.

GORIN, P.A.J. (1981) Advanc. Carbohydr. Chem. Biochem. *38*, 13–104.

GOTTLIEB, C., BAENZIGER, J. and KORNFELD, S. (1975) J. Biol. Chem. *250*, 3303–3309.

GOTTSCHALK, A. (1972) Glycoproteins, 2nd Edn., Parts A and B, Elsevier, Amsterdam.

GRABAR, P. and WILLIAMS, C.A. (1953) Biochim. Biophys. Acta *10*, 193–194.

GRAHAM, E.R.B. and NEUBERGER, A. (1968) Biochem. J. *106*, 593–600.

GREEN, J. (1981) in Techniques in Carbohydrate Metabolism B311, Techniques in Life Sciences, pp. 1–8, Elsevier/North Holland, Shannon.

GREEN, J.R. and NORTHCOTE, D.H. (1978) Biochem. J. *170*, 599–608.

GREENAWAY, P.J. and LEVINE, D. (1973) Nature New Biol. *241*, 191–192.

GREENWOOD, F.C., HUNTER, H.H. and GLOVER, J.S. (1965) Biochem. J. *89*, 114–123.

GREFRATH, S.P. and REYNOLDS, J.A. (1974) Proc. Natl. Acad. Sci. USA *71*, 3913–3916.

GREGORY, J.D., LAURENT, T.C. and RODÉN, L. (1964) J. Biol. Chem. *239*, 3312–3320.

GREIG, R.G. and BROOKS, D.E. (1979) Nature *282*, 738–739.

GROSS, V., TRAN-TRI, T.A., VOSBECK, K. and HEINRICH, P.C. (1983) J. Biol. Chem. *258*, 4032–4036.

GUNTHER, G.R., WANG, J.L., YAHARA, I., CUNNINGHAM, B.A. and EDELMAN,

G.M. (1973) Proc. Natl. Acad. Sci. USA *70*, 1012–1016.
GURD, J.W. (1977) Biochemistry *16*, 369–374.
GURD, J.W. and EVANS, W.H. (1976) Can. J. Biochem. *54*, 477–480.
GURD, J.W. and MAHLER, H.R. (1974) Biochemistry *13*, 5193–5198.
GUSS, J.M., HUKINS, D.W.L., SMITH, P.J.C., WINTER, W.T., ARNOTT, S., MOORHOUSE, R. and REES, D.A. (1975) J. Mol. Biol. *95*, 359–384.
GUTHRIE, R.D. and MCCARTHY, J.F. (1967) Advanc. Carbohydr. Chem. *22*, 11–24.
HAGOPIAN, A., BOSMAN, H.B. and EYLAR, E.H. (1968) Arch. Biochem. Biophys. *128*, 387–396.
HAHN, H., HELLMAN, B., LERMARK, A., SEHLIN, J. and TÜLJEDAL, I. (1974) J. Biol. Chem. *249*, 5275–5284.
HAKOMORI, S.-I. (1964) J. Biochem. (Tokyo) *55*, 205–208.
HAMAGUCHI, H. and CLEVE, H. (1972) Biochim. Biophys. Acta *278*, 271–280.
HAMER, G.K. and PERLIN, A.S. (1976) Carbohydr. Res. *49*, 37–48.
HAMMOND, K.S. and PAPERMASTER, D.S. (1976) Anal. Biochem. *74*, 292–297.
HARBON, S., HERMAN, G. and CLAUSER, J. (1968) Eur. J. Biochem. *4*, 265–272.
HARDINGHAM, T.E. and PHELPS, C. (1968) Biochem. J. *108*, 9–16.
HARDINGHAM, T.E., PERKINS, S.J. and MUIR, H. (1983) Biochem. Soc. Trans. *11*, 128–130.
HARDMAN, K.D. and AINSWORTH, C.F. (1976) Biochemistry *15*, 1120–1128.
HARRIS, E.D. and JOHNSON, C.A. (1969) Biochemistry *8*, 512–518.
HARRISON, R.A. (1983) Biochemistry *22*, 5001–5007.
HART, G.W. (1976) J. Biol. Chem. *251*, 6513–6521.
HART, G.W. (1978) Dev. Biol. *62*, 78–98.
HART, G.W. and LENNARZ, W.J. (1978) J. Biol. Chem. *253*, 5795–5801.
HASCALL, V.C. and SADJERA, S.W. (1970) J. Biol. Chem. *245*, 4920–4030.
HATA, R. and NAGAI, Y. (1972) Anal. Biochem. *45*, 462–468.
HATEFI, Y. and HANSTEIN, W.G. (1969) Proc. Natl. Acad. Sci. USA *62*, 1129–1136.
HATTON, M.W.C., MÄRZ, L. and REGOECZI, E. (1983) Trends Biochem. Sci. *8*, 287–291.
HAYDEN, G.A., CROWLEY, G.M. and JAMIESON, G.A. (1970) J. Biol. Chem. *245*, 5827–5832.
HAYES, M.L. and CASTELLINO, F.J. (1979) J. Biol. Chem. *254*, 8768–8771.
HAYMAN, M.J. and CRUMPTON, M.J. (1972) Biochem. Biophys. Res. Commun. *47*, 923–930.
HEANEY-KIERAS, J., RODÉN, L. and CHAPMAN, D. (1977) Biochem. J. *165*, 1–9.
HELENIUS, A. and SIMONS, K. (1975) Biochim. Biophys. Acta *415*, 29–79.
HENKERT, P. and FISCHER, T. (1975) J. Immunol. *114*, 710–714.
HERBERT, W.J. (1978) in Handbook of Experimental Immunology, 3rd Edn., (Weir, D.M., ed), pp. 20.1–20.20, Blackwell, Oxford.
HESS, H.H. and ROLDE, E. (1964) J. Biol. Chem. *239*, 3215–3220.
HILL, H.D., REYNOLDS, J.A. and HILL, R.L. (1977) J. Biol. Chem. *252*, 3791–3798.
HIRS, C.H.W. (1967) Methods Enzymol. *11*, 59–62.
HOFFMAN, J., LINDBERG, B. and SVENSSON, S. (1972) Acta Chem. Scand. *26*, 661–666.
HONDA, S., YUKI, H. and TAKIURA, K. (1974) J. Biochem. (Tokyo) *76*, 209–211.

Höök, M., Lindahl, U. and Iverius, P. (1974) Biochem. J. *137*, 33–43.
Horowitz, M.I. (ed) (1977) in Glycoconjugates, Vol. 1, pp. 15–34, Academic Press, New York.
Hortin, G. and Boime, I. (1980) J. Biol. Chem. *255*, 8007–8010.
Hortin, G., Stern, A.M., Miller, B., Abeles, R.H. and Boime, I. (1983) J. Biol. Chem. *258*, 4047–4050.
Houdret, N., Le Treut, A., Lhermitte, M., Lamblin, G., Degrand, P. and Roussel, P. (1981) Biochim. Biophys. Acta *668*, 413–419.
Houston, L.L. and Dooley, T.P. (1982) J. Biol. Chem. *257*, 4147–4151.
Howard, I.K. and Sage, H.J. (1969) Biochemistry *8*, 2436–2441.
Howe, C., Lloyd, K.O. and Lee, L.T. (1972) Methods Enzymol. *28*, 236–245.
Huang, C.C., Mayer, H.E. and Montgomery, R. (1970) Carbohydr. Res. *13*, 127–137.
Hubbard, A.L. and Cohn, Z.A. (1972) J. Cell. Biol. *55*, 390–405.
Hubbard, S.C. and Robbins, P.W. (1980) J. Biol. Chem. *255*, 11782–11793.
Huet, C., Longchamp, M., Huet, M. and Bernadec, A. (1974) Biochim. Biophys. Acta *365*, 28–39.
Hughes, R.C. (1976) Membrane Glycoproteins, Butterworths, London.
Hugli, T.E. and Moore, S. (1972) J. Biol. Chem. *247*, 2828–2834.
Hunt, L.A. (1980) Cell *21*, 407–415.
Hunt, R.C. and Brown, J.C. (1975) J. Mol. Biol. *97*, 413–422.
Hunter, G.K., Heersche, J.N.M. and Aubin, J.E. (1983) Biochemistry *22*, 831–837.
Hunter, M.J. (1978) Methods Enzymol. *58*, 23–29.
Hunter, W.M. and Greenwood, F.C. (1962) Nature *194*, 495–496.
Hyman, R., Lacorbiere, M., Stavarek, S. and Nicolson, G. (1974) Proc. Natl. Cancer Inst. *52*, 963–969.
Inbar, M. and Sach, L. (1969) Nature *223*, 710–712.
Irimura, T., Nakajima, M., Di Ferrante, N. and Nicolson, G.L. (1983) Anal. Biochem. *130*, 461–468.
Isemura, M. and Schmid, K. (1971) Biochem. J. *124*, 591–604.
Iwase, H. and Hotta, K. (1977) J. Biol. Chem. *252*, 5437–5443.
Iwase, H., Kato, Y. and Hotta, K. (1981) J. Biol. Chem. *256*, 5638–5642.
Iyer, R.N. and Carlson, D.M. (1971) Arch. Biochem. Biophys. *142*, 101–105.
Jacobs, S. and Cuatrecasas, P. (1976) Biochim. Biophys. Acta *433*, 482–495.
Jaffe, C.L., Ehrlich-Rogozinski, S., Lis, H. and Sharon, N. (1977) FEBS Lett. *82*, 191–196.
Jaffe, C.L., Lis, H. and Sharon, N. (1979) Biochem. Biophys. Res. Commun. *91*, 402–409.
Jansson, P.-E., Kenne, L., Liedgren, L., Lindberg, B. and Lönngren, J. (1976) Chem. Commun. Univ. Stockholm [8].
Juliano, R.L. (1972) Biochim. Biophys. Acta *266*, 301–306.
Juliano, R.L. (1978) Current Topics Membranes Transport *11*, 107–144.
Juliano, R.L. and Li, G. (1978) Biochemistry *17*, 678–683.
Kabat, A (1956) Blood Group Substances, Academic Press, New York.
Kabat, E.A. (1978) J. Supramol. Struct. *8*, 79–88.

KAHANE, I., FURTHMAYR, H. and MARCHESI, V.T. (1976) Biochim. Biophys. Acta *426*, 464–476.

KALYAN, N.K. and BAHL, O.P. (1983) J. Biol. Chem. *258*, 67–74.

KAMERLING, J.P., HEERMA, W., VLIEGENTHART, J.F.G., GREEN, B.W., LEWIS, I.A.S. and SPIK, G. (1983) Biomed. Mass Spectrom. *10*, 420–425.

KAMERLING, J.P., VLIEGENTHART, J.F.G. and VINK, J. (1974) Carbohydr. Res. *33*, 297–306.

KANDUKURI, S.P., BHUSHANA RAO, K.S.P. and MASSON, P.L. (1977) J. Biol. Chem. *252*, 7788–7795.

KÄRKKÄINEN, J. (1970) Carbohydr. Res. *14*, 27–33.

KÄRKKÄINEN, J. (1971) Carbohydr. Res. *17*, 11–18.

KARLSSON, K. (1973) FEBS Lett. *32*, 317–320.

KARLSSON, K. (1974) Biochemistry *13*, 3643–3647.

KARLSSON, K.A., PASCHER, I., SAMUELSSON, B.E., FINNE, J., KRUSIUS, T. and RAUVALA, T. (1978) FEBS Lett. *94*, 413–417.

KATZ, F.N., ROTHMAN, J.E., LINGAPPA, V.R., BLOBEL, G. and LODISH, H.F. (1977) Proc. Natl. Acad. Sci. USA *70*, 2330–2334.

KAWASHIMA, I., KUKUDA, K., TOMITA, M. and HAMADA, A. (1982) J. Biochem. (Tokyo) *91*, 865–872.

KENNEDY, J.F. (1974) Biochem. Soc. Trans. *2*, 54–64.

KENNEDY, J.F. (1979) Proteoglycans — Biological and Chemical Aspects in Human Life, Elsevier, Amsterdam.

KENNEDY, J.F. and FOX, J.E. (1977) Carbohydr. Res. *54*, 13–21.

KENNEDY, J.F. and FOX, J.E. (1980) Methods Carbohydr. Chem. *8*, 3–12.

KHAN, M.A., WOLF, D.P. and LITT, M. (1976) Biochim. Biophys. Acta *444*, 369–373.

KIELY, M.L., MCKNIGHT, G.S. and SCHIMKE, R.T. (1976) J. Biol. Chem. *251*, 5490–5495.

KIM, J.J. and CONRAD, H.E. (1976) J. Biol. Chem. *251*, 6210–6217.

KIMURA, J.H., OSDOBY, P., CAPLAN, A.I. and HASCALL, V.C. (1978) J. Biol. Chem. *253*, 4721–4729.

KITAMIKADO, M., ITO, M. and LI, Y. (1982) Methods Enzymol. *83*, 619–625.

KOBATA, A. (1979) Anal. Biochem. *100*, 1–14.

KOBATA, A. and TAKASAKI, S. (1978) Methods Enzymol. *50*, 560–567.

KOCHETKOV, N.K., CHIZHOW, O.S., KADENSTEV, V.I., SMIROVA, G.P. and ZHUKOVA, I.G. (1973) Carbohydr. Res. *27*, 5–10.

KOHNO, A. and YAMASHINA, M. (1972) Methods Enzymol. *28*, 786–792.

KOIDE, N. and MURAMATSU, T. (1974) J. Biol. Chem. *249*, 4897–4904.

KORNFELD, K., REITMAN, M.L. and KORNFELD, R. (1981) J. Biol. Chem. *256*, 6633–6640.

KORNFELD, R. and KORNFELD, S. (1970) J. Biol. Chem. *245*, 2536–2545.

KORNFELD, R. and KORNFELD, S. (1976) Annu. Rev. Biochem. *45*, 217–237.

KORNFELD, R. and KORNFELD, S. (1980) in The Biochemistry of Glycoproteins and Proteoglycans (Lennarz, W.J., ed), pp. 1–34, Plenum, New York and London.

KORNFELD, R., KELLER, J., BAENZIGER, J. and KORNFELD, S. (1971) J. Biol. Chem. *246*, 3259–3268.

KORNFELD, S. (1982) in The Glycoconjugates, Vol. 3 (Horowitz, M.I., ed), pp. 3–23,

Academic Press, New York.
KRAEMER, P.M. (1971) Biochemistry *10*, 1437–1445.
KRATKY, O., LEOPOLD, H. and STABINGER, H. (1973) Methods Enzymol. *27*, 98–110.
KREIBICH, G. and SABATINI, D. (1974) J. Cell Biol. *61*, 789–807.
KREIBICH, G., DEBEY, P. and SABATINI, D. (1973) J. Cell Biol. *58*, 436–462.
KRISTIANSEN, T. (1975) Methods Enzymol. *34*, 331–341.
KRUPKE, D.W. and BEAMS, J.W. (1972) Methods Enzymol. *26*, 74–107.
KRUPKE, D.W. and DORRIER, T.E. (1978) Methods Enzymol. *48*, 155–162.
KRUSIUS, T. and FINNE, J. (1978) Eur. J. Biochem. *84*, 394–403.
KRUSIUS, T. and FINNE, J. (1981) Carbohydr. Res. *90*, 203–214.
KRUSIUS, T., FINNE, J. and RAUVALA, H. (1978) Eur. J. Biochem. *92*, 289–300.
LAEMMLI, U.K. (1970) Nature *227*, 680–685.
LAGUNOFF, D. and WARREN, G. (1962) Arch. Biochem. Biophys. *99*, 396–400.
LAMPORT, D.T.A. and MILLER, D.H. (1971) Plant Physiol. *48*, 454–456.
LAMPORT, D.T.A., KATONA, L. and ROERIG, S. (1973) Biochem. J. *133*, 125–131.
LAURELL, C.B. (1965) Anal. Biochem. *10*, 358–361.
LAURENT, T.C. and KILLANDER, J. (1964) J. Chromatog. *14*, 317–330.
LAWFORD, G.R. and SCHACHTER, H. (1966) J. Biol. Chem. *241*, 5408–5418.
LAWFORD, G.R. and SCHACHTER, H. (1967) Canad. J. Biochem. *45*, 507–522.
LEACH, B.S., COLLAWN, J.F. and FISH, W.W. (1980a) Biochemistry *19*, 5734–5741.
LEACH, B.S., COLLAWN, J.F. and FISH, W.W. (1980b) Biochemistry *19*, 5741–5747.
LEDEEN, R.W. and YU, R.K. (1976) in Biological Roles of Sialic Acid (Rosenberg, A. and Schengrund, C., eds), pp. 1–57, Plenum Press, New York.
LEE, J.C. and TIMASHEFF, S.N. (1979) Methods Enzymol. *61*, 49–57.
LEE, J.C., GEKKO, K. and TIMASHEFF, S.N. (1979) Methods Enzymol. *61*, 26–49.
LEE, Y.C. (1972) Methods Enzymol. *28*, 63–73.
LEE, Y.C. and LEE, T.R. (1982) in The Glycoconjugates, Vol. 4 (Horowitz, M.I., ed), pp. 57–83, Academic Press, New York.
LEE, Y.C. and SCOCCA, J.R. (1972) J. Biol. Chem. *247*, 5753–5758.
LEHNHARDT, W.F. and WINZLER, R.J. (1968) J. Chromatog. *34*, 471–479.
LEMKIN, M.C. and FARQUHAR, M.G. (1981) Proc. Natl. Acad. Sci. USA *78*, 1726–1730.
LETARTE-MUIRHEAD, M., BARCLAY, A.N. and WILLIAMS, A.F. (1975) Biochem. J. *151*, 685–697.
LETTS, P.J., MEISTRICH, M.L., BRUCE, W.R. and SCHACHTER, H. (1974) Biochim. Biophys. Acta *343*, 192–207.
LI, E., TABAS, I. and KORNFELD, S. (1978) J. Biol. Chem. *253*, 7762–7770.
LI, Y.-H., NAKAGAWA, H., KITAMIKADO, M. and LI, S.-C. (1982) Methods Enzymol. *83*, 610–619.
LIAO, T.-H., GALLOP, P.M. and BLUMENFELD, O.O. (1973) J. Biol. Chem. *248*, 8247–8253.
LIENER, I.E. (1976) in Concanavalin A as a Tool (Bittiger, H. and Schnebli, H.P., eds), pp. 17–31, Wiley, London.
LILJAS, L., LUNDAHL, P. and HJERTEN, S. (1976) Biochim. Biophys. Acta *426*, 526–534.
LINDAHL, U. and RODÉN, L. (1965) J. Biol. Chem. *240*, 2821–2826.

LINDAHL, U. and RODÉN, R. (1966) J. Biol. Chem. *241*, 2113–2119.
LINDBERG, B. (1972) Methods Enzymol. *28*, 178–195.
LINDBERG, B. and LÖNNGREN, J. (1978) Methods Enzymol. *50*, 3–33.
LINDBERG, B., LÖNNGREN, J. and SVENSSON, S. (1975) Advanc. Carb. Chem. Biochem. *31*, 185–240.
LIS, H. and SHARON, N. (1978) J. Biol. Chem. *253*, 3468–3476.
LIS, H. and SHARON, N. (1980) in The Biochemistry of Plants: A Comprehensive Treatise, Vol. 6 (Marcus, A., ed), pp. 371–447, Academic Press, New York.
LIU, T.-Y. and CHANG, Y.H. (1971) J. Biol. Chem. *246*, 2842–2848.
LOHMANDER, S., DE LUCA, S., NILSSON, B., HASCALL, V., CAPUTO, C.B., KIMURA, J.H. and HEINEGARD, D. (1980) J. Biol. Chem. *255*, 6084–6091.
LOTAN, R., BEATTIE, G., HUBBELL, I.W. and NICOLSON, G.L. (1977) Biochemistry *16*, 1787–1794.
LOTAN, R., LIS, H., ROSENWASSER, A., NOVOGRODSKY, A. and SHARON, N. (1973) Biochem. Biophys. Res. Commun. *55*, 1347–1355; correction *56* (1974) 1100.
LOTAN, R., SIEGELMAN, H.W., LIS, H. and SHARON, N. (1974) J. Biol. Chem. *249*, 1219–1224.
LOWE, C.R. (1979) in An Introduction to Affinity Chromatography, Laboratory Techniques in Biochemistry and Molecular Biology (Work, T.S. and Work, E., eds), North-Holland, Amsterdam.
LUCY, J.A., DINGLE, J.T. and FELL, H.B. (1961) Biochem. J. *79*, 500–508.
LUDOWEIG, J.J. and BENMAN, J.D. (1968) Carbohydr. Res. *8*, 185–192.
LUDOWEIG, J. and DORFMAN, A. (1960) Biochim. Biophys. Acta *38*, 212–218.
MCCAIRNS, E. and BEELEY, J.G. (1972) Biochem. J. *128*, 120P.
MCDEVITT, C.A. and MUIR, H. (1971) Anal. Biochem. *44*, 612–622.
MCFARLANE, A.S. (1958) Nature *182*, 53.
MCGINNIS, G.D. and FANG, P. (1980) Advanc. Carbohydr. Chem. *8*, 33–43.
MCKENZIE, G.H., SAWYER, W.H. and Nichol, L.W. (1972) Biochim. Biophys. Acta *263*, 283–293.
MCNEIL, M., DARVILL, A.G., AMAN, P., FRANZEN, L. and ABERSHEIM, P. (1982) Methods Enzymol. *83*, 3–45.
MADDY, A.H., DUNN, M.J. and KELLY, P.G. (1972) Biochim. Biophys. Acta *288*, 263–276.
MAHONEY, W.C. and DUSKIN, D. (1979) J. Biol. Chem. *254*, 6572–6576.
MALAWISTA, I. and SCHUBERT, M. (1958) J. Biol. Chem. *280*, 535–544.
MALMSTRÖM, A., RODÉN, L., FEINGOLD, D.S., JACOBSSON, I., BÄCKSTRÖM, G. and LINDAHL, U. (1980) J. Biol. Chem. *255*, 3878–3883.
MANJUNATH, P. and SAIRAM, M.R. (1982) J. Biol. Chem. *257*, 7109–7115.
MANN, K.G. and FISH, W.W. (1972) Methods Enzymol. *26*, 28–42.
MARCH, S.C., PARIKH, I. and CUATRECASAS, P. (1974) Anal. Biochem. *60*, 149–152.
MARCHALONIS, J.J. (1969) Biochem. J. *113*, 299–305.
MARCHESI, V.T. (1972) Methods Enzymol. *28*, 252–254.
MARCHESI, V.T., FURTHMAYR, H. and TOMITA, M. (1976) Advanc. Enzymol. *45*, 667–698.
MARGOLIS, R.K. and MARGOLIS, R.U. (1973) J. Neurochem. *20*, 1285–1298.
MARIKOVSKY, Y., LOTAN, R., LIS, R., SHARON, N. and DANON, D. (1976) Exp. Cell

Res. *99*, 453–456.

MARSHALL, R.D. (1972) Annu. Rev. Biochem. *41*, 673–702.

MARSHALL, R.D. (1974) Biochem. Soc. Symp. *11*, 17–26.

MARSHALL, R.D. and NEUBERGER, A. (1972a) in Glycoproteins, Their Composition, Structure and Function, 2nd Edn., Part A (Gottschalk, A., ed), pp. 224–299, Elsevier, Amsterdam.

MARSHALL, R.D. and NEUBERGER, A. (1972b) in Glycoproteins, Their Composition Structure and Function, 2nd Edn., Part A (Gottschalk, A., ed), pp. 322–380. Elsevier, Amsterdam.

MARSHALL, R.D. and NEUBERGER, A. (1976) Methods Carbohydr. Chem. *7*, 212–220.

MASON, R.M. and MAYES, R.W. (1973) Biochem. J. *131*, 535–540.

MASON, R.M., KIMURA, J.H. and HASCALL, V.C. (1982) J. Biol. Chem. *257*, 2236–2245.

MATSUMOTO, A., YOSHIMA, H. and KOBATA, A. (1983) Biochemistry *22*, 188–196.

MATTHEWS, M.B. (1976) Methods Carbohydr. Chem. *7*, 116–119.

MATTHEWS, M.B. and DECKER, L. (1971) Biochim. Biophys. Acta *244*, 30–34.

MATTHEWS, M.B. and DECKER, L. (1977) Biochim. Biophys. Acta *498*, 259–263.

MAYES, R.W., MASON, R.M. and GRIFFIN, D.C. (1973) Biochem. J. *131*, 541–553.

MEAGER, A., UNGKITCHANUKIT, A. and HUGHES, R.C. (1976) Biochem. J. *154*, 113–124.

MEITTINEN, T. and TAKKI-LUUKKAINEN, I.T. (1959) Acta Chem. Scand. *13*, 856–858.

MELLIS, S.J. and BAENZIGER, J.U. (1981) Anal. Bioch. *114*, 276–280.

MERRILEES, M.J., MERRILEES, M.A., BIRNBAUM, P.S., SCOTT, P.J. and FLINT, M.H. (1971) Atherosclerosis *27*, 259–264.

METCALFE, D.D., SMITH, J.A., AUSTEN, K.F. and SILBERT, J.E. (1980) J. Biol. Chem. *255*, 11753–11758.

MILLER, D.H., LAMPORT, D.T.A. and MILLER, M. (1972) Science *176*, 918–920.

MIZUOCHI, T., YAMASHITA, K., FUJIKAWA, K., KISIEL, W. and KOBATA, A. (1979) J. Biol. Chem. *254*, 6419–6425.

MOLYNEAUX, R.J. and JAMES, L.F. (1982) Science *216*, 190–191.

MONONEN, I., FINNE, J. and KÄRKKÄINEN, J. (1978) Carbohydr. Res. *60*, 371–375.

MONTGOMERY, R. (1972) in Glycoproteins, 2nd Edn., Part A (Gottschalk, A., ed), pp. 518–528, Elsevier, Amsterdam.

MONTREUIL, J. (1980) Advanc. Carbohydr. Chem. Biochem. *37*, 157–223.

MOORE, S. and STEIN, W.H. (1954) J. Biol. Chem. *211*, 907–913.

MORELL, A.G. and ASHWELL, G. (1972) Methods Enzymol. *28*, 205–208.

MORGAN, W.T.J. (1967) Methods Immunol. Immunochem. *1*, 75–81.

MORRIS, E.R. (1981) in Techniques in Carbohydrate Metabolism, B 309, pp. 1–39, Elsevier/North Holland, Shannon, Ireland.

MORRIS, E.R. and ROSS-MURPHY, S.B. (1981) in Techniques in Carbohydrate Metabolism, B 310, pp. 1–46, Elsevier/North Holland, Shannon, Ireland.

MORRIS, E.R., REES, D.A., SANDERSON, G.R. and THOM, D. (1975) J. Chem. Soc. Trans. Perkin II, 1418–1425.

MORRIS, J.E., CANOY, D.W. and RYND, L.S. (1981) J. Chromatog. *224*, 407–413.

MORT, A.J. and LAMPORT, D.T.A. (1977) Anal. Biochem. *82*, 289–309.

MUIR, H. (1958) Biochem. J. *69*, 195–204.

Muir, H. (1983) Biochem. Soc. Trans. *11*, 613–622.

Muir, H. and Hardingham, T.E. (1975) M.T.P. Int. Rev. Sci. Biochem. Ser. 1, *5*, 153–222.

Muir, L. and Lee, Y.C. (1970) J. Biol. Chem. *245*, 502–509.

Muramatsu, T. and Nathenson, S.G. (1970) Biochemistry *9*, 4875–4883.

Muramatsu, T., Koide, N. and Maeyama, K. (1978) J. Biochem. (Tokyo) *83*, 363–370.

Muramatsu, T., Koide, N., Ceccarini, C. and Atkinson, P.H. (1976) J. Biol. Chem. *251*, 4673–4679.

Murata, K. (1980) Methods Carbohydr. Chem. *8*, 81–88.

Murata, K., Harada, T. and Okubo, K. (1968) J. Atheroscl. Res. *8*, 951.

Nagasawa, K. and Inoue, Y. (1974) Carbohydr. Res. *36*, 265–271.

Nagata, Y., Goldberg, A.R. and Burger, M.M. (1974) Methods Enzymol. *32*, 611–615.

Nakajima, T. and Ballou, C.E. (1974) J. Biol. Chem. *249*, 7679–7684.

Nakazawa, K. and Suzuki, S. (1975) J. Biol. Chem. *250*, 912–917.

Nakazawa, K., Newsome, D.A., Nilsson, B., Hascall, V.C. and Hassell, J.R. (1983) J. Biol. Chem. *258*, 6051–6055.

Narasimhan, S., Wilson, J.R., Martin, E. and Schachter, H. (1979) Canad. J. Biochem. *57*, 83–96.

Narasimhan, S., Harpaz, N., Longmore, G., Carver, J.P., Grey, A.A. and Schachter, H. (1980) J. Biol. Chem. *255*, 4876–4884.

Natowicz, M. and Baenziger, J.U. (1980) Anal. Biochem. *105*, 159–164.

Neuberger, A. and Ratcliffe, W.A. (1972) Biochem. J. *129*, 683–693.

Neuberger, A. and Ratcliffe, W.A. (1973) Biochem. J. *133*, 623–628.

Neufeld, E. and Ashwell, G. (1980) in The Biochemistry of Glycoproteins and Proteoglycans (Lennarz, W.J., ed), pp. 241–266, Plenum, New York.

Ng Ying Kin, N.M.K. and Wolfe, L.S. (1980) Anal. Biochem. *102*, 213–219.

Nicholson, G.I., Lacorbiere, M. and Eckhart, W. (1975) Biochemistry *14*, 172–179.

Nicolson, G.L. (1974) Int. Rev. Cytol. *39*, 89–190.

Nicolson, G.L. (1976a) Biochim. Biophys. Acta *457*, 57–108.

Nicolson, G.L. (1976b) Biochim. Biophys. Acta *458*, 1–72.

Nicolson, G.L., Blaustein, J. and Etzler, M.E. (1974) Biochemistry *13*, 196–204.

Niebes, P. and Schifflers, M.H. (1975) Clin. Chim. Acta *62*, 195–202.

Nilsson, B. and Svensson, S. (1979) Carbohydr. Res. *72*, 183–190.

Nilsson, B. and Zopf, D. (1982) Methods Enzymol. *83*, 46–58.

Nilsson, B., De Luca, S., Lohmander, S. and Hascall, V.S. (1982) J. Biol. Chem. *257*, 10920–10927.

Nilsson, B., Norden, N.E. and Svensson, S. (1979) J. Biol. Chem. 254, 4545–4553.

Nilsson, S.F. and Waxdal, M.J. (1976) Biochemistry *15*, 2698–2705.

Noonan, K. and Burger, M.M. (1973) J. Biol. Chem. *248*, 4286–4292.

Norden, A.G.W. and O'Brien, J.S. (1974) Biochem. Biophys. Res. Commun. *56*, 193–198.

Oegema, T.R., Hascall, V.C. and Dziewiatkowski, D.D. (1975) J. Biol. Chem. *250*, 6151–6159.

OGATA, A.M., MURAMATSU, T. and KOBATA, A. (1977) Arch. Biochem. Biophys. *181*, 353–358.

OGATA, S.-I. and LLOYD, K.O. (1982) Anal. Biochem. *119*, 351–359.

OGATA, S., MURAMATSU, T. and KOBATA, A. (1975) J. Biochem. (Tokyo) *78*, 687–696.

OLDEN, K., PARENT, J.B. and WHITE, S.L. (1982) Biochim. Biophys. Acta *650*, 209–232.

OLSNES, S., SALTVEDT, E. and PIHL, A. (1974) J. Biol. Chem. *249*, 803–810.

ONODERA, K., KOMANA, T. and HIRANO, S. (1964) Biochim. Biophys. Acta *83*, 20–26.

OUCHTERLONY, Ö. (1967) in Handbook of Experimental Immunology (Weir, D.M., ed), pp. 655–706, Blackwell, Oxford.

OZANNE, B. and SAMBROOK, J. (1971) Nature New Biol. *232*, 156–160.

PARODI, A.J., BEHRENS, N.H., LELOIR, L.F. and DANKERT, M. (1972) Biochim. Biophys. Acta *270*, 529–536.

PASTERNACK, S.G., VEIS, A. and BREEN, M. (1974) J. Biol. Chem. *249*, 2206–2211.

PAULSON, J.C., HILL, R.L., TANABE, T. and ASHWELL, G. (1977) J. Biol. Chem. *252*, 8624–8628.

PERKINS, S.J., MILLER, A., HARDINGHAM, T.E. and MUIR, H. (1981) J. Mol. Biol. *150*, 69–95.

PERLIN, A.S. (1976) Methods Carbohydr. Chem. *7*, 94–100.

PERLIN, A.S. (1977) Fed. Proc. *36*, 106–109.

PERLIN, A.S., CASU, B., SANDERSON, G.R. and JOHNSON, L.F. (1970) Canad. J. Chem. *48*, 2260–2268.

PERLIN, A.S., NG YING KIM, M.K. and BHATTACHARJEE, S.S. (1972) Canad. J. Chem. *50*, 2437–2441.

PHILIPS, P.G. and FURMANSKI, P. (1976) in Concanavalin A as a Tool (Bittiger, H. and Schnebli, H.P., eds), pp. 195–199, Wiley, London.

PHILIPS, P.G., FURMANSKI, P. and LUBIN, M. (1974) Exp. Cell Res. *86*, 301–308.

PLANTER, J.J. and CARLSON, D.M. (1972) Methods Enzymol. *28*, 46–48.

PLUMMER, T.H., TARENTINO, A. and MALEY, F. (1968) J. Biol. Chem. *243*, 5158–5164.

POLONOFF, E., MACHADA, C.A. and KABAT, D. (1982) J. Biol. Chem. *257*, 14023–14028.

POUYSSÉGUR, J. and PASTAN, I. (1977) J. Biol. Chem. *252*, 1639–1646.

PRATT, R.S. and COOK, G.M.W. (1979a) Biochem. J. *179*, 299–303.

PRATT, R.S. and COOK, G.M.W. (1979b) Biochem. J. *179*, 305–314.

PRICE, P.A., LUI, T.-Y., STEIN, W.H. and MOORE, S. (1969) J. Biol. Chem. *244*, 917–923.

PRIVAT, J.-P., DELMOTTE, F., MIALONIËR, G., BOUCHARD, P. and MONSIGNY, M. (1974) Eur. J. Biochem. *47*, 5–14.

RADEMACHER, T.W., HOMANS, S.W., FERNANDED, D.L., DWEK, R.A., MIZNOCHI, T., TANIGUCHI, T. and KOBATA, A. (1983) Biochem. Soc. Trans. *11*, 132–134.

RAIZADA, M.K., SCHUTZBACH, J.S. and ANKEL, H. (1975) J. Biol. Chem. *250*, 3310–3315.

RAUVALA, H. (1979) Carbohydr. Chem. *72*, 257–260.

RAUVALA, H. and FINNE, J. (1979) FEBS Lett. *97*, 1–8.

RAUVALA, H., FINNE, J., KRUSIUS, T., KÄRKKÄINEN, J. and JÄRNEFELT, J. (1981) Advanc. Carbohydr. Chem. Biochem. *38*, 389–416.

READING, C.L., PENHOET, E.E. and BALLOU, C.E. (1978) J. Biol. Chem. *253*, 5600–5612.

REBOIS, R.V. and FISHMAN, P.H. (1983) J. Biol. Chem. *258*, 12775–12778.

REES, D.A. (1963) Biochem. J. *88*, 343–345.

REISNER, Y. and SHARON, N. (1980) Trends Biochem. Sci. *5*, 29–31.

REISNER, Y., LIS, H. and SHARON, N. (1976) Exp. Cell Res. *97*, 445–448.

REISSIG, J., STROMINGER, J.L. and LELOIR, L.F. (1955) J. Biol. Chem. *217*, 959–966.

REYNOLDS, J.A. and TANFORD, C. (1976) Proc. Natl. Acad. Sci. USA *73*, 4467–4470.

RICKWOOD, D. (1984) Centrifugation, A Practical Approach (2nd Edn.), IRL Press, Oxford.

RIGHETTI, P.G. (1983) in Isoelectric focusing: theory, methodology and applications, Laboratory Techniques in Biochemistry and Molecular Biology (Work, T.S. and Burdon, R.H., eds), Elsevier, Amsterdam.

RITTENHOUSE, H.G., BENIAN, G., RITTENHOUSE, J.W., HANSEN, E.R. and BOYD, L.E. (1976) in Membranes and Neoplasia (Marchesi, V., ed), pp. 203–213, Alan Liss, New York.

ROARK, D.E. and YPHANTIS, D.A. (1969) Ann. N.Y. Acad. Sci. *164*, 245–278.

ROBBINS, P.W., HUBBARD, S.C., TURCO, S.J. and WIRTH, D.F. (1977) Cell *12*, 893–900.

ROBERTS, B.E. and PATTERSON, B.M. (1973) Proc. Natl. Acad. Sci. USA *70*, 2330–2334.

ROBERTS, D.D. and GOLDSTEIN, I.J. (1982) J. Biol. Chem. *257*, 11274–11277.

ROBERTS, G.P. (1974) Eur. J. Biochem. *50*, 265–280.

ROBERTS, G.P. (1976) Arch. Biochem. Biophys. *173*, 528–537.

ROBINSON, H.C., HORNER, A.A., HÖÖK, M., ÖRGREN, S. and LINDAHL, U. (1978) J. Biol. Chem. *253*, 6687–6693.

ROBINSON, P.J., BULL, F.G., ANDERTON, B.H. and ROITT, I.M. (1975) FEBS Lett. *58*, 330–333.

ROBSON, T., ALLEN, A. and PAIN, R.H. (1975) Biochem. Soc. Trans. *3*, 1105–1107.

RODÉN, L. (1980) in The Biochemistry of Glycoproteins and Proteoglycans (Lennarz, W.J., ed), pp. 267–371, Plenum, New York.

RODÉN, L. and SCHWARTZ, N.B. (1975) in Biochemistry of Carbohydrates, MTP International Review of Science, Biochemistry Series One, Vol. 5 (Whelan, W.J., ed), p. 95, Butterworths, London.

RODÉN, L., BAKER, J.R., CIFONELLI, J.A. and MATTHEWS, M.B. (1972) Methods Enzymol. *28*, 73–140.

RONDLE, C.M.J. and MORGAN, W.T.J. (1955) Biochem. J. *61*, 586–589.

ROSE, M.C., VOTER, W.A., SAGE, H., BROWN, C.F. and KAUFMAN, B. (1984) J. Biol. Chem. *259*, 3167–3172.

ROSEMAN, S., CARLSON, D.M., JOURDIAN, G.W., MCGUIRE, E.J., KAUFMAN, B., BASU, S. and BARTHOLOMEW, B. (1966) Methods Enzymol. *8*, 354–372.

ROSENBERG, L., HELLMANN, W. and KLEINSCHMIDT, A.K. (1975) J. Biol. Chem. *256*, 1877–1883.

ROSENBERG, L., PAL, S., BEALE, R. and SCHUBERT, M. (1970) J. Biol. Chem. *245*,

4112–4122.

ROSNER, M.R., HUBBARD, S.C., IVATT, R.J. and ROBBINS, P.W. (1982) Methods Enzymol. *83*, 399–408.

ROTHMAN, J.E. and FINE, R.E. (1980) Proc. Natl. Acad. Sci. USA *77*, 780–784.

ROTHMAN, J.E. and LODISH, H. (1977) Nature *269*, 775–780.

ROVIS, L., ANDERSON, B., KABAT, E.A., GRUEZO, F. and LIAO, J. (1973) Biochemistry *12*, 1955–1961.

ROYAL, P.D., SPARKS, K.J. and GOETINCK, P.F. (1980) J. Biol. Chem. *255*, 9870–9878.

RYLEY, H.C. and BROGAN, T.D. (1968) Brit. J. Exp. Pathol. *49*, 625–633.

SADLER, J.E., BEYER, T.A., OPPENHEIMER, C.L., PAULSON, J.C., PRIEELS, J.-P., REARICK, J.I. and HILL, R.L. (1982) Methods Enzymol. 83, 458–514.

SAEED, M.S. and WILLIAMS, J.M. (1980) Carbohydr. Res. *84*, 83–94.

SAGE, H.J. and GREEN, R.W. (1973) Methods Enzymol. *28*, 332–339.

SAITO, H., YAMAGATA, T. and SUZUKI, S. (1968) J. Biol. Chem. *243*, 1536–1542.

SAJDERA, S.W. and HASCALL, V.C. (1969) J. Biol. Chem. *244*, 77–87.

SAKURA, J.D. and REITHEL, F.J. (1972) Methods Enzymol. *26*, 107–119.

SANDSON, J., ROSENBERG, S. and WHITE, D. (1966) J. Exp. Med. *123*, 817–828.

SANDVIG, K., OLSNES, S. and PIHL, A. (1976) J. Biol. Chem. *251*, 3977–3984.

SANDVIG, K., OLSNES, S. and PIHL, A. (1978) Eur. J. Biochem. *82*, 13–23.

SATO, C.S. and GYORKEY, F. (1974) Anal. Biochem. *61*, 305–310.

SAUNIER, B., KILKER, R.D., TRACZ, J.S., QUARONI, A. and HERSKOVICS, A. (1982) J. Biol. Chem. *257*, 14155–14161.

SCATCHARD, G. (1949) Ann. N.Y. Acad. Sci. *51*, 660–672.

SCHACHMAN, H.K. (1968) Methods Enzymol. *4*, 32–103.

SCHACHTER, H. (1975) Methods Enzymol. *41*, 3–10.

SCHACHTER, H. (1978) in The Glycoconjugates, Vol. 2 (Horowitz, M.I., ed), pp. 88–181, Academic Press, New York.

SCHACHTER, H. and ROSEMAN, S. (1980) in The Biochemistry of Glycoproteins and Proteoglycans (Lennarz, W.J., ed), pp. 85–160, Plenum Press, New York.

SCHAUER, R. (1978) Methods Enzymol. *50*, 64–89.

SCHAUER, R., BUSCHER, H.-P. and CASALS-STENZEL, J. (1974) Biochem. Soc. Symp. *40*, 87–116.

SCHAUER, R., WEMBER, M., WIRTZ-PIETZ, G. and FERREIRA DO AMARAL, C. (1971) Hoppe-Seyler's Z. Physiol. Chem. *352*, 1073–1080.

SCHEETZ, M.P. (1979) Biochim. Biophys. Acta *557*, 122–134.

SCHMID, K., BINETTE, J.P., DORLAND, L., VLIEGENTHART, J.F.G., FOURNET, B. and MONTREUIL, J. (1979) Biochim. Biophys. Acta *581*, 356–359.

SCHMID, K., NIMBERG, R.B., KIMURA, A., YAMAGUCHI, H. and BINETTE, J.B. (1977) Biochim. Biophys. Acta *492*, 291–302.

SCHMIDT-ULLRICH, R., WALLACH, D.F.H. and HENDRICKS, J. (1975) Biochim. Biophys. Acta *382*, 295–310.

SCHNEBLI, H.B. (1976) in Concanavalin A as a Tool (Bittiger, H. and Schnebli, H.P., eds), pp. 249–255, Wiley, London.

SCHOOP, H.J. and FAILLARD, H. (1967) Hoppe-Seyler's Z. Physiol. Chem. *348*, 1509–1517.

SCHUBERT, D. (1970) J. Mol. Biol. *51*, 287–301.
SCHUR, B.D. (1982) in The Glycoconjugates, Vol. 3 (Horowitz, M.I., ed), pp. 146–185, Academic Press, New York.
SCHWARTZ, N.B. (1975) FEBS Lett. *49*, 342–345.
SCHWARTZ, N.B. (1977) J. Biol. Chem. *252*, 6316–6321.
SCHWARTZ, N.B. and DORFMAN, A. (1975) Arch. Biochem. Biophys. *171*, 136–144.
SCHWARTZ, N.B. and RODÉN, L. (1975) J. Biol. Chem. *250*, 5200–5207.
SCHWARTZ, N.B., GALLIGANI, L., HO, P.-L. and DORFMAN, A. (1974a) Proc. Natl. Acad. Sci. USA *71*, 4047–4051.
SCHWARTZ, N.B., RODÉN, L. and DORFMAN, A. (1974b) Biochem. Biophys. Res. Commun. *56*, 717–724.
SCHWARTZ, R.T. and DATEMA, R. (1982a) Methods Enzymol. *83*, 432–443.
SCHWARTZ, R.T. and DATEMA, R. (1982b) Adv. Carbohydr. Chem. Biochem. *40*, 287–379.
SCHWARTZ, R.T. and DATEMA, R. (1984) Trends Biochem. Sci. *9*, 32–34.
SCHWYTZER, M. and HILL, R.L. (1977a) J. Biol. Chem. *252*, 2338–2345.
SCHWYTZER, M. and HILL, R.L. (1977b) J. Biol. Chem. *252*, 2346–2355.
SCOPES, R. (1982) Protein Purification, Springer-Verlag, New York.
SCOTT, J.E. (1960) Methods Biochem. Anal. *8*, 145–197.
SCOTT, J.E. (1968) Biochim. Biophys. Acta *170*, 471–473.
SCOTT, J.E. and HARBINSON, R.J. (1968) Histochemie *14*, 215–220.
SEGREST, J.P. and JACKSON, R.L. (1972) Methods Enzymol. *28*, 54–63.
SEGREST, J.P., JACKSON, R.L., ANDREWS, E.R. and MARCHESI, V.T. (1971) Biochem. Biophys. Res. Commun. *44*, 390–395.
SENO, N., ANNO, K., KONDO, K., NAGASE, S. and SAITO, S. (1970) Anal. Biochem. *37*, 197–202.
SERAFFINI-FACASSINI, A., PETERS, T.J. and FLOREANI, L. (1967) Biochem. J. *105*, 569–575.
SHARMA, C.B., BABCZINSKI, P., LEHLE, L. and TANNER, W. (1974) Eur. J. Biochem. *46*, 35–41.
SHARON, N. and LIS, H. (1975) Methods Membrane Biol. *3*, 147–200.
SHARON, N. and LIS, N. (1982) in The Proteins, 3rd Edn. Vol. 5 (Neurath, H. and Hill, R.L., eds), pp. 1–144, Academic Press, New York.
SHEEHAN, J.K., NIEDUSZYNSKI, I.A., SHEEHAN, J.K., PHELPS, C.F., MUIR, H. and HARDINGHAM, T.E. (1978) Biochem. J. *171*, 109–114.
SHIVELY, J.E. and CONRAD, H.E. (1976) Biochemistry *15*, 3932–3942.
SHIVELY, J.E. and CONRAD, H.E. (1977) Fed. Proc. *36*, 28–31.
SILBERT, J.E. (1967) J. Biol. Chem. *242*, 5153–5157.
SIMPSON, D.L., HRANISAVLJEVIC, J. and DAVIDSON, E.A. (1972) Biochemistry *11*, 1849–1856.
SIMPSON, D.L., THORNE, D.R. and LOH, H.H. (1978) Life Sciences *22*, 727–748.
SLAYTER, H.S. (1981) in Electron Microscopy of Proteins, Vol. 1 (Harris, J.R., ed), pp. 197–254, Academic Press, London.
SMITH, R.L., GILKDERSON, E., KOHATSU, N., MERCHANT, T. and SCHURMAN, D.J. (1980) Anal. Biochem. *103*, 191–200.
SNARY, D. and ALLEN, A. (1971) Biochem. J. *123*, 845–853.

SNARY, D. and ALLEN, A. (1972) Biochem. J. *127*, 577–587.
SNARY, D., ALLEN, A. and PAIN, R.H. (1974) Biochem. J. *141*, 641–646.
SNOW, L.D., COLTON, D.G. and CARRAWAY, K.L. (1977) Arch. Biochem. Biophys. *179*, 690–697.
SO, L.-L. and GOLDSTEIN, I.J. (1967) J. Biol. Chem. *242*, 1617–1622.
SO, L.-L. and GOLDSTEIN, I.J. (1969) Carbohydr. Res. *10*, 231–244.
SOMOGYI, M. (1952) J. Biol. Chem. *195*, 19–23.
SPIRO, M.J., SPIRO, R.G. and BHOYROO, V.D. (1976) J. Biol. Chem. *251*, 6420–6425.
SPIRO, R.G. (1962) J. Biol. Chem. *235*, 2860–2869.
SPIRO, R.G. (1966a) Methods Enzymol. *8*, 3–25.
SPIRO, R.G. (1966b) Methods Enzymol. *8*, 26–52.
SPIRO, R.G. (1972) Methods Enzymol. *28*, 3–43.
SPIRO, R.G. (1973) Advanc. Protein Chem. *27*, 349–467.
SPIRO, R.G. (1976) Methods Carbohydr. Chem. *7*, 205–211.
SPIRO, R.G. and BHOYROO, V.D. (1971) Fed. Proc. *30*, 1223.
SPIRO, R.G. and BHOYROO, V.D. (1974) J. Biol. Chem. *249*, 5704–5717.
SPOONER, E., FUKUDA , M., KLOCK, J.C., OATES, J.E. and DELL, A. (1984) J. Biol. Chem. *259*, 4792–4801.
SPRINGER, G.F., NAGAI, Y. and TEGTMEYER, H. (1966) Biochemistry *5*, 3254–3272.
STANLEY, P (1980) in The Biochemistry of Glycoproteins and Proteoglycans (Lennarz, W.J., ed), pp. 161–189, Plenum Press, New York.
STANLEY, P. (1983) Methods Enzymol. *96*, 157–184.
STARKEY, B.J., SNARY, D. and ALLEN, A. (1974) Biochem. J. *141*, 633–639.
STECK, T.L. and DAWSON, G. (1974) J. Biol. Chem. *249*, 2135–2142.
STECK, T.L. and WALLACH, D.F.H. (1965) Biochim. Biophys. Acta *97*, 510–522.
STEIN, M.D., HOWARD, I.K. and SAGE, H.J. (1971) Arch. Biochem. Biophys. *146*, 353–355.
STELLNER, K., SAITO, H. and HAKOMORI, S. (1973) Arch. Biochem. Biophys. *155*, 464–472.
STONE, A.L. (1976) Methods Carbohydr. Chem. *7*, 120–138.
STONE, W.L. and REYNOLDS, J.A. (1975) J. Biol. Chem. *250*, 8045–8048.
STRAUS, J.H., BURGE, B.W. and DARNELL, J.E. (1970) J. Mol. Biol. *47*, 437–448.
STRECKER, G. and MONTREUIL, J. (1979) Biochimie *61*, 1199–1246.
STRECKER, G., PIECE-CRETEL, A., FOURNET, B., SPIK, G. and MONTREUIL, J. (1981) Anal. Biochem. *111*, 17–26.
SUTTON, B.J. and PHILIPS, D.C. (1983) Biochem. Soc. Trans. *11*, 130–132.
SUZUKI, S. (1972a) Methods Enzymol. *28*, 911–917.
SUZUKI, S. (1972b) Methods Enzymol. *28*, 917–921.
SVEDBERG, T. and PEDERSON, K.O. (1940) The Ultracentrifuge, Oxford University Press, London.
SVENNERHOLM, E. and SVENNERHOLM, L. (1958) Nature *181*, 1154–1155.
SVENNERHOLM, L. (1957) Biochim. Biophys. Acta *24*, 604–611.
SWANN, D.A., SLAYTER, H.S. and SILVER, F.H. (1981) J. Biol. Chem. *258*, 5921–5925.
SWEET, D.P., SHAPIRO, R.H. and ALBERSHEIM, P. (1975) Carbohydr. Res. *40*, 217–225.

TABAS, I. and KORNFELD, S. (1982) Methods Enzymol. *83*, 416–429.
TABAS, I., SCHLESSINGER, S. and KORNFELD, S. (1978) J. Biol. Chem. *253*, 716–722.
TAKASAKI, S., YAMASHITA, K., SUSUKI, K., IWANAGA, S. and KOBATA, A. (1979) J. Biol. Chem. *254*, 8548–8553.
TAKASAKI, S. and KOBATA, A. (1976) J. Biol. Chem. *251*, 3610–3615.
TAKASAKI, S. and KOBATA, A. (1978) Methods Enzymol. *50*, 50–54.
TAKASAKI, S., MIZUOCHI, T. and KOBATA, A. (1982) Methods Enzymol. *83*, 263–268.
TANAKA, K. (1978) J. Biochem. (Tokyo) *83*, 655–659.
TANAKA, K. and PIGMAN, W. (1965) J. Biol. Chem. *240*, PC 1487–1488.
TANFORD, C. and REYNOLDS, J.A. (1976) Biochim. Biophys. Acta *457*, 133–170.
TANFORD, C., NOZAKI, Y., REYNOLDS, J.A. and MAKINO, S. (1974) Biochemistry *13*, 2369–2375.
TANNER, M.J.A. (1979) Current Topics Membranes Transport *12*, 1–51.
TANNER, M.J.A. and ANSTEE, D.J. (1976) Biochem. J. *153*, 265–270.
TARENTINO, A.L. and MALEY, F. (1972) Methods Enzymol. *28*, 782–786.
TARENTINO, A.L., PLUMMER, T.H. and MALEY, F. (1974) J. Biol. Chem. *249*, 818–824.
TARENTINO, A.L., TRIMBLE, R.B. and MALEY, F. (1978) Methods Enzymol. *50*, 574–580.
TAYLOR, R.L. and CONRAD, H.E. (1972) Biochemistry *11*, 1383–1388.
TAYLOR, R.L., SHIVELY, J.E., CONRAD, H.F. and CIFONELLI, J.A. (1973) Biochemistry *12*, 3633–3637.
TETAERT, D., TAKAHASHI, N. and PUTNAM, F.W. (1982) Anal. Biochem. *123*, 430–437.
THOMAS, D.B. and WINZLER, R.J. (1969) J. Biol. Chem. *244*, 5943–5946.
THORPE, S.R. and BAYNES, J.W. (1982) in Glycoconjugates, Vol. 3 (Horowitz, M.I., ed), pp. 113–132, Academic Press, New York.
TODA, N., DOI, A., JIMBO, A., MATSUMOTO, I. and SENO, N. (1981) J. Biol. Chem. *256*, 5345–5349.
TOMITA, M., FURTHMAYR, H. and MARCHESI, V.T. (1978) Biochemistry *17*, 4756–4770.
TORCHIA, D.A., HASSON, M.A. and HASCALL, V.C. (1977) J. Biol. Chem. *252*, 3617–3625.
TOWBIN, H., STAEHELIN, T. and GORDON, J. (1979) Proc. Natl. Acad. Sci. USA *76*, 4350–4354.
TREVELYAN, W.E., PROCTER, D.P. and HARRISON, J.S. (1950) Nature *166*, 444–445.
TRIMBLE, R.B. and MALEY, F. (1977a) J. Biol. Chem. *252*, 4409–4412.
TRIMBLE, R.B. and MALEY, F. (1977b) Biochem. Biophys. Res. Commun. *78*, 935–944.
TRIMBLE, R.B., TARENTINO, A.L., AUMICK, G.E. and MALEY, F. (1982) Methods Enzymol. *83*, 603–610.
TSAO, D. and KIM, Y.S. (1981) J. Biol. Chem. *256*, 4847–4950.
TSIGONAS, C.P. and MUIR, H. (1966) Anal. Biochem. *17*, 495–501.
TSUJI, T., IRAMURA, T. and OSAWA, T. (1980) Biochem. J. *187*, 677–686.
TULSIANI, D.R.P., HARRIS, T.M. and TOUSTER, O. (1982) J. Biol. Chem. *257*, 7936–7939.

TURVEY, J.R. (1965) Advanc. Carbohydr. Chem. *20*, 183–218.

UHR, J.W. (1970) Cell Immunol. *1*, 228–255.

UI, N. (1981) J. Chromatog. *215*, 289–294.

UTTERMAN, G. and SIMONS, K. (1974) J. Mol. Biol. *85*, 569–587.

VAN HALBECK, H., DORLAND, L., VLIEGENTHART, J.F.G., FIAT, A.-M. and JOLLÈS, P. (1980a) Biochim. Biophys. Acta *623*, 295–300.

VAN HALBECK, H., DORLAND, L., VLIEGENTHART, J.F.G., SCHMID, K., MONTREUIL, J., FOURNET, B. and HULL, W.E. (1980b) FEBS Lett. *114*, 11–16.

VAN LENTEN, L. and ASHWELL, G. (1972) Methods Enzymol. *28*, 209–211.

VESTERGAARD, B.F. and BØG-HANSEN, T.C. (1975) Scand. J. Immunol. *4*, Suppl. 2, 211.

VLIEGENTHART, J.F.G., DORLAND, L. and VAN HALBECK, H. (1983) Adv. Carbohydr. Chem. Biochem. *41*, 209–374.

VLIEGENTHART, J.F.G., VAN HALBECK, H. and DORLAND, L. (1981) Pure Appl. Chem. *53*, 45–77.

VLIEGENTHART, J.F.G., VAN HALBECK, H. and DORLAND, (1980) I.U.P.A.C. 27th International Congress of Pure and Applied Chemistry, Helsinki, 1979 (Varmavuori, A., ed), pp.253–262, Pergamon Press, Oxford.

WAECHTER, C.J., LUCAS, J.J. and LENNARZ, W.J. (1973) J. Biol. Chem. *248*, 7570–7579.

WAGNER, R.R. and CYNKIN, M.A. (1969a) Arch. Biochem. Biophys. *129*, 242–247.

WAGNER, R.R. and CYNKIN, M.A. (1969b) Biochem. Biophys. Res. Commun. *35*, 139–143.

WARREN, L. (1959) J. Biol. Chem. *234*, 1971–1977.

WARREN, L., FUHRER, J.P., TUSZYNSKI, G.P. and BUCK, C.A. (1974) Biochem. Soc. Symp. *40*, 147–157.

WASTESON, A. (1971a) Biochem. J. *122*, 477–485.

WASTESON, A. (1971b) J. Chromatog. *59*, 87–97.

WEDGEWOOD, J.F. and STROMINGER, J.L. (1980) J. Biol. Chem. *255*, 1120–1123.

WELLS, G.B., KONTOYIANNIDOU, V., TURCO, S.J. and LESTER, R.L. (1982) Methods Enzymol. *83*, 132–137.

WELLS, G.B., TURCO, S.J., HANSON, B.A. and LESTER, R.L. (1982) Methods Enzymol. *83*, 137–139.

WENN, R.V. (1975) Biochem. J. *145*, 281–285.

WESSLER, E. (1968) Anal. Biochem. *26*, 439–444.

WESSLER, E. (1971) Anal. Biochem. *41*, 67–69.

WEST, C.M. and MCMAHON, D. (1977) J. Cell Biol. *74*, 264–273.

WHEAT, R.W. (1966) Methods Enzymol. *8*, 60–78.

WILSON, I.A., LADNER, R.C., SKEHEL, J.J. and WILEY, D.C. (1983) Biochem. Soc. Trans. *11*, 145–147.

WILSON, I.A., SKEHEL, J.J. and WILEY, D.C. (1981) Nature *289*, 366–373.

WINTER, W.T. (1982) Methods Enzymol. *83*, 87–104.

WINTER, W.T., SMITH, P.J.C. and ARNOTT, S. (1975) J. Mol. Biol. *99*, 219–235.

WINZLER, R.J. (1955) Methods Biochem. Anal. *2*, 279–311.

WINZLER, R.J., HARRIS, E.D., PEKAS, D.J., JOHNSON, C.A. and WEBER, P. (1967) Biochemistry *6*, 2195–2202.

WOLFROM, M.L. and JULIANO, B.O. (1960) J. Amer. Chem. Soc. *82*, 1673–1677.
WONG, C.G., SUNG, S.J. and SWEELEY, C.C. (1980) Methods Carbohydr. Chem. *8*, 55–65.
WOOD, D.L. (1965) Methods Enzymol. *4*, 104–126.
WOODWARD, C.B., HRANISAVLJEVIC, J. and DAVIDSON, E.A. (1972) Biochemistry *11*, 1168–1176.
WRIGHT, C.S. (1980a) J. Mol. Biol. *139*, 53–60.
WRIGHT, C.S. (1980b) J. Mol. Biol. *141*, 267–291.
WURMSER, R. and FILITTE-WURMSER, S. (1957) Progr. Biophys. Biophys. Chem. *7*, 87–113.
YAMAGISHI, T., MATSUDA, K. and WATANABE, T. (1976) Carbohydr. Res. *50*, 63-74.
YAMASHITA, K., MIZUOCHI, T. and KOBATA, A. (1982) Methods Enzymol. *83*, 105–126.
YAMASHITA, K., TACHIBANA, Y. and KOBATA, A. (1977) Arch. Biochem. Biophys. *182*, 546–555.
YAMASHITA, K., HITOI, A. and KOBATA, A. (1983) J. Biol. Chem. *258*, 14753–14755.
YAMASHITA, K., MIZUOCHI, T. and KOBATA, A. (1982) Methods Enzymol. *83*, 105–126.
YAMASHITA, K., TACHIBANA, Y. and KOBATA, A. (1978) J. Biol. Chem. *253*, 3862–3869.
YANAGISHITA, M. and HASCALL, V.C. (1983a) J. Biol. Chem. *258*, 12847–12856.
YANAGISHITA, M. and HASCALL, V.C. (1983b) J. Biol. Chem. *258*, 12857–12864.
YOSHIMA, H., NAKANISHI, M., OKADA, Y. and KOBATA, A. (1981) J. Biol. Chem. *256*, 5355–5361.
YOSIZAWA, A., SATO, T. and SCHMID, K. (1966) Biochim. Biophys. Acta *121*, 417–420.
YPHANTIS, D.A. (1964) Biochemistry *3*, 297–317.
YU, J. and STECK, T.L. (1975) J. Biol. Chem. *250*, 9170–9175.
YU, R.K. and LEDEEN, R.W. (1970) J. Lipid Res. *11*, 506–516.
YURCHENCO, P.D. and ATKINSON, P.H. (1977) Biochemistry *16*, 944–953.
YURCHENCO, P.D., CECCARINI, C. and ATKINSON, P.H. (1978) Methods Enzymol. *50*, 175–204.
ZAMYATNIN, A.A. (1972) Prog. Biophys. Mol. Biol. *24*, 109–123.
ZANETTA, Z.P., BRECKENRIDGE, W.C. and VINCENTON, G. (1972) J. Chromatog. *69*, 291–304.
ZEHAVI, U. and JEANLOZ, R.W. (1971) Biochem. Prep. *13*, 14–18.
ZILVERSMIT, D.B., ENTENMAN, C. and FISHLER, M.C. (1942) J. Gen. Physiol. *26*, 325–331.
ZINN, A.B., MARSHALL, J.B. and CARLSON, D.M. (1978) J. Biol. Chem. *253*, 6761–6767.

Subject index